Intelligent Building Health Monitoring and Assessment

Intelligent Building Health Monitoring and Assessment

Editors

Jingzhou Xin
Yan Jiang
Bo Wu
Simon X. Yang

MDPI • Basel • Beijing • Wuhan • Barcelona • Belgrade • Manchester • Tokyo • Cluj • Tianjin

Editors
Jingzhou Xin
Chongqing Jiaotong
University
Chongqing
China

Yan Jiang
Southwest University
Chongqing
China

Bo Wu
Chongqing Jiaotong
University
Chongqing
China

Simon X. Yang
University of Guelph
Guelph
Canada

Editorial Office
MDPI
St. Alban-Anlage 66
4052 Basel, Switzerland

This is a reprint of articles from the Special Issue published online in the open access journal *Buildings* (ISSN 2075-5309) (available at: https://www.mdpi.com/journal/buildings/special_issues/95O46H6PY7).

For citation purposes, cite each article independently as indicated on the article page online and as indicated below:

LastName, A.A.; LastName, B.B.; LastName, C.C. Article Title. *Journal Name* **Year**, *Volume Number*, Page Range.

ISBN 978-3-0365-8448-5 (Hbk)
ISBN 978-3-0365-8449-2 (PDF)

Contents

Preface

Buildings play an indispensable role in urban development. As typical structures of transportation buildings, bridges serve as crucial nodes in connecting different regions, promoting economic growth, and ensuring social security. However, with the extension of their service life, the performance of bridges will inevitably decline. Performance monitoring and evaluation are crucial during the life cycle of bridges. The accelerating convergence of civil engineering, materials science, and artificial intelligence has sparked the interest of researchers from different disciplines in the emerging field of bridge state perception.

This book covers topics on condition monitoring and assessment of bridges, featuring 13 papers. These studies provide some novel methods, models, and technological applications for bridge condition monitoring and assessment, which are of great significance for the design, construction, and assessment of bridges.

Jingzhou Xin, Yan Jiang, Bo Wu, and Simon X. Yang
Editors

buildings

MDPI

Editorial

Intelligent Bridge Health Monitoring and Assessment

Jingzhou Xin [1], Yan Jiang [2,*], Bo Wu [1] and Simon X. Yang [3]

[1] State Key Laboratory of Mountain Bridge and Tunnel Engineering, Chongqing Jiaotong University, Chongqing 400074, China
[2] College of Engineering and Technology, Southwest University, Chongqing 400715, China
[3] School of Engineering, University of Guelph, Guelph, ON N1G 2W1, Canada
* Correspondence: xnjtjiangyan@163.com

Citation: Xin, J.; Jiang, Y.; Wu, B.; Yang, S.X. Intelligent Bridge Health Monitoring and Assessment. *Buildings* **2023**, *13*, 1834. https://doi.org/10.3390/buildings13071834

Received: 17 July 2023
Accepted: 19 July 2023
Published: 20 July 2023

Buildings play an indispensable role in urban development. As transportation structures, bridges serve as crucial nodes in connecting different regions, promoting economic growth, and ensuring social security. However, with the extension of their service life, the performance of bridges will inevitably decline. Performance monitoring and evaluation are crucial during the life cycle of bridges [1,2] because they can provide vital scientific research significance and engineering application value for ensuring the safety of bridges and keeping road networks unblocked. The accelerating convergence of civil engineering, materials science, and artificial intelligence has sparked the interest of researchers from different disciplines in the emerging field of bridge state perception [3,4]. This Special Issue is devoted to new research and development activities with regard to the intelligent monitoring and assessment of bridges. This Special Issue on Intelligent Building Health Monitoring and Assessment features 12 papers. All these contributions effectively address the main topics of this Special Issue in a targeted effort.

According to the research topic, the published 12 papers can be divided into two major categories, i.e., monitoring and assessment. Within the monitoring category, there are two subcategories. The first is direct field monitoring. Paper [5] investigated the effect of pile foundation cutting and under-pinning processes on the stability of bridge substructures during shield tunneling. They utilized numerical simulations and on-site monitoring to study how the active underpinning process of shield tunneling pile foundations affected bridge substructure deformations. Paper [6] proposed a working stress monitoring method for prestressed rebars based on magnetic resonance, which can provide a new perspective on working stress measurements of vertical prestressed rebars. The second subcategory is indirect monitoring based on reverse deduction of monitoring data. Paper [7] proposed a novel method to perform moving load identification using MobileNetV2 and transfer learning. MobileNetV2 has a faster identification speed and requires fewer computational resources in moving load identification. Paper [8] identified moving forces by utilizing the acceleration response of a moving instrumented vehicle. This method exhibited a high recognition accuracy and a good robustness and reliability even amidst substantial environmental noise interference.

Within the assessment category, there are three subcategories. The first is experimental assessments. Paper [9] constructed a test system for the bearing characteristics of a cantilever anti-slide pile based on similarity theory. The distribution laws of internal forces and the deformation of a cantilever anti-slide pile were revealed, and an optimized calculation method for the internal force of a cantilever anti-slide pile was proposed by considering the elastoplastic characteristics of steel bars and concrete. Paper [10] designed an array-type, self-balancing pulley group loading system for a 1:10 scale model of the world's largest span arch bridge. The system can automatically calculate the required load at each loading point using ANSYS and optimize the load points at different construction stages. The second subcategory is numerical-simulation-based assessments. Paper [11] investigated the appropriate range of the inclination angle of arch ribs. A spatial finite element model was established to investigate the effect of different arch rib inclination angles under static loads and then verified using real bridge data. Paper [12] focused on a super-long-span concrete-filled steel tube arch bridge and conducted a nonlinear time history analysis and seismic verification calculations. The third subcategory is algorithm-based assessments. Paper [13] investigated the debonding failure between concrete and steel reinforcements. Correlation and gray correlation analyses were used to establish the index system, from which two debonding prediction models were established. Paper [14] proposed a response surface method to further improve the accuracy of alignment prediction for large-span steel–concrete composite beam track cable-stayed bridges. This method can accurately predict the alignment of rail cable-stayed bridges, thereby providing technical support for alignment control and ensuring the safe and comfortable operation of rail transportation. Paper [15] investigated the surface temperature of the steel box girders of a long-span suspension bridge via a statistical analysis. The time and space distributions of the three-dimensional temperature were studied, and a contour map was created. Based on a machine learning method and the analytical solution of Fick's second law, paper [16] analyzed the diffusion model of corrosion factors and the influence of surface concentrations.

The papers included in this Special Issue cover a wide range of topics, including the utilization of different techniques for monitoring and assessing the health of bridges. These studies provide some novel methods, models, and technological applications for bridge health monitoring, which are of great significance for the design, construction, and assessment of bridges. By employing these methods and technologies, the safe and stable operation of bridges can be effectively ensured.

Conflicts of Interest: The authors declare no conflict of interest.

References

1. Liu, S.; Jiang, Y.; Qiao, K.; Peng, L.; Liu, D. Record-Based Simulation of Three-Component Long-Period Ground Motions: Hybrid of Surface Wave Separation and Multivariate Empirical Mode Decomposition. *Soil Dyn. Earthq. Eng.* **2023**, *172*, 108037. [CrossRef]
2. Tao, T.; Wang, H. Efficient Buffeting Analysis of Long-Span Bridges under Non-Stationary Winds: A 2D Interpolation Enhanced Approach. *J. Sound Vib.* **2023**, *559*, 117754. [CrossRef]
3. Xin, J.; Zhou, C.; Jiang, Y.; Tang, Q.; Yang, X.; Zhou, J. A Signal Recovery Method for Bridge Monitoring System Using TVFEMD and Encoder-Decoder Aided LSTM. *Measurement* **2023**, *214*, 112797. [CrossRef]
4. Tang, Q.; Xin, J.; Jiang, Y.; Zhou, J.; Li, S.; Fu, L. Fast Identification of Random Loads Using the Transmissibility of Power Spectral Density and Improved Adaptive Multiplicative Regularization. *J. Sound Vib.* **2022**, *534*, 117033. [CrossRef]
5. Zheng, F.; Jiang, Y.; Wang, N.; Geng, D.; Xu, C. Investigation on the Influence of Active Underpinning Process on Bridge Substructures during Shield Tunnelling: Numerical Simulation and Field Monitoring. *Buildings* **2023**, *13*, 241. [CrossRef]
6. Xia, J.; Zhang, S.; Liao, L.; Liu, H.; Sun, Y. Working Stress Measurement of Prestressed Rebars Using the Magnetic Resonance Method. *Buildings* **2023**, *13*, 1416. [CrossRef]
7. Qin, Y.; Tang, Q.; Xin, J.; Yang, C.; Zhang, Z.; Yang, X. A Rapid Identification Technique of Moving Loads Based on MobileNetV2 and Transfer Learning. *Buildings* **2023**, *13*, 572. [CrossRef]
8. Liu, D.; Liu, B.; Li, X.; Shi, K. Theoretical and Numerical Examination of a Novel Method for Identifying Bridge Moving Force Using an Instrumented Vehicle. *Buildings* **2023**, *13*, 1481. [CrossRef]
9. Jin, H.; Ren, Q.; Xiao, S. Study on the Evolution Law of Internal Force and Deformation and Optimized Calculation Method for Internal Force of Cantilever Anti-Slide Pile under Trapezoidal Thrust Load. *Buildings* **2023**, *13*, 322. [CrossRef]
10. Fan, Y.; Zhou, J.; Luo, C.; Yang, J.; Xin, J.; Wang, S. Research on Loading Scheme for Large-Scale Model Tests of Super-Long-Span Arch Bridge. *Buildings* **2023**, *13*, 1639. [CrossRef]
11. Liu, Z.; Wu, Y.; Wang, C.; Fan, Y.; Luo, C.; Wang, S. Research on Optimal Arch Rib Inclination of Large Span Highway CFST through Arch Bridge. *Buildings* **2023**, *13*, 1415. [CrossRef]
12. Ye, D.; Tong, Y.; Gan, L.; Tang, Z.; Zhang, R. Improvement in the Seismic Performance of a Super-Long-Span Concrete-Filled Steel-Tube-Arch Bridge. *Buildings* **2023**, *13*, 1811. [CrossRef]
13. Hu, T.; Zhang, H.; Zhou, J. Prediction of the Debonding Failure of Beams Strengthened with FRP through Machine Learning Models. *Buildings* **2023**, *13*, 608. [CrossRef]
14. Li, X.; Luo, H.; Ding, P.; Chen, X.; Tan, S. Prediction Study on the Alignment of a Steel-Concrete Composite Beam Track Cable-Stayed Bridge. *Buildings* **2023**, *13*, 882. [CrossRef]

15. Ma, W.; Wu, B.; Qin, D.; Zhao, B.; Yang, X. Statistical Analyses of the Non-Uniform Longitudinal Temperature Distribution in Steel Box Girder Bridge. *Buildings* **2023**, *13*, 1316. [CrossRef]
16. Li, S.; Yao, G.; Wang, W.; Yu, X.; He, X.; Ran, C.; Long, H. Research on the Diffusion Model of Cable Corrosion Factors Based on Optimized BP Neural Network Algorithm. *Buildings* **2023**, *13*, 1485. [CrossRef]

buildings

MDPI

Article

Investigation on the Influence of Active Underpinning Process on Bridge Substructures during Shield Tunnelling: Numerical Simulation and Field Monitoring

Fengqu Zheng [1], Yalong Jiang [1,2,*], Ning Wang [1], Daxin Geng [1] and Changjie Xu [1]

1 School of Civil Engineering and Architecture, East China Jiaotong University, Nanchang 330013, China
2 State Key Laboratory of Performance Monitoring and Protecting of Rail Transit Infrastructure, East China Jiaotong University, Nanchang 330013, China
* Correspondence: yalongjiang@whu.edu.cn

Abstract: The pile foundation cutting and underpinning process during shield tunnelling significantly impacts the stability of bridge substructures. In this paper, the shield tunnel area from Hongguzhong Avenue Station to Yangming Park Station of Nanchang Metro Line 2 was taken as the research subject, which crosses the pile foundation underpinning project of the south approach section of Bayi Bridge. Through numerical simulation and on-site monitoring analysis, the influence of the active underpinning process of shield tunnelling pile foundation on the deformation of bridge substructure was studied. First, through analyzing on-site conditions and comparing technical solutions, an active gantry bridge pile foundation underpinning technology was proposed, and the specific construction steps were determined. On this basis, for the C15 pile foundation with the most complex working conditions, ABAQUS software was applied to simulate the jack-up, unloading and pile-cutting process during the pile foundation underpinning construction, and the displacement development of the bridge pier, underpinning beam and new pile during the whole construction process were analyzed. Finally, through on-site monitoring data analysis, the technology's feasibility and safety were further verified. At the same time, according to the analysis of the monitoring results of the bridge piers, underpinning beams and new piles, the results from the finite element software were nearly the same as the trend shown by the monitoring results, and the displacement of the main structures of the lower part of the bridge was small and within the control range. The above research work verified the applicability of the active gantry type bridge pile foundation underpinning technology in the pile foundation underpinning condition of the single-column single-pile bridge in the narrow space curved bridge section, and is worthy of further promotion and application.

Keywords: shield tunnelling; active underpinning; bridge substructure; 3D FEM modelling; field verification

Citation: Zheng, F.; Jiang, Y.; Wang, N.; Geng, D.; Xu, C. Investigation on the Influence of Active Underpinning Process on Bridge Substructures during Shield Tunnelling: Numerical Simulation and Field Monitoring. *Buildings* **2023**, *13*, 241. https://doi.org/10.3390/buildings13010241

Academic Editor: Bartolomeo Pantò

Received: 5 December 2022
Revised: 9 January 2023
Accepted: 11 January 2023
Published: 15 January 2023

1. Introduction

With the rapid urban development in China, an increasing number of metro tunnels are being constructed in major cities to alleviate traffic pressure and improve transportation efficiency [1–3]. According to statistics offered by China Association of Metros, there are more than 50 metro cities in China, and the total length of metro lines in operation exceeded 7209.7 km up to December 2021. Benefiting from unique advantages, such as high automation in facilities, fast construction speed, and little disturbances to ground traffic [4–8], the shield technique has been widely used in metro tunnelling. However, since metro lines usually go across densely populated urban areas, shield tunnel excavation inevitably disturbs the surrounding environment, leading to ground uplifting or settlement [9–13], further inducing deformation and even damage of adjacent structures, including buildings, bridges, buried pipelines and existing tunnels [14–17]. Hence, how to minimize the interference of shield tunnel construction process on adjacent structures has been a key issue in the field of geotechnical engineering, which has attracted more and more attention from engineers and scholars [18–21].

As one of the most typical engineering cases, once the existing pile foundations of the adjacent buildings and structures intrude into the designed shield tunnel line, the intruded piles are usually cut off and new piles are constructed, acting as an underpinning foundation to continuously support the upper structures [19,22]. In this underpinning process, the overlying load of the upper structures should be smoothly transferred to the underpinning pile foundation to ensure the stability of the buildings and the safety of shield digging [16]. It is a very complex construction process involving complicated interactions between the tunnel, soil, and pile foundation [23], which merit detailed investigation.

In recent years, many efforts have been made by researchers to study the mechanical responses during pile foundation underpinning and the optimization of underpinning schemes through laboratory experiments, numerical simulation, and field tests [16,24]. Stulgis et al. [25] revealed that the pile foundation should be replaced before the destructive settlement of the building by installing grouting micro-piles in the replacement area to ensure the safety of the structures based on the case study of expansion project of St. Joseph Mercy Hospital in Georgetown, Guyana. Ma and Wang [26] theoretically analyzed the bearing capacity of a single pile and the joist deformation according to the structural design and found that the pile foundation underpinning method in the Xi'an Metro shield tunnel project was valid. Yan et al. [27] obtained the formula of shear force of a pile foundation underpinning structure through theoretical analysis and field model tests, which enabled calculation of the shear bearing capacity. Taking Beijing Metro Line 8 as an engineering case, Yao et al. [27] numerically analyzed the pile foundation underpinning scheme using FLAC^{3D} software to improve the effectiveness of the underpinning in reducing the deformation of structures and isolating piles. Xu et al. [22] proposed a pile underpinning technology for shield tunnel cross through group pile foundation in a shield tunnel interval in Shanghai Metro Line 10, and improved the rationality of the scheme by theoretical and numerical analysis as well as field monitoring tests. Park et al. [28] proposed and verified the application of a modified underpinning method in the new subway #9 line in Seoul Metropolitan in South Korea, which was able to reduce construction period by 1.5 times and the construction cost by 1.2 times compared with conventional methods. In addition to the aforementioned studies on pile foundation underpinning schemes, many scholars have focused on the pile cutting process [19,29,30]. Fu [31] numerically investigated the feasibility of the direct shield cutting pile foundation construction technology in cutting plain concrete, glass fiber concrete, and reinforced concrete, based on the mechanical response of the shield cutterhead as well as the tunnelling parameters are analyzed. Chen et al. [32] analyze the effect and mechanism of large-diameter pile cutting process during shield tunnelling and obtained the characteristics of cutting parameters and the damage law of cutting tools through field tests. To achieve stable and effective pile cutting process, a new cutterhead configuration as well as pile cutting scheme were proposed by [33], suggesting that the shield advance rate should not exceed 2 mm/min, and the rotation speed should be controlled at a relatively low level.

In the pile foundation underpinning–cutting scheme, dynamic control of the displacement of superstructure and the underpinning foundation is crucial to the safety and efficiency of this process [16], which still remains to be explored in detail. Moreover, complex geological conditions and differences in the surroundings often make it hard to determine an appropriate underpinning scheme. Considering that several single-column and single-piles in the upper Bayi Bridge invaded the tunnel, the shield tunnel section pile foundations from Hongguzhong Avenue Station to Yangming Park Station of Nanchang Metro Line 2 need to be underpinned, which will inevitably have a great impact on the substructure of the bridge and the surrounding engineering environment, and the construction risk is extremely high. Therefore, in this paper, based on the analysis of on-site working conditions and the comparison of technical solutions, an active gantry type bridge pile foundation underpinning technology was proposed, and the corresponding specific construction steps were designed. On this basis, for the pile foundation C15 with the most complex working conditions, the 3D finite element ABAQUS numerical simulation was used to analyze the underpinning process. Moreover, the influence of three key stages of the underpinning scheme, i.e., jack lifting, unloading and pile cutting on the bridge pier, underpinning beam and new underpinning pile were studied in detail, and the feasibility and safety of the underpinning technology were preliminarily verified. Finally, by monitoring and analyzing the three important stages of jacking, unloading and pile cutting of the C15 pile, it was further verified that the technical solution is reliable and worthy of further promotion and application.

2. Engineering Background

2.1. Project Introduction

The left line from Hongguzhong Avenue Station to Yangming Park Station of Nanchang Metro Line 2 is 2335 m long, and the minimum plane curve radius is 600 m. In this section, the mud–water balance shield construction technique was adopted, and several bridge piles of the south approach bridge of Bayi Bridge invaded the tunnel section, as shown in Figure 1. For the invaded the subway tunnel line, there are seven pile foundations in ramp C and ramp F, of which ramp C has a deck width of 11 m, and the upper structure of the bridge is a two-box multi-span reinforced concrete continuous box girder bridge. The ramp F has a deck width of 7 m, and the superstructure of the bridge is a single-box multi-span reinforced concrete continuous box girder bridge. The pile foundations intruded into the tunnel are all single-column and single piles, so these invaded pile foundations need to be underpinned. The Bayi Bridge spans the Ganjiang River, and is an important transportation hub in Nanchang, hence there are strict requirements on structural deformation and stability. However, in this project, performing the single-column and single-pile underpinning is extremely risky in the narrow space of the curved and unclosed bridge section, and it is extremely difficult to control the stability of the underpinning pile foundation and bridge structure during the construction process.

2.2. Engineering Geological Conditions

The upper part of the stratum from Hongguzhong Avenue Station to Yangming Park Station is mainly artificial fill <Q4al>, Quaternary Holocene shock layer <Q4al>, Pleistocene alluvial <Q3al>, and the underlying bedrock is mainly the third Xinyu Group (Ex) argillaceous siltstone, etc. According to lithology and engineering geological characteristics, the site stratum can be divided into miscellaneous fill <1-1>, plain fill <1-2>, silty clay <2-1>, silt <2-2>, fine sand <2-3>, medium sand <2-4>, coarse sand <2-5>, gravel sand <2-6>, round gravel <2-7>, silty clay layer <3-1>, fine sand <3-3>, fine sand layer <3-4>, coarse sand <3-5>, gravel sand layer <3-6>, round gravel layer <3-7>, pebble <3-8>, argillaceous siltstone layer <5-1>, glutenite <5-2>, and mudstone <5-3>. In the section between Hongguzhong Avenue Station and Yangming Park Station, shield tunnels mostly pass through the Quaternary overburdened soil layer, the surrounding rock types are VI to IV which mainly consists of weathered argillaceous siltstone, as shown in Table 1.

2.3. Position of Bridge Pile Foundation and Shield Tunnel

The structures corresponding to the pile foundations that need to be underpinned for the south approach bridge of the Bayi Bridge are all multi-span reinforced concrete box girder structures. The pavement of the bridge deck consists of 6 cm expanded metal mesh on the lower layer, C40 concrete and 4 cm thick asphalt concrete on the upper layer, and the horizontal width of the guardrail on both sides is 0.44 m. The lower structure is a single circular column and single pile structure. Among them, the upper structure of ramp C is a double-boxed reinforced concrete continuous girder bridge with a bridge deck width of 11.88 m. The bridge deck layout and related size parameters are shown in Figure 2. The bridge piles on ramp C are relatively complex, and there are different distances between new underpinning piles and the old piles, resulting in eccentricity. In this paper, the bridge pile C15 with the most complex working conditions was selected for research. The column diameter is 1.5 m and the column height is 9.7 m. The pile diameter is 1.5 m, the pile length is 24.9 m, and its intrusion length is 3.5 m.

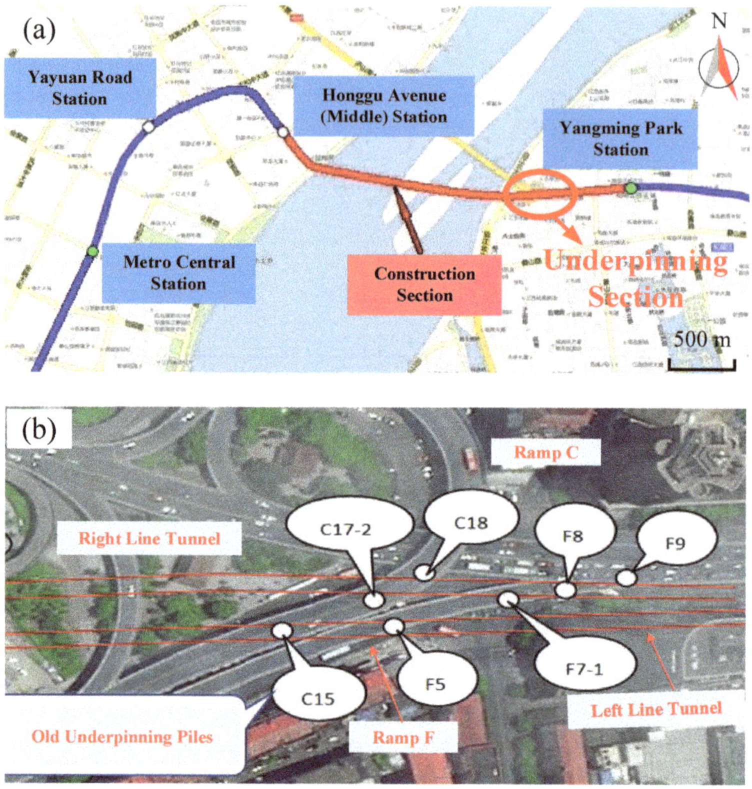

Figure 1. Engineering background of the shield tunnel of Nanchang Metro Line 4 passing through the bridge pile foundation of Bayi bridge: (**a**) location of the underpinning section; and (**b**) layout of the underpinning piles.

Table 1. Classification of surrounding rock in shield tunnel section.

Line	Classification of Surrounding Rock			Surrounding Rock Classification	Stratum Passed by Tunnel
	Tunnel Top	Tunnel Wall	Tunnel Bottom		
Left line	VI	VI~IV	IV	IV	Cave top: <2-1>, <2-2>, <2-3>, <2-4>, <2-5>, <2-6>, <2-7> Cave wall: <2-4>, <2-6>, <2-7> Cave bottom: <2-4>, <2-6>, <2-7>, <5-1-1>, <5-1-2>
Right line	VI	VI~IV	IV	IV	Cave top: <2-1>, <2-2>, <2-3>, <2-4>, <2-5>, <2-6>, <2-7> Cave walll: <2-4>, <2-6>, <2-7> Cave bottom: <2-4>, <2-6>, <2-7>, <5-1-1>, <5-1-2>

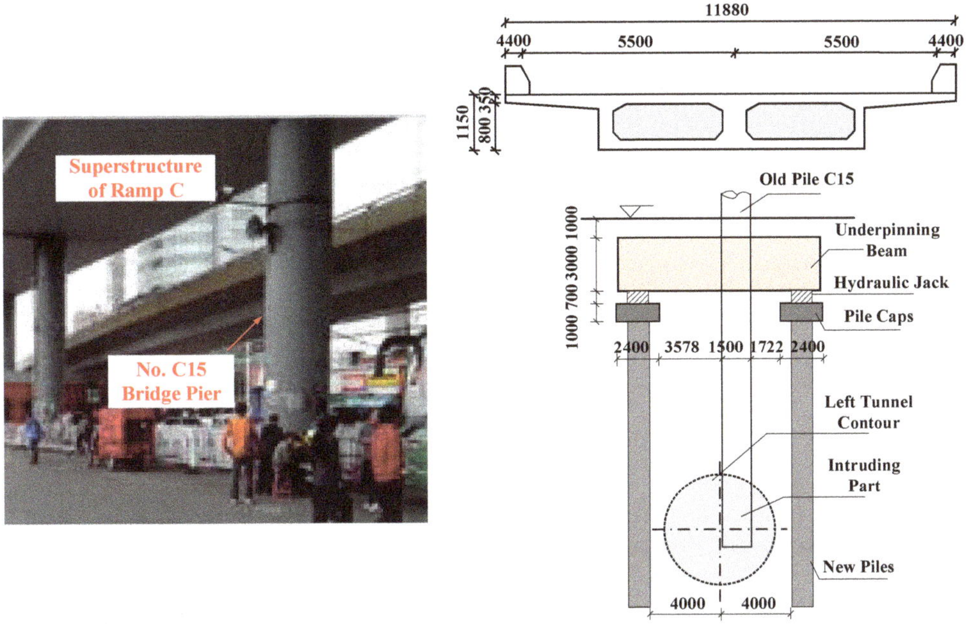

Figure 2. Pile underpinning scheme of C15 bridge pier and the corresponding parameters.

3. The Underpinning Scheme of Bridge Pile

In order to ensure that the Bayi Bridge can still operate normally during the construction, an active underpinning method was adopted to deal with the problem of pile foundation intruding into the tunnel so as to minimize the impact of the construction on the upper bridge structure. The underpinning structure selected was a gantry-type underpinning system consisting of new underpinning piles and underpinning beams. Before the pile foundation underpinning, the new underpinning pile and the foundation pit enclosure structure were to be constructed sequentially. The underpinning beam construction and pile underpinning operation were carried out in the foundation pit. The underpinning steps are as follows:

(1) A jack is set on the underpinning platform, the underpinning beam and the underpinning pile are constructed independently, and a rigid overall structure is formed after the underpinning force of the pile foundation is converted. When the underpinning pile, underpinning beam and underpinning platform concrete reach the design strength, the loading underpinning construction will be carried out;
(2) During underpinning, a jack is set up between the underpinning beam and the underpinning cap to transfer the load of the superstructure to the new underpinning pile and make most of the displacement of the new pile to offset the preload of the jack simultaneously, making the new underpinning pile can replace the force of the original pile through active loading;
(3) During the underpinning process of pile foundation, the PLC hydraulic synchronous control system is used to jack up the underpinning beam and preload the underpinning pile. At the same time, the load and displacement changes of the underpinning pile are monitored to ensure the settlement of underpinned pile does not exceed the control value. The loading is divided 10 times, with 10% of the total jacking force as the interval;
(4) After the end of the jacking stage, the unloading stage begins, and the unloading process is divided 5 times, with 20% as the interval;

(5) Considering that the structure above the underpinning beam needs to remain balanced, the original pile foundation is cut off using a wire saw. The cutting height of the pile is controlled within the range of 300~500 mm, and the cutting gap is 5~10 mm. After the first cutting is completed, it is necessary to ensure continuous monitoring for more than 1 h, and the second cutting can only be carried out after the settlement becomes stable. During the pile cutting process and after the pile cutting is completed, the jacking force should be monitored and adjusted to ensure that the original column elevation remains unchanged. The construction process of the underpinning pile foundation is shown in Figure 3.

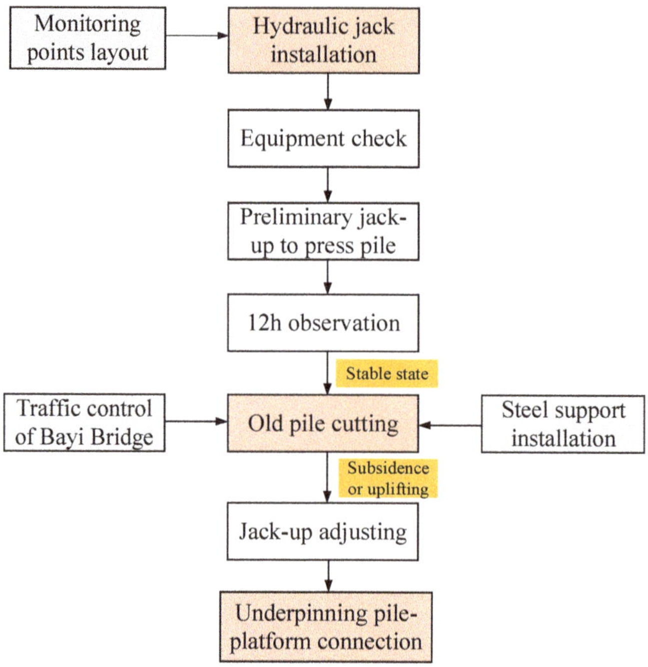

Figure 3. Construction process of underpinning pile foundation.

4. Numerical Simulation of Pile Foundation Underpinning Process

During the underpinning construction, the stress and deformation of the new underpinning pile, the underpinning beam and the underpinning pile are relatively complex, and at the stage of jack loading, unloading and pile cutting, the deformation control of the new pile and underpinning beam, such as settlement, inclination and horizontal displacement of pile top is the key to pile underpinning technology [16]. In many previous studies in the literature, finite element method software ABAQUS 3D has been used in the numerical simulation of shield tunnelling and pile foundation underpinning process, which has proved the validity in capturing the mechanical behaviors of the tunnels and the surrounding pile foundations [16,19]. Hence, in this part, ABAQUS 3D finite element simulation was used to analyze the deformation of the bridge substructure of the C15 pile foundation in the three stages of jacking, unloading and pile cutting so as to verify the feasibility and safety of the underpinning technology.

4.1. The Establishment of Model

As shown in Figure 4, the model size is set as 50 m (length) × 80 m (width) × 50 m (height) to avoid boundary effects [19], and the size of the foundation pit is 14 m (length) × 5.2 m (width) × 5.9 m (height). The dimensions of other components, such as old pile, cap, underpinning beam, and new underpinning pile are also the same as those of the engineering case. The two new underpinning piles are 1.2 m in diameter and 27 m in length. The model unit is C3D8R, and the total number of units is 32,700.

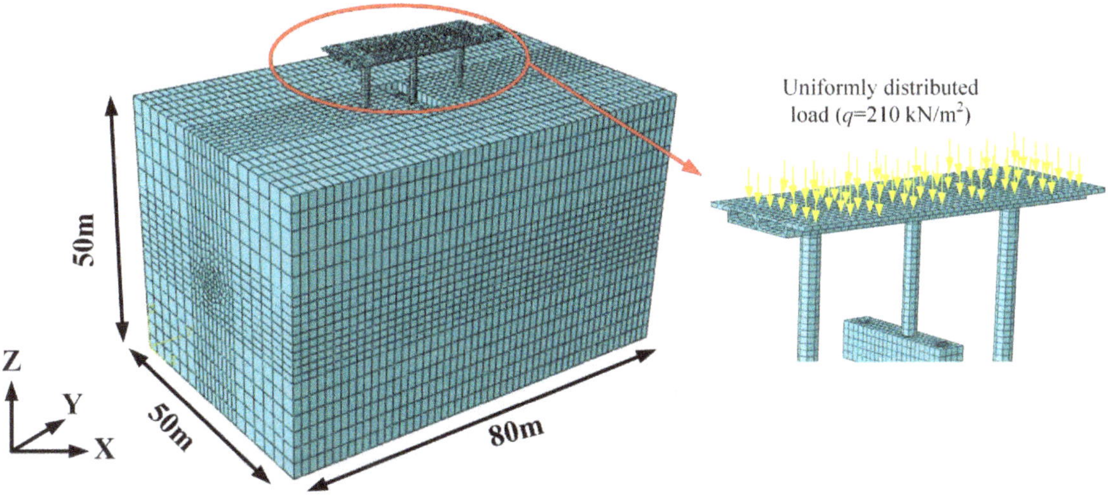

Figure 4. 3D-FEM modelling configuration and boundary conditions.

The left and right boundaries of the model are constrained in the X direction, the front and rear boundaries are constrained in the Y direction, the bottom boundary is fixed, and the top boundary is free. The mechanical behavior of the soil obeys the Mohr–Coulomb criterion, and the underpinning beams, caps and piles are assumed to be homogeneous and linear elastic materials [19]. The corresponding model material parameters are listed in Table 2. A parametric study was conducted for the variability that the surrounding terrain could bring to the structural analysis, and the results improve that the model parameters adopted in the simulation can effectively avoid boundary effects. Considering that only the impact of the construction process on the substructure of the bridge can be analyzed in numerical simulation, continuous mechanical change cannot be monitored in the same way as the engineering. Therefore, in the modelling process, the "jack" was directly regarded as the concrete block after the pile sealing has been completed. The details of the underpinning are shown in Figure 5a, and the relationship between the underpinning pile and the tunnel is shown in Figure 5b.

Table 2. Material parameters used in numerical modelling.

Material Type	Density (ρ) /kg/m^3	Young's Modulus (E)/MPa	Poisson's Ratio (ν)	Friction Angle (φ)/°	Cohesion (c)/kPa	Thickness (h)/m
Miscellaneous fill	1780	7	0.32	10	5	4
Silty clay	1870	8	0.34	15	25	3.4
Coarse sand	1900	20	0.32	29	0	3.9
Pebble	2000	30	0.33	30	0	7
Strongly weathered argillaceous siltstone	2100	140	0.32	28	38	3.7
Moderately weathered argillaceous siltstone	2450	4×10^3	0.3	38	1000	28
Concrete (underpinning beam, platform, piles)	2400	3×10^4	0.23	—	—	—

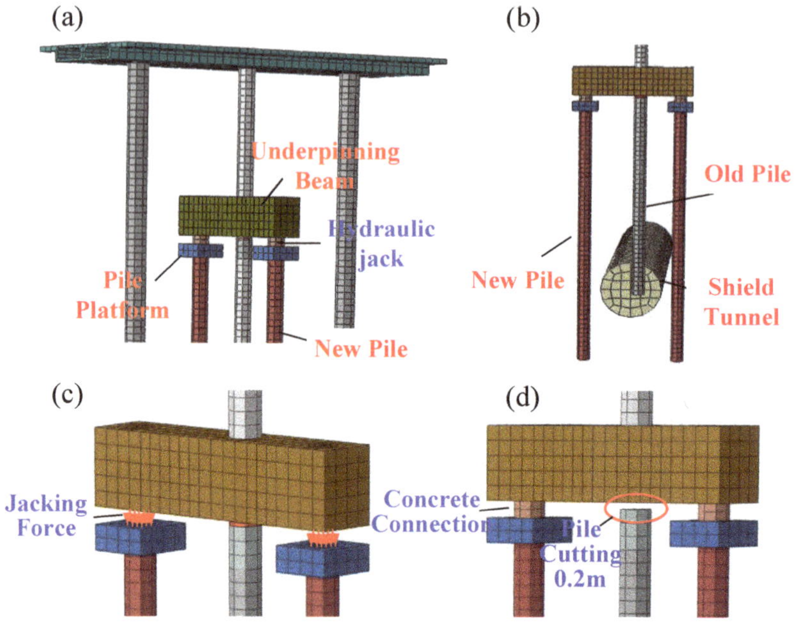

Figure 5. Details of the 3D-FEM modelling: (**a**) underpinning diagram; (**b**) location of the underpinning pile and shield tunnel; (**c**) application of jacking force; and (**d**) old pile cutting and new pile sealing.

4.2. Simulation Procedure

There are 17 steps in the model analysis, and the specific settings are as follows:

(1) Step 0: the balance of earth stress;
(2) Before the underpinning, because the "jack" does not work between the underpinning beam and the bearing platform, the "jack" concrete block is not activated first;
(3) According to the "General Code for Design of Highway Bridges and Culverts" (JTG D60-2015) [34], the standard value of the uniformly distributed load of the I-level lane is q_k = 10.5 kN/m, and the span is 20 m, hence a uniformly distributed load of 210 kN is applied to the bridge deck to simulate the traffic load;
(4) Step 1–10 jacking stage: as shown in Figure 5c, jacking force is subjected to the bottom of underpinning beams and the upper part of the cap at two sides. According to the construction scheme, the loadings applied to the left and right pile platforms are 3466 kN and 4588 kN, respectively, which are converted into equivalent pressure in Table 3. The loading is divided 10 times, and 10% of the total jacking force is applied each time;
(5) Step 11–15 unloading stage: after the lifting is completed, unloading is carried out. The unloading interval is 20% of the total lifting force, and the unloading is carried out 5 times;
(6) Step 16 Pile cutting and sealing stage: In the actual project, the truncation is carried out in two steps. In this simulation, it is simplified to activate the "jack" concrete block to simulate the completion of the pile sealing and truncate the old pile to be separated by 0.2 m, as shown in Figure 5d.

Table 3. Equivalent jacking pressures applied on the pile platforms.

Pile No.	Total Pressure on Left Pile Platform/kPa	Pressure on Left Pile Platform Each Loading Step/kPa	Pressure on Left Pile Platform Each Unloading Step/kPa	Total Pressure on Right Pile Platform/kPa	Pressure on Right Pile Platform Each Loading Step/kPa	Pressure on Left Pile Platform Each Unloading Step/kPa
C15	3066	306.6	613.2	4058	405.8	811.6

4.3. The Analysis of Simulation Result

In the process of pile foundation underpinning, the displacement changes of bridge piers, underpinning beams and new piles are the key monitoring contents. The data extraction points of numerical simulation are shown in Figure 6. Because the whole construction process is simulated, the displacement changes of each extraction point in the whole construction process were analyzed. During the jacking process of pile foundation underpinning, the new underpinning pile will experience settlement due to the different engineering environments, such as insufficient foundation bearing capacity, pile bottom sediment, etc., which will induce additional stress to the upper structure to damage the beam body, which is very important in the study of the settlement law of new piles.

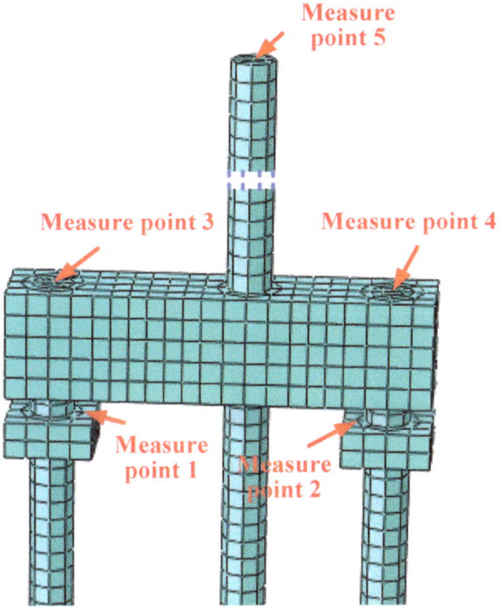

Figure 6. Data extraction points of the numerical model.

Figure 7 shows the changing trend of the new pile displacement during the pile foundation underpinning process, and it is the same as the changing trend of the displacement of the bridge pier and the underpinning beam. In the first five times of jacking, this new pile displacement has a relatively stable increase, and it starts to increase from the sixth jacking, especially the settlement of new piles near the old pile side are increased by 40% compared to the previous in the last jacking. The maximum settlement of the new pile on the side far from the old pile is 2.12 mm, and the maximum settlement value of the new pile is 3.48 mm, which has reached the warning value during monitoring but not the control value. Therefore, during the jack-up process of the pile foundation underpinning, it is necessary to focus on the settlement changes of the new piles near the old piles [19].

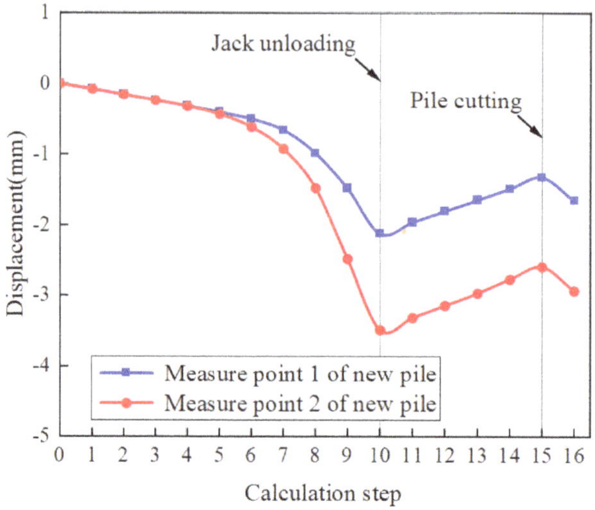

Figure 7. Displacement of the new piles calculated by numerical modelling.

Figure 8 shows the displacement change curve of the underpinning beam. It can be seen that the displacement changes of the underpinning beam on the side away from the old pile is small, and the corresponding growth rate is small, while the displacement of the underpinning beam close to the old pile side changes greatly, and the maximum displacement is 0.88 mm, but it is still within the early warning. The displacement of the underpinning beams on both sides changes smoothly during the unloading stage, and it has no sudden change when the old pile is truncated.

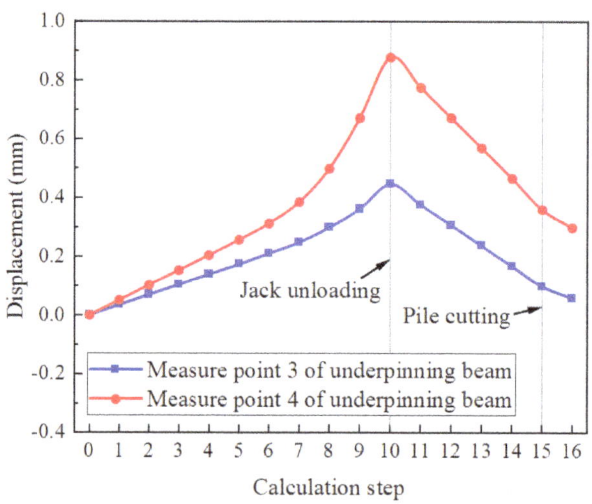

Figure 8. Displacement of the underpinning beam calculated by numerical modelling.

The difference in settlement of the new piles on both sides is caused by the difference in the lifting force. However, with the end of the jacking, this difference becomes small, indicating that the deformation of the superstructure is well controlled in the underpinning process, and the design of the jacking force in the underpinning construction is feasible.

The displacement of the bridge pier calculated by numerical modelling is given in Figure 9. It can be seen from the figure that in the first five times of the jack lifting process, the displacement increased relatively stable. However, its growth rate starts to increase from the 6th jacking to the end, and the maximum displacement is 0.68 mm. The displacement is gradually decreased by 0.43 mm when unloading, indicating that the load is well transferred. In the final jacking, the old pile is cut off, and the displacement continues to drop by 0.06 mm, indicating that the load transfer is completed in the underpinning construction, and the new pile is able to sustain the load safely. In addition, there is no sudden change in displacement during the whole simulation process, and all the displacement changes are also within the warning value. In general, the underpinning technical solution was preliminarily verified to be safe and feasible.

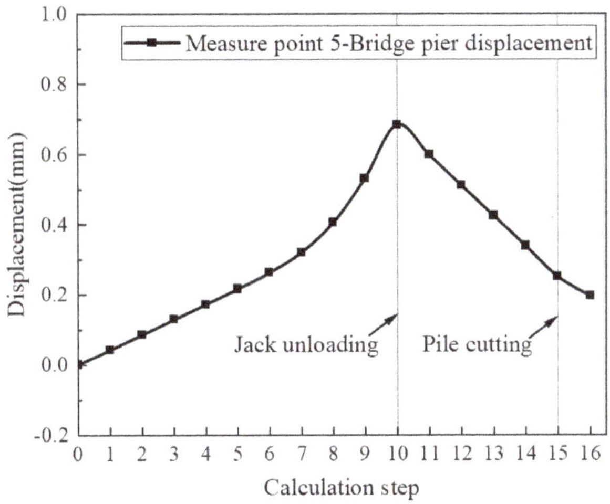

Figure 9. Displacement of the bridge pier calculated by numerical modelling.

5. Field Verification of Pile Foundation Underpinning Scheme

5.1. Monitoring Content and Methods

Pile foundation underpinning is very difficult and has a high construction risk; construction monitoring is the key to ensuring construction safety, and the setting of construction monitoring control values and early warning values is crucial. The monitored data are the deformation of the bridge pier and the underpinning beam during the underpinning process. The monitoring content and warning control value are shown in Table 4. The settlement of the bridge piers is monitored by a crystalline silicon static level instrument, and each bridge pier is installed with a static level measuring point installed on the upper part of the pier column. The inclination of the bridge pier is measured by a dual-axis inclinometer that is installed on the side of the pier parallel to the bridge axis at the top of the pier column. During the monitoring, each pier is installed with a dual-axis inclinometer. Two static level monitoring points are set at the top of the two new piles to monitor the settlement using the static level instrument. By monitoring the relative settlement of the new pile top and the pier body, the absolute settlement of the new pile top can be obtained. Thus, the settlement change of the new pile and the difference in the displacement of new piles on both sides of the main beam during the jacking process. During the underpinning process, due to the inconsistency of the jacking force on the left and right sides, the underpinning beam may be inclined to induce additional loads on the pier body and the main beam, affecting the safety of underpinning construction and bridge structure. Therefore, the monitorization of settlement of the underpinning beam is very necessary. The settlement of the underpinning beam is monitored by a static levelling instrument, and two static levelling points are set at two ends of the underpinning beam (the corresponding displacement of the new pile). The main monitored parameters include absolute settlement and differential settlement at both ends of the joist beam.

Table 4. Field monitoring variables and corresponding alarming and controlling values.

No.	Monitoring Variables	Monitoring Instruments	Precision	Alarming Value	Controlling Value
1	Pier displacement	Hydrostatic level	0.1 mm	2.0 mm	3.0 mm
2	Pier inclination	Inclinometer	0.001°	0.06°	0.12°
3	New pile displacement	Hydrostatic level	0.1 mm	3.5 mm	5 mm
4	Underpinning displacement	Hydrostatic level	0.1 mm	3.5 mm	5 mm

5.2. Monitoring Data Analysis

During construction, the monitoring in the jacking process is the most critical stage of the project. The underpinning jacking adopts the graded loading, this process is divided into 10 stages, and the interval is 10% of the predetermined limit of loading. Each loading stage is held for 10 min until the tested displacement parameters are stable and meet the limit requirements (controlled by 0.1 mm/h), and then the next stage of loading can be carried out. After the static compression of the new pile is completed (reaching the design control value of the jacking force), the monitorization continues until the settlement is stable (consecutively controlled by 0.1 mm/h for 12 h). During the pile-cutting stage, the settlement of the pier and underpinning beam are continuously monitored (including the difference in settlement change), and sudden displacement of the underpinning structure is not allowed. At the same time, the pier settlement and underpinning beam settlement (including differential settlement changes) are continuously monitored, and sudden displacement of the underpinning structure is not allowed. In order to focus on the study of the deformation law of the main structure in the lower part of the bridge during the three important stages of jacking, unloading and pile cutting in the underpinning project, the change of the monitoring data of the key nodes of the C15 pile foundation construction was analyzed. As shown in Figure 10, the data of monitoring points are the same as that of the extraction points in the numerical simulation. Measuring points 6 and 7 are the new pile settlement monitoring points, and measuring points 8 and 9 are the underpinning beam settlement monitoring points. The inclinometer is placed on the top of the pier column (measurement point 10).

5.2.1. Jacking Stage of Underpinning Pile

As shown in Figure 11, during the jacking process of C15, both the pier and the underpinning beam experience upward displacement within the control range under the action of jacking force. When the jacking is completed, the change of displacement gradually becomes stable without significant fluctuation. The maximum change rate of the pier displacement is −0.25 mm/h, and the cumulative change is 0.23 mm. The maximum change rate of the underpinning beam displacement is 1.28 mm/h, and the cumulative change is 1.32 mm. The maximum change rate of the pier inclination is 0.006°/h, the cumulative change value is 0.016°. During the jacking process, the displacement and inclination of the piers are all within the normal range. Among them, the measured value of seven measuring points (new piles near the F ramp) exceeds the control value after the completion of the 7th level of loading, but after the completion of the 10th level of loading, the differential settlement of the new piles is 1.74 mm, and other monitoring items have no abnormality. According to the monitoring results after reaching the jacking design load for 30 h, it is shown that the monitoring data are gradually stable, the final differential settlement of the new pile is 1.66 mm, and the overall monitoring cumulative value and change rate are both small. In the jacking stage, although the settlement of the new pile exceeds the control value, the deformation of the superstructure is stable, which is also the same as the conclusion of the numerical simulation, indicating that the design of the jacking force is reasonable, and the construction plan is feasible.

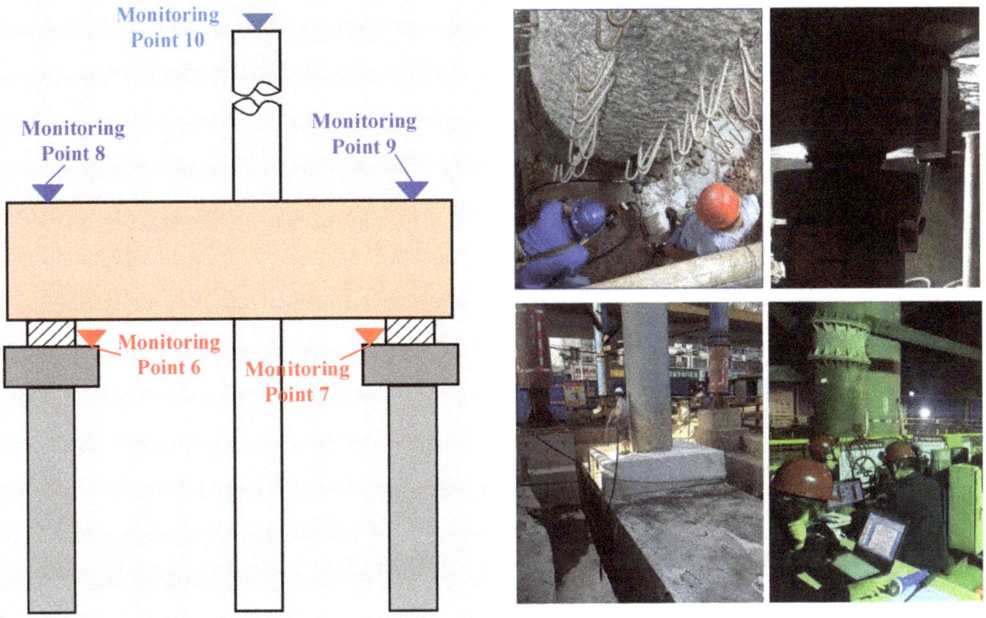

Figure 10. Layout of the monitoring points during underpinning construction (**left**) and the field monitoring (**right**).

5.2.2. Underpinning Pile Unloading Stage

As shown in Figure 12, during the unloading process of C15, the force from the jack becomes smaller, and the load transfer is gradually realized. The settlement of the bridge pier and the underpinning beam gradually increases, but the change of the displacement tends to be stable after the unloading of the jack is completed. There is no sudden change in displacement during the process, indicating that the construction plan is reasonable in the unloading stage. During this period, the maximum change rate of pier settlement is -0.01 mm/h, and the final cumulative maximum value is −0.04 mm. The maximum change rate of pier inclination angle is 0.003°/h, and the final cumulative maximum value is 0.007°. The maximum change rate of underpinning beam settlement is −0.04 mm/h and the final cumulative maximum value is −0.09 mm. The settlement and inclination of the bridge superstructure are within the safe range.

5.2.3. Underpinning Pile Cutting Stage

As shown in Figure 13, in the process of two piles cutting of C15, the maximum change rate of the bridge pier displacement is 0.06 mm/h, and the cumulative maximum value is 0.18 mm, showing an upward trend. This is because in the jacking stage, due to the existence of the old piles, the jacking force will generate prestress between the old piles and the bridge piers. When the old piles are truncated, the stress is gradually released, increasing the displacement of the bridge piers, but the displacement of the bridge piers tends to be stable at the end, and the magnitude of the change is also within a safe range. The maximum change rate of the pier inclination angle is 0.038°/h, and the cumulative maximum value is 0.046°. The maximum change rate of the displacement of the underpinning beam is −0.1 mm/h, and the cumulative maximum value is −0.42 mm, which are both within the safe range. Combined with the analysis results of numerical simulation and field measurement data, it can be seen that the active gantry type bridge pile foundation underpinning technology scheme is highly feasible, and it is worth popularizing in pile foundation underpinning projects of single-column and single-pile bridges in similar narrow-space curved bridge sections.

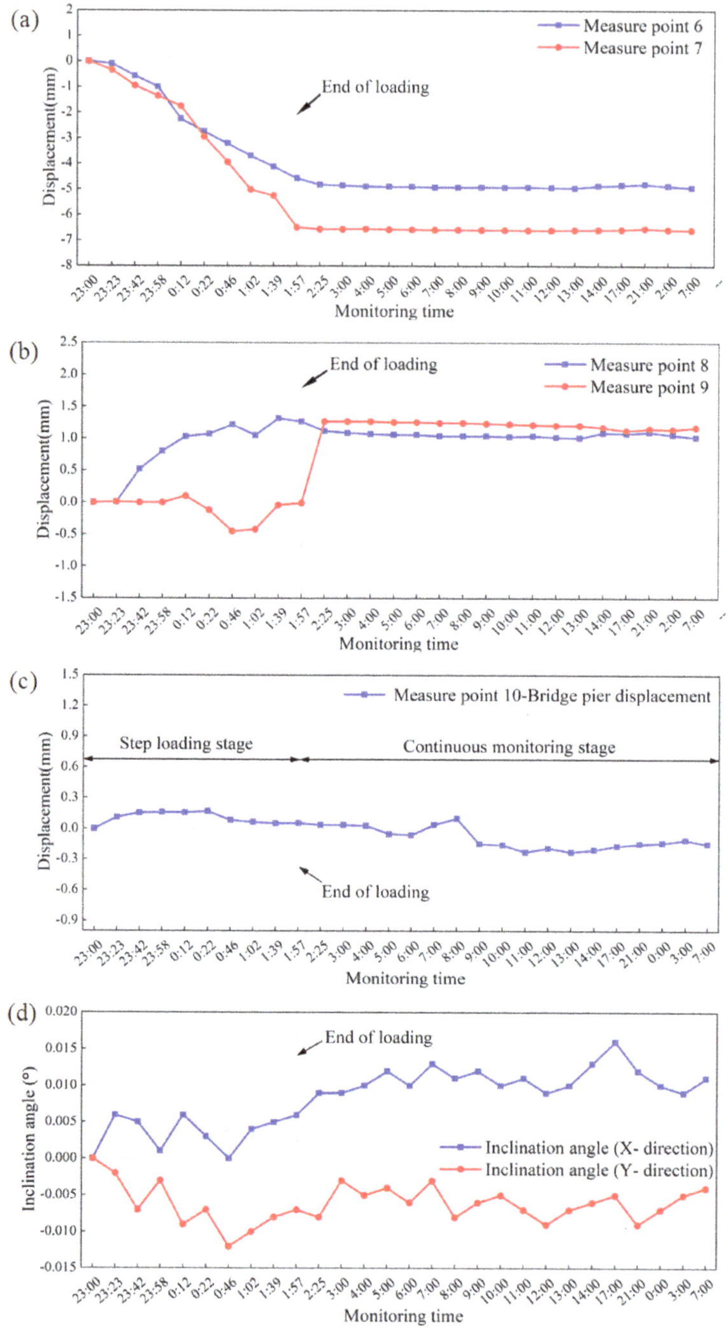

Figure 11. Displacement monitored during jacking-up stage: (**a**) displacement of new pile; (**b**) displacement of underpinning beam; (**c**) displacement of bridge pier; and (**d**) pier inclination.

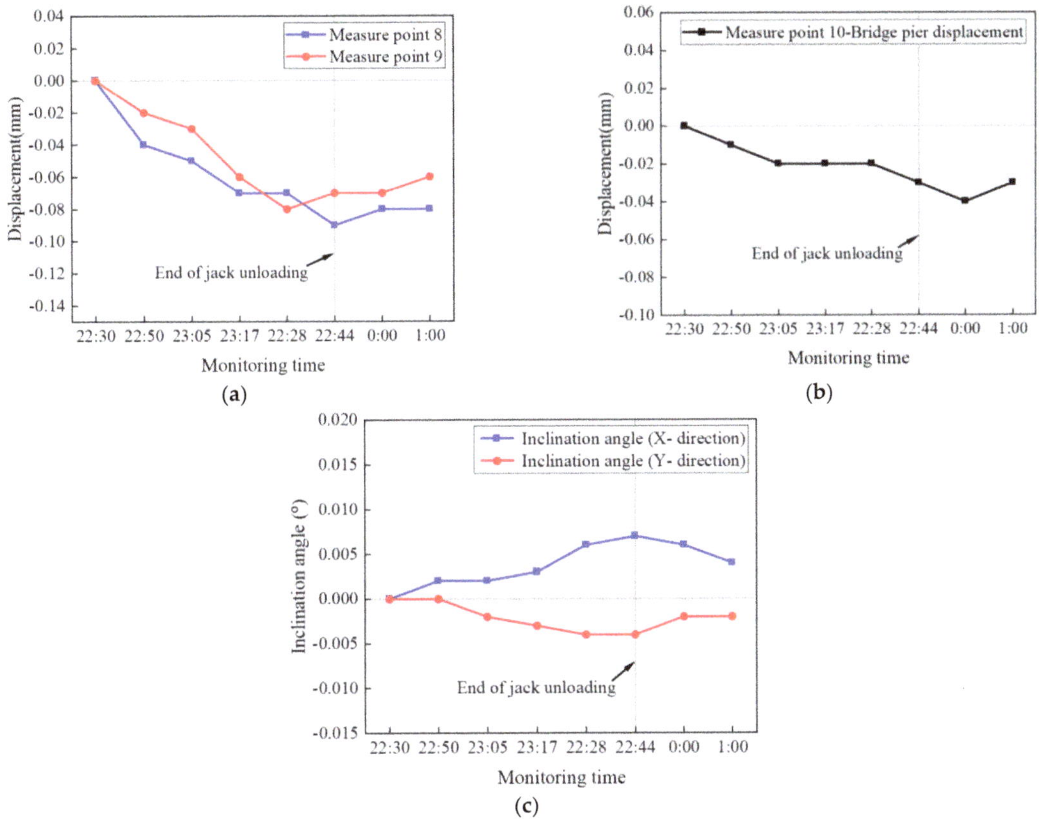

Figure 12. Displacement monitored during unloading stage: (**a**) displacement of underpinning beam; (**b**) displacement of bridge pier; and (**c**) pier inclination.

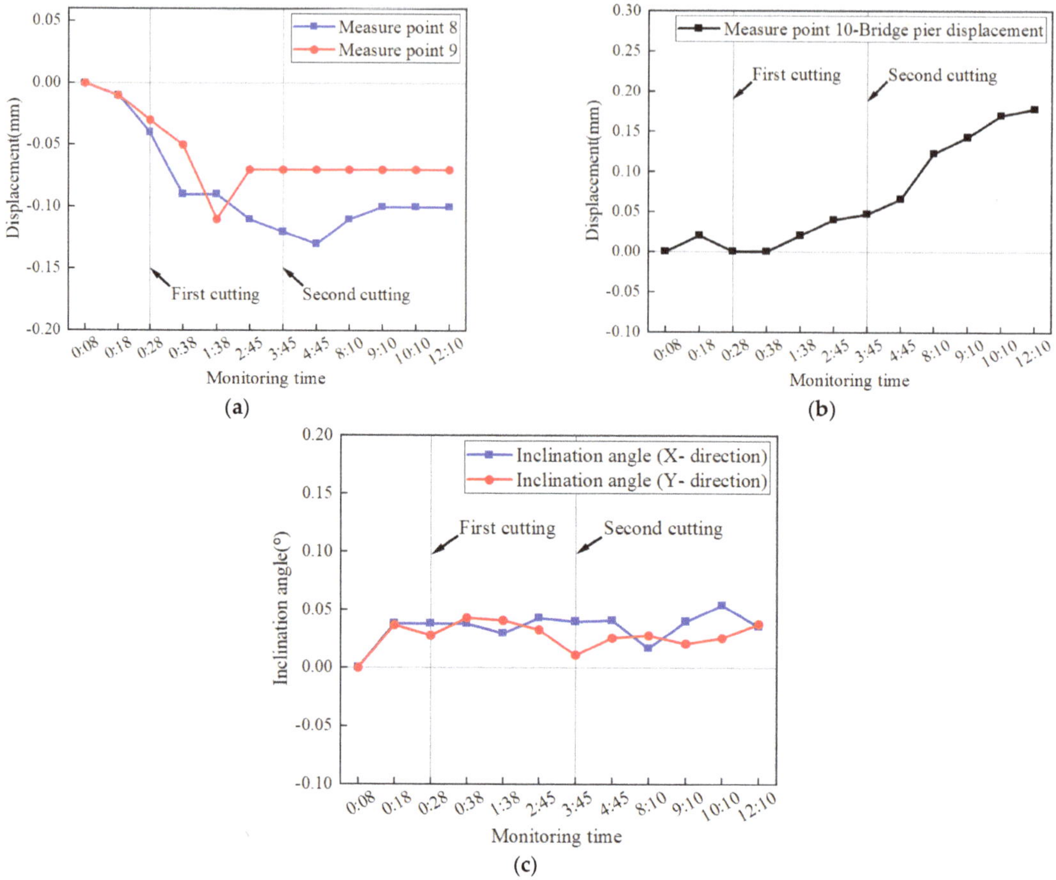

Figure 13. Displacement monitored during pile cutting stage: (**a**) displacement of underpinning beam; (**b**) displacement of bridge pier; and (**c**) pier inclination.

6. Conclusions

In this paper, the shield tunnel between Nanchang Metro No. 2 Hongguzhong Avenue Station and Yangming Park Station was taken as the engineering background, and an active gantry-type bridge pile foundation underpinning technology and its specific construction sequence were developed based on the technology comparison and analysis of in-situ engineering conditions. On this basis, for the C15 pile foundation with the most complex working conditions, ABAQUS finite element software was used to simulate the pile foundation's jacking, unloading and pile cutting process, and the displacement variation law of the bridge pier, underpinning beam and new pile during the whole construction process were analyzed. The application of this newly proposed technology was verified through the analysis of the practical construction process on-site monitoring data, and the following main conclusions were drawn:

(1) It can be seen from the changing trend of the displacement of the new piles through numerical simulation that were the same as the bridge pier and the underpinning beam, that there is a relatively stable growth of the displacement in the first five times of jacking, and after the 6th jacking, especially at the new pile near the old pile side, the settlement value in the last jacking was increased by 40% compared with the previous one. The maximum settlement value of the new pile on the side away from the old pile was 2.12 mm, while the maximum settlement value of the new pile was

3.48 mm, which reached the early warning value during monitoring, but did not reach the control value; thus, the focus should be on monitoring the settlement changes of the new piles near the old piles during the jacking process of the jack underpinning the pile foundation;

(2) When the jack was unloaded in the numerical simulation, the piers, underpinning beams and new piles had relatively stable displacement changes. In the actual monitoring data, the same displacement change was within the safe range. When the old piles were cut off, there was no sudden change in the displacement of the bridge piers, underpinning beams and new piles, which proved that the pile foundation underpinning construction scheme is reasonable and feasible. In the actual monitoring of the pile cutting stage, there was no displacement exceeding the early warning value in the two cutting-off old piles;

(3) In the monitoring of the two piles, the displacement of the bridge piers had an upward trend. This is because in the jacking stage, due to the existence of the old piles, the jacking force will generate prestress between the old piles and the bridge piers. When the old piles are cut off, the stress is gradually released, resulting in an increase in the displacement of the bridge piers. The displacement tends to be stable, and the magnitude of the change is also within a safe range. The above phenomena indicate more attention should be paid to the monitoring of the substructure of the bridge during the pile-cutting stage;

(4) According to the analysis of the monitoring results of the bridge piers, underpinning beams and new piles, the data calculated by the finite element software were roughly the same as the monitoring trend. The main structure of the lower part of the bridge did not have a large displacement, indicating that the finite element calculation results are credible.

However, it should be noted that the above conclusions are case-based findings, which are directly related to the engineering conditions. The shield tunnelling parameters and the underpinning schemes should be evaluated and optimized according to the actual engineering conditions of certain cases when the proposed active gantry bridge pile foundation underpinning technology is adopted.

Author Contributions: Y.J.: Conceptualization, funding acquisition and writing; N.W.: Data curation, formal analysis and funding acquisition; D.G.: Conceptualization and funding acquisition; F.Z.: Data curation, formal analysis and writing. C.X.: Conceptualization and funding acquisition. All authors have read and agreed to the published version of the manuscript.

Funding: The research work is supported by the National Natural Science Foundation of China (Grant No. 42267022, 51768021), the Natural Science Foundation of Jiangxi Province (No. 20202ACBL214016), and the State Key Laboratory of Performance Monitoring and Protecting of Rail Transit Infrastructure Foundation (No. HJGZ2021102).

Data Availability Statement: The general data are included in the article. Additional data are available on request.

Conflicts of Interest: The authors declare no conflict of interest.

References

1. Xiao, C.; Yang, Y.; Chi, G. Subway development and obesity: Evidence from China. *J. Transp. Health* **2021**, *21*, 101065. [CrossRef]
2. Ding, L.; Jie, X.U. A review of metro construction in China: Organization, market, cost, safety and schedule. *Front. Eng. Manag.* **2017**, *16*, 4–19. [CrossRef]
3. Wang, T.; Tan, L.; Xie, S.; Ma, B. Development and applications of common utility tunnels in China. *Tunn. Undergr. Space Technol.* **2018**, *76*, 92–106. [CrossRef]
4. Chang, X.S. Underground Tunnel Development Survey of Shield Tunnelling Machine. *Adv. Mater. Res.* **2011**, *299–300*, 1111–1114. [CrossRef]
5. Koyama, Y. Present status and technology of shield tunneling method in Japan. *Tunn. Undergr. Space Technol.* **2003**, *18*, 145–159. [CrossRef]

6. Mao, J.-H.; Yuan, D.-J.; Jin, D.-L.; Zeng, J.-F. Optimization and application of backfill grouting material for submarine tunnel. *Constr. Build. Mater.* **2020**, *265*, 120281. [CrossRef]
7. Cao, L.; Kalinski, M. Risk Analysis of Subway Shield Tunneling. In *Geo-Risk 2017*; The American Society of Civil Engineers: Reston, VA, USA, 2017; pp. 309–319.
8. Liu, C.; Li, J.; Zhang, Z.; Li, P.; Cui, J.; Liu, H.; Yang, Y. Model tests on tail-grouting process during URUP shield tunneling in soft soil. *Tunn. Undergr. Space Technol.* **2020**, *103*, 103451. [CrossRef]
9. Hajjar, M.; Hayati, A.N.; Ahmadi, M.M.; Sadrnejad, S.A. Longitudinal Settlement Profile in Shallow Tunnels in Drained Conditions. *Int. J. Geomech.* **2015**, *15*, 04014097. [CrossRef]
10. Fargnoli, V.; Boldini, D.; Amorosi, A. TBM tunnelling-induced settlements in coarse-grained soils: The case of the new Milan underground line 5. *Tunn. Undergr. Space Technol.* **2013**, *38*, 336–347. [CrossRef]
11. Loganathan, N.; Poulos, H.G.; Stewart, D.P. Centrifuge model testing of tunnelling-induced ground and pile deformations. *Géotechnique* **2000**, *50*, 283–294. [CrossRef]
12. Wu, C.; Zhu, Z. Analytical Method for Evaluating the Ground Surface Settlement Caused by Tail Void Grouting Pressure in Shield Tunnel Construction. *Adv. Civ. Eng.* **2018**, *2018*, 3729143. [CrossRef]
13. Xie, X.; Yang, Y.; Ji, M. Analysis of ground surface settlement induced by the construction of a large-diameter shield-driven tunnel in Shanghai, China. *Tunn. Undergr. Space Technol.* **2016**, *51*, 120–132. [CrossRef]
14. Li, X.-G.; Chen, X.-S. Using Grouting of Shield Tunneling to Reduce Settlements of Overlying Tunnels: Case Study in Shenzhen Metro Construction. *J. Constr. Eng. Manag.* **2012**, *138*, 574–584. [CrossRef]
15. Mirhabibi, A.; Soroush, A. Effects of surface buildings on twin tunnelling-induced ground settlements. *Tunn. Undergr. Space Technol.* **2012**, *29*, 40–51. [CrossRef]
16. Li, Z.; Chen, Z.; Wang, L.; Zeng, Z.; Gu, D. Numerical simulation and analysis of the pile underpinning technology used in shield tunnel crossings on bridge pile foundations. *Undergr. Space* **2021**, *6*, 396–408. [CrossRef]
17. Zhang, K.; Chavez Torres, J.L.; Zang, Z. Numerical Analysis of Pipelines Settlement Induced by Tunneling. *Adv. Civ. Eng.* **2019**, *2019*, 4761904. [CrossRef]
18. Lin, C.-G.; Zhang, Z.-M.; Wu, S.-M.; Yu, F. Key techniques and important issues for slurry shield under-passing embankments: A case study of Hangzhou Qiantang River Tunnel. *Tunn. Undergr. Space Technol.* **2013**, *38*, 306–325. [CrossRef]
19. Wang, N.; Jiang, Y.; Geng, D.; Huang, Z.; Ding, H. Numerical Investigation of the Combined Influence of Shield Tunneling and Pile Cutting on Underpinning Piles. *Front. Earth Sci.* **2022**, *10*, 896634. [CrossRef]
20. Zumsteg, R.; Langmaack, L. Mechanized Tunneling in Soft Soils: Choice of Excavation Mode and Application of Soil-Conditioning Additives in Glacial Deposits. *Engineering* **2017**, *3*, 863–870. [CrossRef]
21. Soomro, M.A.; Ng, C.W.W.; Liu, K.; Memon, N.A. Pile responses to side-by-side twin tunnelling in stiff clay: Effects of different tunnel depths relative to pile. *Comput. Geotech.* **2017**, *84*, 101–116. [CrossRef]
22. Xu, Q.; Zhu, H.; Ma, X.; Ma, Z.; Li, X.; Tang, Z.; Zhuo, K. A case history of shield tunnel crossing through group pile foundation of a road bridge with pile underpinning technologies in Shanghai. *Tunn. Undergr. Space Technol.* **2015**, *45*, 20–33. [CrossRef]
23. Yan, L.; Wang, G.; Chen, M.; Yue, K.; Li, Q. Experimental and Application Study on Underpinning Engineering of Bridge Pile Foundation. *Adv. Civ. Eng.* **2018**, *2018*, 5758325. [CrossRef]
24. Liu, C.; Zhang, Z.; Regueiro, R.A. Pile and pile group response to tunnelling using a large diameter slurry shield—Case study in Shanghai. *Comput. Geotech.* **2014**, *59*, 21–43. [CrossRef]
25. Stulgis, R.P.; Barry, B.E.; Francis, S.; Harvey, J. Foundation Underpinning with Mini-Piles: "A First" in Guyana, South America. In Proceedings of the GeoSupport Conference 2004, Orlando, FL, USA, 29–31 December 2004; pp. 700–711.
26. Ma, L.; Wang, J. Technology of Pile Foundation Underpinning in Shield Tunnel of Xi'an Subway. *Appl. Mech. Mater.* **2012**, *204–208*, 1445–1448. [CrossRef]
27. Yao, A.; Yang, X.; Dong, L. Numerical Analysis of the Influence of Isolation Piles in Metro Tunnel Construction of Adjacent Buildings. *Procedia Earth Planet. Sci.* **2012**, *5*, 150–154. [CrossRef]
28. Park, Y.-H.; Kim, J.-P.; Cho, K.-H. Stability analysis of subway box structure supported by modified underpinning method. *Tunn. Undergr. Space Technol.* **2015**, *50*, 199–208. [CrossRef]
29. Wang, F.; Yuan, D.; Cai, R.; Mu, Y.; Wang, M. Field Test Study on Cutting Obstacle Piles Directly by Shield Cutters. *Appl. Mech. Mater.* **2013**, *353–356*, 1433–1439. [CrossRef]
30. Peng, F.; Ma, S.; Li, M.; Fu, K. Stress Performance Evaluation of Shield Machine Cutter Head during Cutting Piles under Masonry Structures. *Adv. Civ. Eng.* **2022**, *2022*, 4111637. [CrossRef]
31. Fu, D.-M. Model Test on Concrete Cutting Directly by Shield and Pile Foundation Cutting Technology. *Tunn. Constr.* **2014**, *34*, 472–477.
32. Chen, H.-F.; Yuan, D.-J.; Wang, F.; Wang, M.-S. Study on shield cutting parameters when cutting big diameter piles. *China Civ. Eng. J.* **2016**, *49*, 103–109. (In Chinese)

33. Yuan, D.-J.; Wang, F.; Dong, C.-W.; Han, B.; Wang, M.-S. Study on New-style Cutter for Shield Cutting Large-diameter Reinforced Concrete Pile. *China J. Highw. Transp.* **2016**, *29*, 89–97.
34. Institute CHPaD. *General Code for Design of Highway Bridges and Culverts*; People's Transpotation Press; China Communications Press: Beijing, China, 2015.

buildings

MDPI

Article

Study on the Evolution Law of Internal Force and Deformation and Optimized Calculation Method for Internal Force of Cantilever Anti-Slide Pile under Trapezoidal Thrust Load

Honghua Jin [1,2], Qingyang Ren [1,2,*] and Songqiang Xiao [1,2]

[1] State Key Laboratory of Mountain Bridge and Tunnel Engineering, Chongqing Jiaotong University, Chongqing 400074, China
[2] School of Civil Engineering, Chongqing Jiaotong University, Chongqing 400074, China
* Correspondence: qyren@cqjtu.edu.cn; Tel.: +86-13983042719

Abstract: The evolution law of internal force and deformation of an anti-slide pile affects the slope stability and prevention design in a significant way. Based on the similarity theory, a test system for the bearing characteristics of a cantilever anti-slide pile was constructed, and the physical model test for the bearing characteristics of a cantilever anti-slide pile under trapezoidal thrust load was carried out. The distribution laws of internal force and deformation of a cantilever anti-slide pile were revealed, and the optimized calculation method for internal force of a cantilever anti-slide pile was proposed by taking the elastoplastic characteristics of steel bars and concrete into consideration. Furthermore, a numerical model was employed to conduct a parametric analysis of a cantilever anti-slide pile. The results show that the whole process of stress and deformation of a cantilever anti-slide pile can be classified as the uncracked stage, the cracks emerging and developing stage, and the steel bars yielding–failing stage. In the uncracked stage, the bending moment of the cantilever anti-slide pile calculated by the traditional method is smaller than that calculated by the optimized calculation method established in this paper. The traditional calculation method is no longer applicable in the stage of cracks emerging and developing. The lateral displacement and bending moment of the cantilever anti-slide pile are negatively and positively correlated with the strength of the pile material, respectively, and the influence of the deterioration of steel bars' strength on the ultimate bearing performance of the anti-slide pile is more obvious than that of the deterioration of concrete strength. The bearing capacity of the anti-slide pile could not be significantly improved by increasing the length of the anchored section when the strength of the rock stratum embedded in anchored section was large enough. As the thrust load behind the pile increased, the difference of the bearing performances of the cantilever anti-slide pile under the uniform load and trapezoidal load increased gradually. The research results can provide guidance for the evaluation of the service performance of the cantilever anti-slide pile and the slope stability.

Keywords: cantilever anti-slide pile; trapezoidal thrust load; bearing characteristics; optimized calculation method of internal force; evolution law of internal force and deformation

Citation: Jin, H.; Ren, Q.; Xiao, S. Study on the Evolution Law of Internal Force and Deformation and Optimized Calculation Method for Internal Force of Cantilever Anti-Slide Pile under Trapezoidal Thrust Load. *Buildings* **2023**, *13*, 322. https://doi.org/10.3390/buildings13020322

Academic Editor: Binsheng (Ben) Zhang

Received: 24 December 2022
Revised: 16 January 2023
Accepted: 17 January 2023
Published: 21 January 2023

1. Introduction

Due to the advantages of a flexible structure and strong resistance, anti-slide pile has become the most widely used slope prevention technology in the field of geotechnical disaster prevention and mitigation since the 1930s [1]. However, with the anti-slide pile structures gradually entering the middle or even aging stage of their service life, how to evaluate their service status in the whole life cycle has become one of the most important key issues in this field. The service performance of an anti-slide pile depends heavily on its initial design level and service conditions. However, at present, the structural design of the anti-slide pile depends more on engineering experience and the industry standard, and the relevant theoretical system is still far behind the engineering application. There are many cases of economic waste caused by an over-conservative design of anti-slide pile structures, and more and more cases of early instability and failure due to insufficient control of the structure itself and geological environment conditions. With the temporal and spatial evolution of the service environment, the working state of the anti-slide pile will always be in a dynamic process during the service period. It is particularly urgent to strengthen the monitoring of the stress and deformation of the anti-slide pile to obtain the bearing characteristics and then evaluate the preventive effect or service status.

The research methods of internal force and deformation of the anti-slide pile mainly include field monitoring or numerical simulation for outdoor prototype piles and physical model tests or numerical simulations on indoor scale piles. Zhang et al. [2] and Zhou et al. [3] used a vibrating wire reinforcement meter or strain gauges to monitor the stress state of steel bars in anti-slide piles for a long time and obtained the stress characteristics of piles. Zhang et al. [4] and Zhang et al. [5] conducted long-term dynamic monitoring on the target anti-slide pile by laying optical-fiber-sensing equipment and carried out the inverse analysis on the internal force of the pile. However, due to the lack of monitoring equipment at the initial stage of the construction of some in-service prototype piles and the complexity of on-site monitoring, the physical model test or numerical simulation on indoor scaled piles have become a very effective method to analyze the bearing performance of anti-slide piles [6]. Hu et al. [7] and Zhang et al. [8] obtained the distribution law of the bending moment of the test anti-slide pile at the initial stage of loading by monitoring the strain evolution process of the pile concrete. Wei et al. [9] and Xie [10] obtained the internal force distributions at different sections of the large test pile by welding reinforcement meters on the steel bars. Cao [11], Li et al. [12], and Li et al. [13] monitored the evolution law of internal force in the whole loading process of test piles by laying steel strain gauges. As indicated above, all studies deduced the distribution of internal force of the test pile by monitoring the stress and strain states of concrete and steel bars. However, the internal force of the test pile in the linear elastic stage was only analyzed [7,8], and the anti-slide pile was considered as a homogeneous elastomer in the whole calculation process of internal force [9–13]. In fact, the concrete and steel bars of the piles are all elastic-plastic materials, and ignoring the plastic characteristics of the concrete and steel bars will significantly affect the calculation results of the internal force. Moreover, the monitoring methods are only applicable to the stress monitoring of outdoor anti-slide piles due to the large size of reinforcement meter [9,10]. In conclusion, the calculation of internal force of the anti-slide pile in laboratory model test lacks reasonable theoretical analysis, the whole process analysis of internal force and deformation under external load has not been carried out, and the nonlinear characteristics and internal force calculation method of anti-slide piles need to be further studied.

In view of this, based on a cantilever anti-slide pile treatment project in Nan'an District of Chongqing City, China, an indoor physical model test on the bearing performance of the cantilever anti-slide pile under trapezoidal load was carried out in this study. Considering the evolution process of the stress–strain constitutive relationship of the pile materials, the optimized calculation method for internal force of the cantilever anti-slide pile was proposed, and the whole process of the stress and deformation and the nonlinear characteristics of the internal force of the pile was explored. It is expected to provide a reasonable evaluation standard for the working state of cantilever anti-slide piles during the service period, and further optimize the structural design.

2. Model Test

2.1. Engineering Background

The prototype pile of this test is derived from a cantilever anti-slide pile treatment structure located in the inner ring of Nan'an District, Chongqing, China, the upper part of the slope treated by the target anti-slide pile is covered with 0~1 m Quaternary eluvial slope gravelly soil and rock block soil, and the lower bedrock is 56~269 m purple mudstone with quartz sandstone of the Lower Jurassic Zhenzhuchong Formation (J_{1z}). The attitude of the rocks is 289°∠69°, and its strike is roughly the same as that of the slope direction, as shown in Figure 1. A total of about 50 cantilever anti-slide piles were designed for the slope protection project (only 15 are shown in Figure 1c). The sectional width and sectional height of the cantilever anti-slide pile section were 1.8 m and 2.7 m, respectively. The pile length was 16.5 m, of which the loaded section and the anchored section were 10.5 m and 6.0 m, respectively, and the pile spacing was 6.0 m. C30 concrete was adopted for the anti-slide pile, and HRB400 steel bars were used for the steel bars of the pile. The tensile steel bars at the rear side of the pile were 42 C(Φ) 32, and the reinforcement ratio was 0.69%. The structural steel bars at the front side and the steel bars at both sides of the pile were 7 C(Φ) 25, and the stirrup was 1 C(Φ) 14 with a spacing of 100 mm, as described in Table 1.

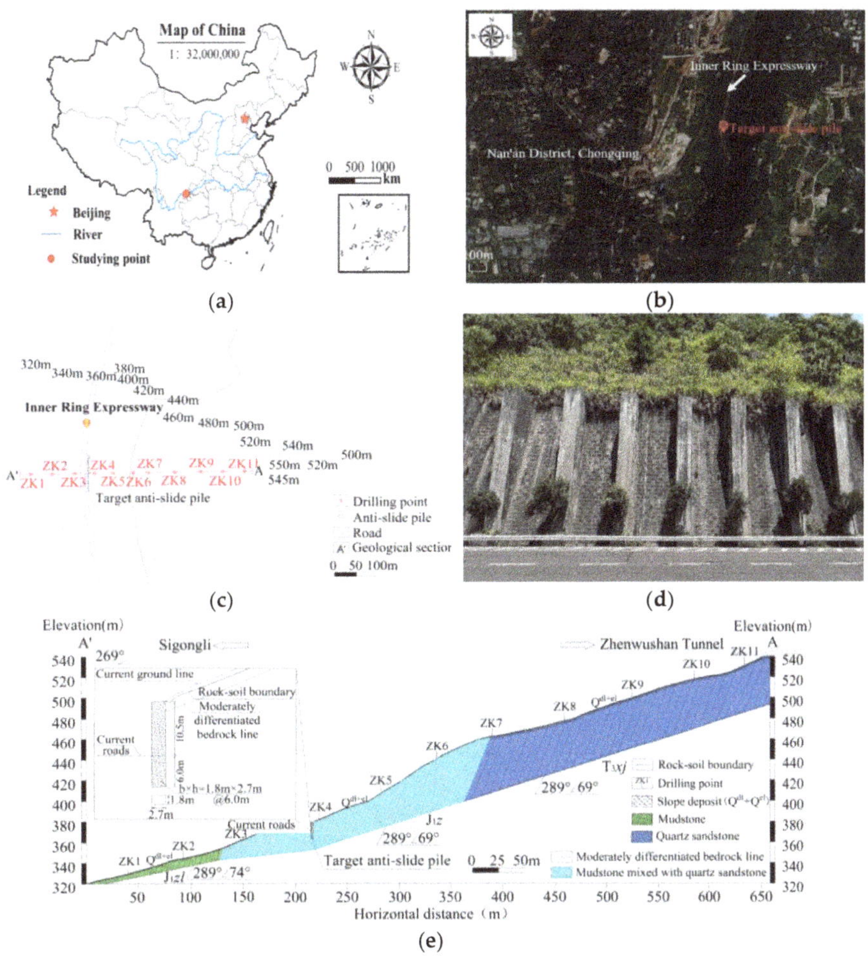

Figure 1. Basic information of the target anti-slide pile. (**a**) Location; (**b**) geographic location; (**c**) topographical map; (**d**) realistic picture; (**e**) engineering geological profile.

Table 1. Basic parameters of the actual cantilever anti-slide pile.

Sectional Width (m)	Sectional Height (m)	Length (m)	Length of Loaded Section (m)	Length of Anchored Section (m)	Spacing between Piles (m)	Concrete	Steel Bars	Ratio of Tensile Steel Bars (%)
1.8	2.7	16.5	10.5	6.0	6.0	C30	HRB400	0.69

2.2. Experimental Design

2.2.1. Determination of Model Test Similarity

Relying on the above slope as the prototype and considering the conditions of the indoor model field, the model test on the bearing performance of the cantilever anti-slide pile under lateral load was conducted based on the similarity theory. The geometric dimension and the elastic modulus of the anti-slide pile were selected as the control quantities. The similarity relation of geometric dimensions (C_l), elastic modulus (C_E), and the steel ratio (C_ρ) of the test pile were selected as the basic similarity ratio, and C_l, C_E, C_ρ were defined as 15, 1, and 1, respectively. The dimensional analysis method was used to determine the similarity of other physical quantities, as listed in Table 2.

Table 2. Similarity ratio of model test.

Parameter Type	Physical Quantities	Similarities Relationship	Similarity Constants
Geometric parameters	Length (L)	C_l	15
	Displacement (y)	$C_y = C_l \cdot C_\varepsilon$	15
	Area (A)	$C_A = {C_l}^2$	225
Material parameters	Strain (ε)	C_ε	1
	Modulus of elasticity (E)	C_E	1
	Stress (σ)	$C_\sigma = C_E$	1
	Poisson's ratio (μ)	C_μ	1
	Reinforcement rate (ρ)	C_ρ	1

2.2.2. Geometric Dimensions and Material Properties of the Test Pile

According to the similarity constants of individual physical quantities in Table 2, the dimensions and reinforcement information of the test pile are presented in Figure 2 and Table 3. The concrete grade of the test pile was determined to be C30, and the thickness of concrete cover was 20 mm. The pile length (L) is 1.1 m, in which the loaded section and the anchored section were 70 cm and 40 cm, respectively. The sectional width (b) and the sectional height (h) were 10 cm and 15 cm, respectively. The tensile steel bars at the rear side of the test pile were composed of two steel bars with a diameter of 8 mm (N1), and the steel ratio (ρ) was 0.67%. The structural steel bars at the front side of the test pile were composed of two steel bars with a diameter of 6 mm (N2), and the stirrup was composed of one steel bar with a diameter of 6 mm and a spacing of 106 mm (N3).

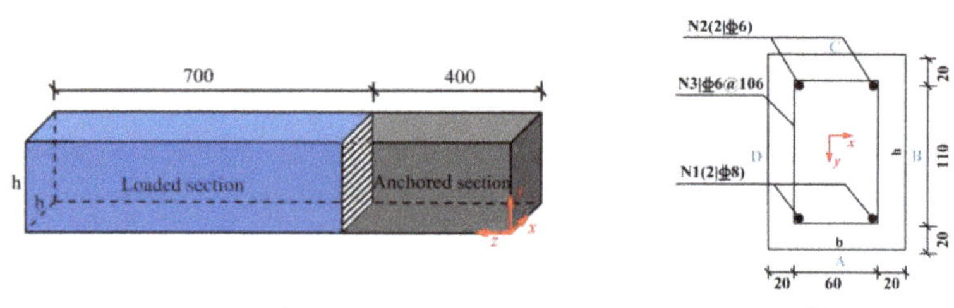

(a) (b)

Figure 2. Diagrammatic sketch of the dimensions and section reinforcement of the test piles (unit: mm). (**a**) Dimensions; (**b**) section reinforcement.

Table 3. Parameters of the dimensions and reinforcement of the test pile.

Sectional Width (cm)	Sectional Height (cm)	Length (cm)	Length of Loaded Section (cm)	Length of Anchored Section (cm)	Thickness of Concrete Cover (cm)	Concrete	Steel Bars	Ratio of Tensile Steel Bars (%)
1.8	2.7	16.5	10.5	6.0	6.0	C30	HRB400	0.69

2.2.3. Distribution Form of Thrust Load of the Test Pile

(1) Selection of the distribution form of the thrust load

The distribution form of the landslide thrust behind the cantilever anti-slide pile is complex, and it does not change regularly along the pile depth. According to the Chinese specification (GB/T 38509-2020) [14], the distribution of landslide thrust in the loaded section of anti-slide pile can be divided into three forms: triangle, trapezoid, and rectangle, and the specific distribution form should be determined according to the nature and geometric characteristics of the sliding mass. The determination of the actual distribution form of landslide thrust is mostly based on an indoor model test, field test, and numerical simulation. Liu et al. [15] established a function model of the thrust distribution form based on the analysis of field measured data at home and abroad, indicating that the thrust distribution form of a sand and clay sliding mass should be considered as triangular or parabolic, and the distribution form between them should be trapezoidal. In the first section of this article, the rock slope behind the target anti-slide pile was covered with 0~1 m eluvial soil layer and 2~3 m strongly weathered rock layer. Therefore, it is more consistent with the actual working conditions to assume that the thrust distribution form behind the actual anti-slide pile is trapezoidal, as illustrated in Figure 3. The upper load and lower load of trapezoidal load are q_1 and q_2, respectively, and the length of the loaded section is l.

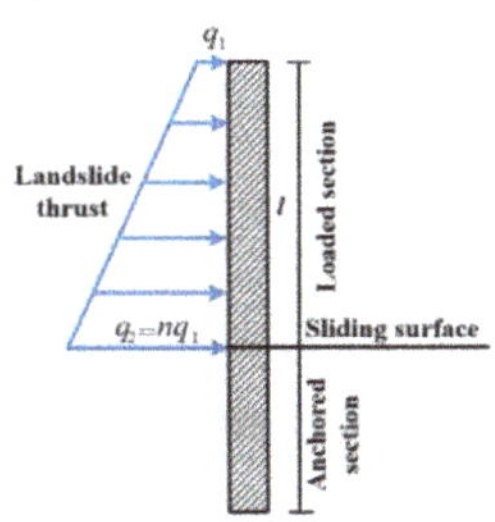

Figure 3. Distribution of trapezoidal thrust load behind an anti-slide pile.

(2) Application mode of thrust

The distribution form of the trapezoidal thrust load behind the anti-slide pile mentioned above is a theoretical assumption, and the idealized trapezoidal thrust load should be simplified as a concentrated force in the indoor test. According to the mechanical characteristics of a cantilever anti-slide pile, it is regarded as a cantilever beam subjected to trapezoidal load. Based on the principle of equivalence (the same load location, the same total load value, and the economic and reasonable equivalent error), the trapezoidal load of an anti-slide pile is equivalently replaced by the concentrated force. In the process of replacement, the accuracy of the result of load equivalence is directly determined by the number, value, and position of the concentrated force. The existing research on the bearing performance of the anti-slide pile mainly concentrates on the bending moment and deformation, where the maximum bending moment of the pile and the maximum deflection at pile top are the main criteria. Therefore, the cantilever beam structure was taken as an example in this paper, and the maximum bending moment and maximum deflection of the cantilever beam were taken as the indicators to discuss the optimum equivalent scheme, so as to finally obtain the distribution form, number, location, and value of the corresponding concentrated load. The specific equivalent calculation method is as follows.

(1) According to the distribution of trapezoidal load in Figure 3, the proportion of the upper load and lower load of the trapezoidal load (n) were set as 1:3, 1:4, 1:5, and 1:6, respectively (i.e., if $q_1 = q$, $q_2 = nq$, then n = 3, 4, 5, 6).
(2) According to the trapezoidal load distribution form in step (1), the centroid position (y_c) of different trapezoidal loads was determined by Equation (1), which is the resultant point of the landslide thrust. Then, the total shear force (F) and the maximum bending moment (M) under different trapezoidal loads were calculated, as shown in Equations (2) and (3).
(3) The arrangement scheme of concentrated force in the process of the loading test was proposed as follows. (I) Assuming the number of concentrated forces was m, three layout schemes (m = 3, 4, 5) were calculated in this paper. (II) Assuming i was the serial number of the concentrated force (i = 1, 2, m, numbered from top to bottom), then F_i is the size of the i-th concentrated force. The ratio of i:1 was determined to distribute F_i to equal the total shear force (F) of the trapezoidal load. Taking m = 4 as an example, then $F_2 = 2F_1$, $F_3 = 3F_1$, $F_4 = 4F_1$, and $F_1 + F_2 + F_3 + F_4 = F$, and the others are similar. (III) Taking m = 3 as an example in the process of solving the concentrated force position, the distance from F_1 to the free end and the distance from F_3 to the fixed end should be made equal firstly and made to be d_1. Then, the distance between F_2 and F_1 and between F_2 and F_3 should be guaranteed to be equal and made to be d_2, where $2(d_1 + d_2) = l$. Finally, the location of each concentrated force (x_i) was calculated through the principle of $M = M_m$, and the others are similar. (IV) The maximum deflections of the cantilever beam under different trapezoidal loads (ω) and different concentrated forces (ω_m) were calculated by using Equations (5) and (6), respectively, for deflection comparison. The final calculation results are listed in Table 4.

Table 4. Calculation results of the bending moment and deflection under trapezoidal load and concentrated load.

q:nq	F	y_c	M	ω/ω_0	m = 3		m = 4		m = 5	
					ω_3/ω_0	D_3	ω_4/ω_0	D_4	ω_5/ω_0	D_5
1:3	$2ql$	$5l/12$	$5ql^2/6$	0.192	0.168	12.14%	0.165	14.03%	0.163	14.99%
1:4	$5ql/2$	$2l/5$	ql^2	0.225	0.208	7.59%	0.201	10.52%	0.198	12.01%
1:5	$3ql$	$7l/18$	$7ql^2/6$	0.258	0.249	3.52%	0.239	7.32%	0.234	9.26%
1:6	$7ql/2$	$8l/21$	$4ql^2/3$	0.292	0.292	0.02%	0.278	4.55%	0.272	6.86%

Note: $\omega_0 = ql^4/(EI)$.

It can be seen from Table 4 that the greater the proportion of the upper load and lower load of the trapezoidal load (n), and the smaller the number of concentrated forces (m), the smaller the error of equivalent schemes (D), and D_3 is the smallest when n = 6 and m = 3. Therefore, three concentrated forces are finally loaded in this study, where F_1, F_2, and F_3 are located at 0.86l, 0.50l, and 0.14l of the loaded section of the test pile, respectively, and $F_1 = F/6$, $F_2 = F/3$, and $F_3 = F/2$, respectively.

$$y_c = \frac{n+2}{3(n+1)} l \tag{1}$$

$$F = \frac{n+1}{2} ql \tag{2}$$

$$M = \frac{n+2}{6} ql^2 \tag{3}$$

$$Mm = \sum_{i=1}^{m} F_i * x_i \tag{4}$$

$$\omega = \frac{ql^4}{EI}(\frac{n-1}{30} + \frac{1}{8}) \quad (5)$$

$$\omega_m = \sum_{i=1}^{m} \frac{F_i x_i^2}{6EI}(3l - x_i) \quad (6)$$

where q is the size of the upper load, and E and I are the elastic modulus and inertia moment of the cantilever beam, respectively.

2.3. Raw Materials

Ordinary Portland cement, sand, gravel, tap water and admixture were used to prepare concrete. Ordinary Portland cement with the strength grade 42.5 (P.O 42.5) was used to prepare concrete with the strength grade of C30, and its parameters met the requirements of Chinese specification (GB 175-2020) [16]. The fine aggregate was the ordinary river sand produced locally in Chongqing, and the fineness modulus was 2.78. The coarse aggregate was continuous graded limestone gravels, and the particle size range was 5~20 mm. The water was tap water, and GJ-1 superplasticizer with a water reduction of 22% was selected as the additive. The mix proportion of the pile concrete was calculated according to the Chinese specification (JGJ 55-2011) [17], and is listed in Table 5. The measured performances of concrete and steel bars are displayed in Tables 6 and 7, respectively.

Table 5. Mix Proportions.

Concrete Strength Grade	Water-Binder Ratio	Cement (kg/m^3)	Water (kg/m^3)	Gravel (kg/m^3)	Sand (kg/m^3)	SP (kg/m^3)
C30	0.57	244 (1)	139 (0.57)	1291 (5.29)	726 (2.98)	2.44

Table 6. Measured performances of concrete.

Grade	Density (kg/m^3)	Cube Compressive Strength (MPa)	Axial Compressive Strength (MPa)	Axial Tensile Strength (MPa)	Elastic Modulus (GPa)
C30	2400	35.3	23.6	2.24	31.6

Table 7. Measured performances of steel bars.

Type	Grade	Diameter (mm)	Density (kg/m^3)	Yield Strength (MPa)	Tensile Strength (MPa)	Elastic Modulus (GPa)	Poisson's Ratio	Elongation (%)
N1	HRB400	8	7850	432	558	201	0.3	19.4
N2, N3		6		411	526	198		21.3

2.4. Device System for the Model Test

In order to simulate the cantilever anti-slide pile subjected to the trapezoidal load, a test system for the bearing performance of the cantilever anti-slide pile under the trapezoidal load was independently developed and fabricated, as shown in Figures 4 and 5. The system is mainly composed of the test pile fixing system, thrust loading system, and data acquisition system. These systems are introduced as follows.

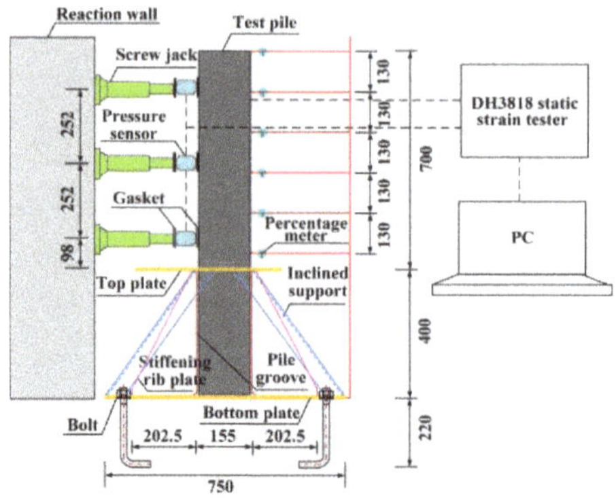

Figure 4. Diagrammatic sketch of test device (unit: mm).

Figure 5. Load test on the test pile.

2.4.1. Fixing System

The test pile fixing system was mainly constructed by connecting the high-strength bolt embedded in the rigid foundation (the rigid foundation was mainly made of in-situ cast concrete and steel bars) with the steel support, which was used to fix the test pile to simulate the anchored section of the actual cantilever anti-slide pile. The steel support was welded by channel steel, angle steel, and steel plate, and its grade was Q235. The fixed part of the test pile was the pile groove formed by channel steel. Considering the dimensional error in the pouring process of the test piles, the size of the pile groove should be slightly larger than the section size of the test piles to facilitate the placement of the test piles, the size of the pile groove was designed as 105 mm × 155 mm, and the thickness of the pile groove was 10 mm. The stiffening rib plate was welded on the side of pile groove and the inclined supports (6 mm thick, mainly made of angle steel) were set at the four corners of the bottom plate to stabilize the pile groove. A rectangular working platform (top plate) was installed on the top of the pile groove, and its dimensions and thickness were 455 mm × 305 mm and 6 mm, respectively. The dimensions of the bottom plate were 750 mm × 500 mm, and its thickness was 10 mm. In order to ensure the stability of the steel support, the pre-embedded bolt component was set at the front and rear sides of the bottom plate of the steel support, which were anchored with the rigid foundation, and then the pile groove transmitted the thrust load to the pre-embedded bolt through the stiffened rib plate and inclined support to obtain the reaction force. The pre-embedded bolts adopted M20 bolts (matched with flat washer and nut), and the embedded length was about 230 mm. The three-dimensional concept diagram of the test pile fixing system is illustrated in Figure 6.

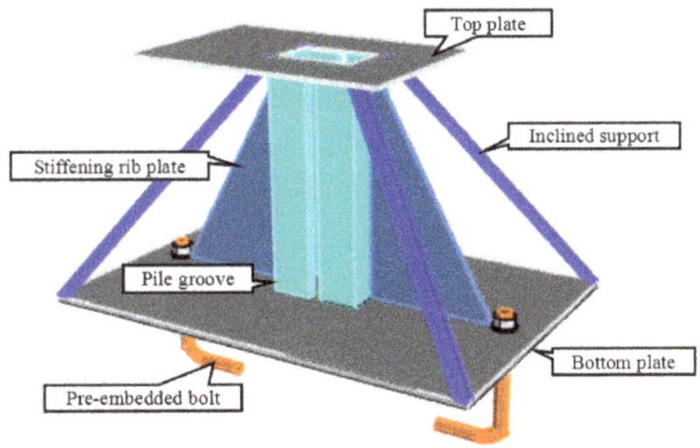

Figure 6. Three-dimensional model diagram of the fixing system of test pile.

2.4.2. Thrust Loading System

The thrust loading system was mainly composed of reaction wall, screw jack, and pressure sensor. The basic principle is that the trapezoidal load is applied by three jacks and reacting on the reaction wall and transferred to the test pile through the pressure sensor, as shown in Figure 5. The jacks were all screw jacks with a maximum lifting capacity of 10 t produced by Shanghai Baoshan Hugong Group, and were arranged at 9.8 cm, 35 cm, and 60.2 cm of the loaded section, as illustrated in Figure 7. The pressure sensor only played the role of load transfer in the thrust loading system, and its parameters are introduced in the following data acquisition system.

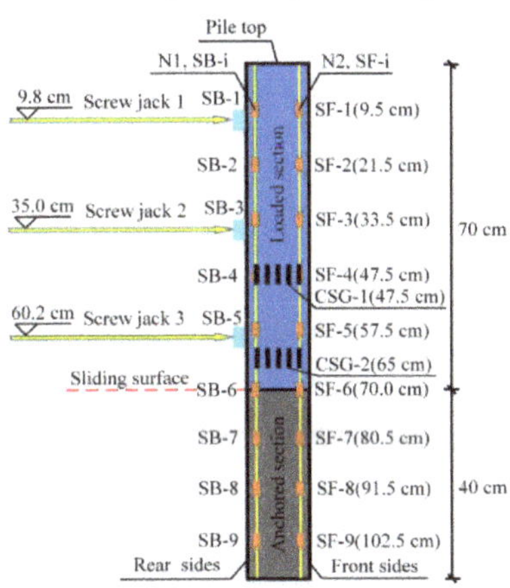

Figure 7. Layout diagram of screw jacks and steel strain gauges.

The monotonic and graded method was applied in the loading process of the trapezoidal load, and the single-stage loading process mainly includes the loading stage and the holding stage. The loading magnitude (F_l), the loading step length (ΔF), the loading rate (V_l), the loading time (t_l) and the holding time (t_h) should be controlled during the loading process. He et al. [18], Zhou et al. [19] and Chen et al. [20] adopted the same loading method in the loading test of test pile, which kept ΔF, V_l, t_l and t_h constant in the whole loading process. This method is only applicable to a type of test where the test pile does not produce cracks during the whole loading process. Therefore, the bearing characteristics experiments of the test pile were carried out according to Chinese specification (GB/T 50152-2012) [21].

(1) Loading magnitude (F_l)

Before the formal loading of the test pile, it was necessary to conduct a pre-loading experiment on the test pile to examine the stability of the loading system and check whether the various original components can operate normally, and then the formal loading of the test pile could only be carried out after the debugging. According to Chinese specification (GB/T 50152-2012) [21], the load size during the pre-loading was about 10% of the estimated ultimate load ($0.1F_u$, F_u = 60 kN), and it was loaded three times in stages, and 2 kN in a single stage. The loading experiment was loaded from 0 kN until the test pile was damaged after the pre-loading.

(2) Loading step length (ΔF)

The loading step length (ΔF) in the loading process of the test pile should be determined by the loading magnitude (F_l), ultimate bearing capacity ($F_{\max}$), and loading rate (V_l). The loading step lengths of each level before and after the cracking of the test pile should be about 5% F_u and 10% F_u, respectively. When the load reached 90% F_u, the loading step length of each stage was loaded at 5% F_u until the test pile was damaged. Therefore, before the test pile cracked, the loading step length of the jack 1, jack 2, and jack 3 (ΔF_1, ΔF_2, ΔF_3) in the loaded section were designed to be 0.5 kN, 1.0 kN, and 1.5 kN, and increased to 1.0 kN ($\Delta F_1{}'$), 2.0 kN ($\Delta F_2{}'$), and 3.0 kN ($\Delta F_3{}'$) after the cracking of the test pile. When the load reached 90%F_u, ΔF_1, ΔF_2, and ΔF_3 were designed to be 0.5 kN, 1.0 kN, and 1.5 kN until the end of the test.

(3) Loading rate (V_l), loading time (t_l), and holding time (t_h)

Referring to the loading method of indoor concrete members [22,23], it was found that the reasonable loading rate (V_l) in the loading process is 10~30 N/s. In order to avoid possible adverse effects caused by excessive loading rate, the loading rates of jack 1, jack 2, and jack 3 proposed in this paper are 10 N/s, 20 N/s, and 30 N/s, respectively. The loading time (t_l) is determined by the loading step length (ΔF) and the loading rate (V_l), and the single-stage loading times before and after cracking was determined to be 50 s and 100 s, respectively. When the load reached 90% F_u, the single-stage loading time was 50 s. After the thrust load was applied, the loading state was maintained for 100 s (t_h) to ensure that the applied trapezoidal load was fully transferred to the test pile, the data of all monitoring elements were stable, and then the next loading step was performed. The loading method is illustrated in Figure 8.

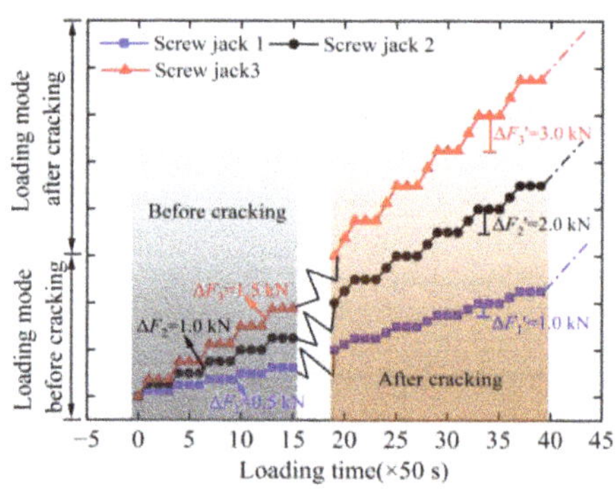

Figure 8. Loading design of the test pile.

2.4.3. Data Acquisition System

The data acquisition system mainly contains the percentage meter, strain gauge, pressure sensor, DH3818 static strain tester, and a laptop computer equipped with DHDAS dynamic signal acquisition system. The percentage indicator was applied to obtain the lateral displacement of the test pile, and a total of six percentage indicators (DIs) were arranged along the loaded section of the test pile in this experiment, which were 0 cm, 13 cm, 26 cm, 39 cm, 52 cm, and 65 cm from the pile top, respectively, as illustrated in Figure 9. The strain gauges on the steel bars were applied to measure the steel bars' strains of the test pile and the type was BE120-3AA. It was arranged in two rows of 18 places on the tensile steel bars and the structural steel bars of the test pile, which were 9.5 cm, 21.5 cm, 33.5 cm, 47.5 cm, 57.5 cm, 70.0 cm, 80.5 cm, 91.5 cm, and 102.5 cm from the pile top, respectively. The concrete strain gauges were applied to measure the concrete strains of the test pile and the type was BX120-50AA. It was arranged in two rows of 10 places on Side B of the test pile (CSG-1,CSG-2), which were 47.5 cm and 65.0 cm from the pile top, respectively (Figure 7). The pressure sensors were arranged between the screw jacks and the test pile to obtain the applied loads. The type was CZLYB-1A and the measuring range was 50 kN.

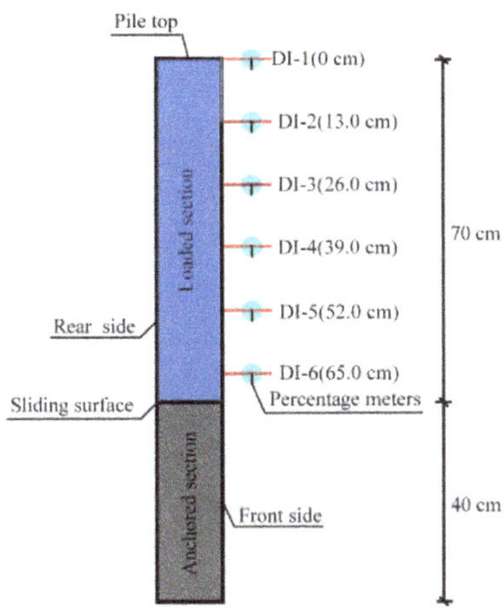

Figure 9. Arrangement of percentage meters.

2.5. Experimental Procedures

Figure 10 shows the main steps of this test. In Step 1, the steel strain gauges were posted according to the steel strain gauge arrangement scheme shown in Figure 7. Then the tension steel, structural steel, and the hoop steel were tied into a steel cage according to Figure 2b. The mold was made with wood boards according to the pile size and finally the pile concrete was poured and cured for 28 days. Step 2 was to set up the experimental system, including fixing the steel support, placing the test pile, laying screw jacks and pressure sensors, and connecting strain gauges and strain tester. Step 3 was to load the test pile with trapezoidal load by referring to the loading method in Section 2.4.2. Step 4 was to collect experimental data, including the pile displacement, residual bearing capacity, steel strain and crack development. Step 5 was to dismantle the experimental system and end the experiment.

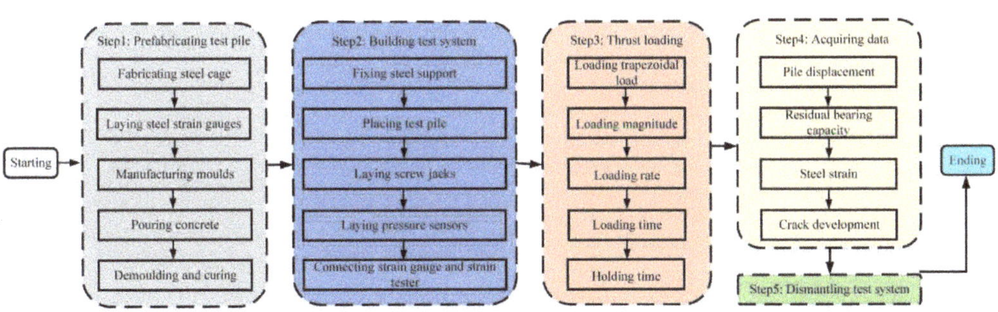

Figure 10. Experimental procedures.

3. Results and Analysis

3.1. The Whole Measuring Process of Stress and Deformation of the Anti-Slide Pile

Figure 11 illustrates the lateral displacement distribution of the test pile under different trapezoidal load levels, and the ordinate in Figure 12 represents the resultant force (F_T)

of the concentrated forces F_1, F_2, F_3. The relationship curve between the pile top lateral displacement and the resultant force of trapezoidal load was extracted in Figure 12. It can be seen that the pile lateral displacement increased linearly along the pile length as trapezoidal load increases, and the deformation process of the cantilever anti-slide pile included three stages. In Stage I, the pile lateral displacement increased approximately proportionally with the resultant force of the trapezoidal thrust load (F_T), but the rate of increase was relatively slow. In this stage, the cantilever anti-slide pile worked in full section, and the concrete and the steel bars in the tension area bore the tensile force together, both of which were in an elastic stage. When F_T increased to 10.2 kN, the maximum lateral displacement at pile top was 1.75 mm, and the tensile area near the sliding surface of side A of the test pile first cracked (Figures 12 and 13a), which was named as the uncracked stage of the cantilever anti-slide pile. In Stage II, due to the concrete cracking in the tension area, the tensile force originally born by the concrete was gradually transferred to the tensile steel bars, and the cracks on side A, side B, and side D of the test pile gradually expanded and extended (Figures 12 and 13b,d). As the trapezoidal load increased, the tensile force at the cracking area was basically born by the tensile steel bars. In this stage, the lateral displacement of the pile increases more obviously due to the presence of cracks, and the lateral displacement at pile top increases in a non-proportional manner with F_T. When F_T increased to 36.6 kN, the maximum lateral displacement at pile top was 17.5 mm, and the tensile steel bars began to yield. Since then, the maximum lateral displacement at the pile top increased sharply, which was named as the crack emerging and developing stage of cantilever anti-slide pile. In Stage III, when F_T gradually increased to 40.2 kN, the concrete strain in the compression area reached the ultimate compressive strain, and the concrete in the compression area crushed and bulged (Figures 12 and 13c). The trapezoidal thrust load gradually rebounded, the maximum lateral displacement at pile top was 63.7 mm, and the anti-slide pile subsequently failed (Figures 12 and 13e), which was named as the steel bar yielding-failing stage of cantilever anti-slide pile.

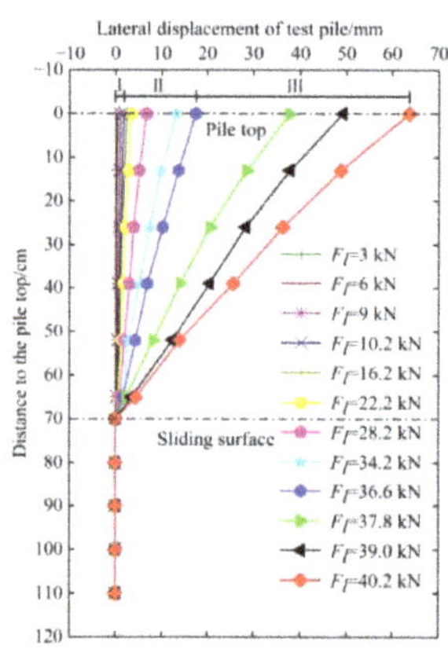

Figure 11. Displacement distributions along the test pile.

Further investigation on the cracks and failure characteristics of the test pile indicated that the initial crack was located 3 cm above the sliding surface of side A. The crack was a transverse crack parallel to the pile width, and its width was about 0.1 mm. When the anti-slide pile entered Stage II, inclined cracks began to appear on side B and side D of the test pile and extended to the compression area (Figure 13b,d). At the same time, the number and width of transverse cracks on side A of the test pile gradually increased, the number of cracks reached 5 and the maximum transverse crack width was about 8 mm (Figure 13a). When the trapezoidal load increased to the ultimate bearing capacity of the test pile, the inclined cracks on side B and side D of the test pile gradually penetrated into the compression area, and the concrete of side C of the test pile showed compression-tension cracks and bulged outward (Figure 13c), resulting in bending failure (Figure 13e). The above phenomena indicate that the cantilever anti-slide pile under trapezoidal load sustains typical plastic failure and ductile failure characteristics, and belongs to under-reinforced beam damage. The cracking load and yield load of the test pile were 25.37% and 91.04% of the failure load, and the cracks were distributed within about 10 cm (0.09 times the pile length) above and below the sliding surface.

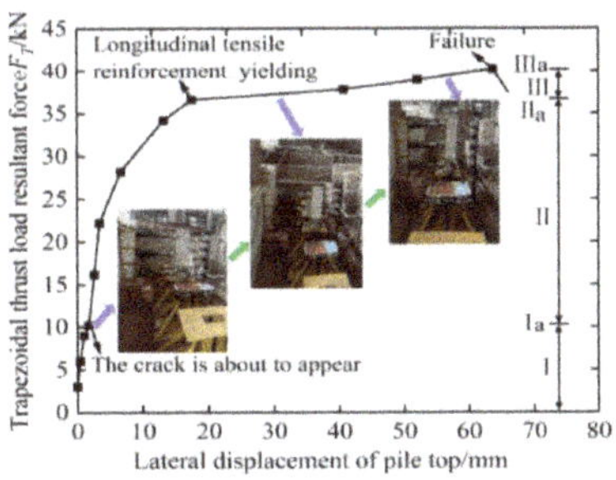

Figure 12. Variation curve of the thrust load with the pile top displacement.

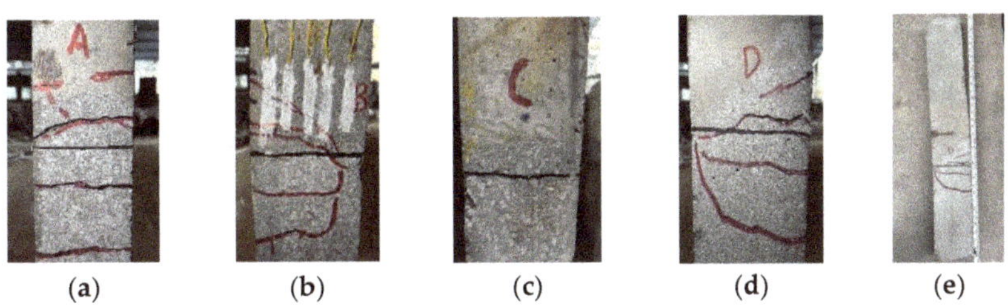

Figure 13. Crack distributions along the test pile. (**a**) Side A; (**b**) Side B; (**c**) Side C; (**d**) Side D; (**e**) Pile.

Figure 14 illustrates the strain distributions along different heights of concrete in side B of test pile under various loads. Due to the wide crack after the cracking of the anti-slide pile section, some strains on the CSG-2 row could not be accurately collected. The conclusion could be drawn that the concrete strains at different heights of two sections of the test pile were basically proportional to the distances from the points to the neutral axis when the load was fixed, and the concrete strains were linear, which indicated that the plane section assumption was available.

3.2. Calculation Method of Internal Force of Anti-Slide Pile

At present, the internal force calculation of the cantilever anti-slide pile generally adopts the traditional calculation method. This section first introduces its calculation principle, and then proposes the optimized calculation method by analyzing its shortcomings.

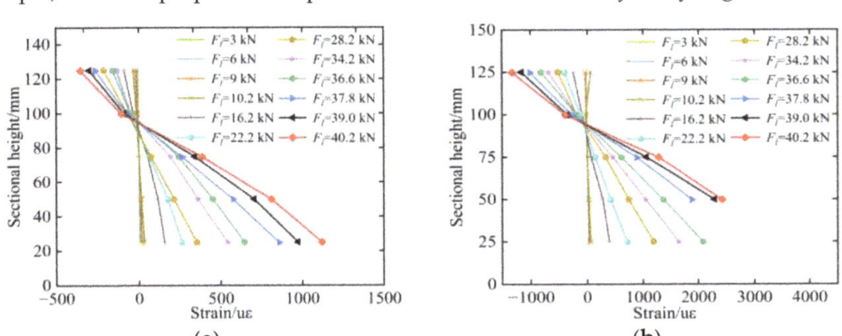

Figure 14. Section strain distribution along the test pile. (**a**) CSG-1; (**b**) CSG-2.

3.2.1. Traditional Calculation Method of Internal Force

The traditional method for calculating the bending moment of the cantilever anti-slide pile was mostly based on the Euler–Bernoulli beam theory, which assumed that the elastic assumption and plane section assumption were satisfied in the whole loading process of the cantilever anti-slide pile [24], and the bending moment $M(y)$ can be obtained according to the following Equations (7)–(11).

$$\sigma(x) = \frac{M}{I_x} d = E\varepsilon \tag{7}$$

$$M(y) = \frac{E\Delta\varepsilon}{d} I_x \tag{8}$$

$$d = h - (a_s + a'_s) \tag{9}$$

$$I_x = \frac{bh^3}{12} \tag{10}$$

$$\Delta\varepsilon = \varepsilon_s - \varepsilon'_s \tag{11}$$

where $\sigma(x)$ is the section stress of the pile; I_x is the inertia moment of pile section; b and h are the pile sectional width and sectional height, respectively; d is the distance between the tensile steel bars and structural steel bars (i.e., N1 and N2 reinforcement); a_s and a'_s are the thickness of concrete cover of reinforcement N1 and N2, respectively; E is the overall elastic modulus of the pile; ε_1 and ε_2 are the strains of reinforcements N1 and N2, respectively.

It can be found that the traditional method always regards the cantilever anti-slide pile as an elastic body in the calculation of the bending moment. However, in the indoor experiment of Section 3.1, the concrete or steel bar exhibited plastic deformation characteristics when the pile reached Stage II and Stage III, and the pile also presented plastic failure characteristics when destroyed. Therefore, the traditional method is not reasonable when the pile reaches Stage II and Stage III. It is imperative to analyze the stress and deformation characteristics of each stage of the pile separately and then establish a reasonable calculation method for the bending moment.

3.2.2. Optimized Calculation Method of Internal Force

(1) Stage I

When the trapezoidal thrust load behind the cantilever anti-slide pile is small, the concrete and steel bars in the tension area bear the tensile force together, and both are in an elastic state. The concrete strain at the edge of the tensile area of the pile is less than the concrete's ultimate tensile strain, and there are no cracks on the pile. Based on the assumption of elastic body, the cross-sectional stress and strain state in the tensile area of the pile at this stage meets Equation (12), and Figure 15 shows the stress and strain distributions of the pile section.

$$\begin{aligned} \varepsilon_s = \varepsilon_t \leq \varepsilon_{tu} \\ \sigma_s \leq \varepsilon_{tu} E_s \end{aligned} \tag{12}$$

where ε_{tu} is 0.0001; and σ_s and E_s are the stress and elastic modulus of the reinforcement N1, respectively.

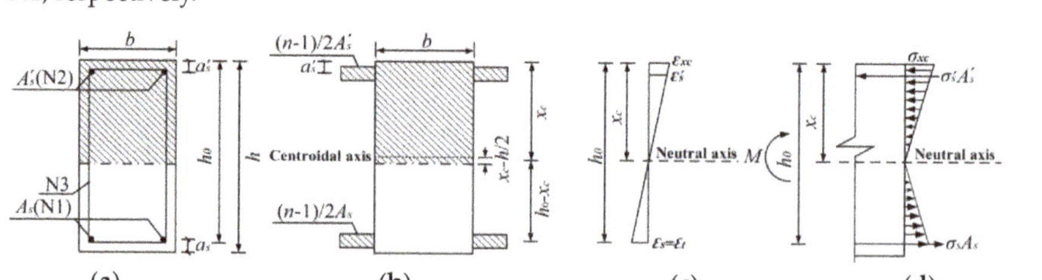

Figure 15. Stress and strain diagrams of the pile normal section in stage I. (**a**) Pile section; (**b**) transformed section; (**c**) distribution of average strain; (**d**) distribution of normal stress.

The height of the compression area (x_c) is determined using the plane section assumption, and it is given as follows:

$$\frac{x_c - a'_s}{\varepsilon'_s} = \frac{h_0 - a'_s}{\varepsilon_s + \varepsilon'_s} \tag{13}$$

$$x_c = \frac{\varepsilon'_s}{\varepsilon_s + \varepsilon'_s} h_0 + \frac{\varepsilon_s}{\varepsilon_s + \varepsilon'_s} a'_s \tag{14}$$

where h_0 is the effective height of the pile section, $h_0 = h - a_s$.

Making $\varepsilon'_s / \varepsilon_s = \zeta$ and simplifying Equation (14), then x_c is:

$$x_c = \frac{\zeta}{1+\zeta} h_0 + \frac{1}{1+\zeta} a'_s \tag{15}$$

In Section 3.2.1, the contribution of reinforcements to the inertia moment of the pile (I_c) section was ignored in the traditional calculation method for pile bending moment. In fact, as a composite material, the pile inertia moment (I_c) is larger than that of plain rectangular concrete beam. Therefore, based on the conversion section method, the reinforcements N1 and N2 with the cross-sectional areas of A_s and A'_s were converted into concrete with the areas of nA_s and nA'_s at the same position respectively. Then, I_c is:

$$I_c = (n-1)A_s(h_0 - x_c)^2 + (n-1)A'_s(x_c - a'_s)^2 + \frac{bh^3}{12} + bh(x_c - h/2)^2 \tag{16}$$

where $n = E_s/E_c$, and E_c is the elastic modulus of pile concrete.

At this time, the tensile stress of the concrete at the same level of the reinforcement N1 (σ_t) is:

$$\sigma_t = E_c \varepsilon_s = \sigma_s \frac{E_c}{E_s} = \frac{M_1(h_0 - x_c)}{I_c} \tag{17}$$

The bending moment of the test pile (M_1) was further obtained as:

$$M_1(h_0 - x_c) = \sigma_s \frac{E_c I_c}{E_s} \tag{18}$$

Similarly, the stress of the concrete at the same level of the reinforcement N2 (σ'_s) should meet Equation (19), namely

$$M_1(x_c - a'_s) = \sigma'_s \frac{E_c I_c}{E_s} \tag{19}$$

Combining Equations (18) and (19), and noting $\sigma_s = E_s \varepsilon_s$ and $\sigma'_s = E_s \varepsilon'_s$, M_1 in Stage I can be obtained as:

$$M_1 = \frac{E_c I_c}{d}(\varepsilon_s + \varepsilon'_s) \tag{20}$$

It can be seen that the structure of Equation (20) is similar to that of Equation (8), and the difference between the two is the inertia moment of the pile section. Equation (20) is more objective and reasonable to consider the increasing effect of the steel bars on the inertia moment of the pile section.

(2) Stage II

When the concrete stress in the pile tensile area exceeds its tensile strength, cracks will occur and develop. At this moment, the concrete in the tension area basically no longer bears the tensile stress, the steel bar stress increases rapidly, and the concrete stress in the compression area increases statistically, which finally leads to the redistribution of the pile section stress, as shown in Figure 16. At this stage, the concrete in the pile compression area starts to enter the plastic compression stage, and the stress–strain relationship of the concrete changes from linear to nonlinear.

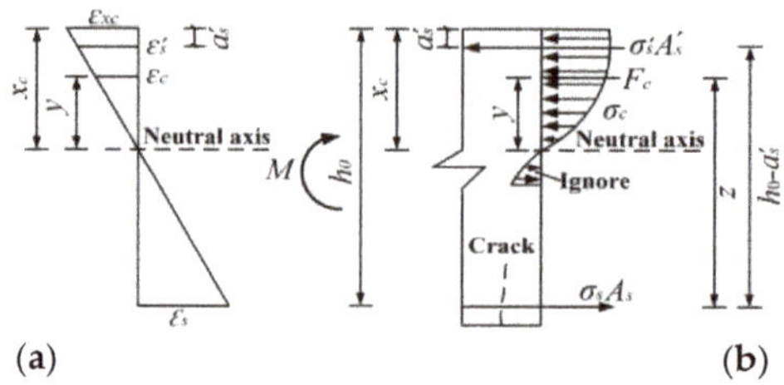

Figure 16. Stress and strain diagrams of the anti-slide pile in normal section Stage II. (**a**) Distribution of average strain; (**b**) distribution of normal stress.

At this stage, the concrete in the pile compression area does not conform to the elastic body assumption, and its stress–strain constitutive relationship should be adopted according to Chinese specification (GB 50010-2019) [25], as shown in Figure 17 and Equation (21).

$$\begin{aligned} \sigma_c &= \sigma_0\left[1-(1-\varepsilon_c/\varepsilon_0)^2\right],\ 0<\varepsilon_c\le\varepsilon_0=0.002 \\ \sigma_c &= \sigma_0=\text{const},\ 0.002\le\varepsilon_c\le\varepsilon_{cu}=0.033 \end{aligned} \tag{21}$$

where σ_c and ε_c are the concrete stress and strain in the pile compression area, respectively; ε_{cu} is the concrete ultimate compressive strain; σ_0 is the concrete peak compressive stress, which is taken as the design value of the axial compressive strength.

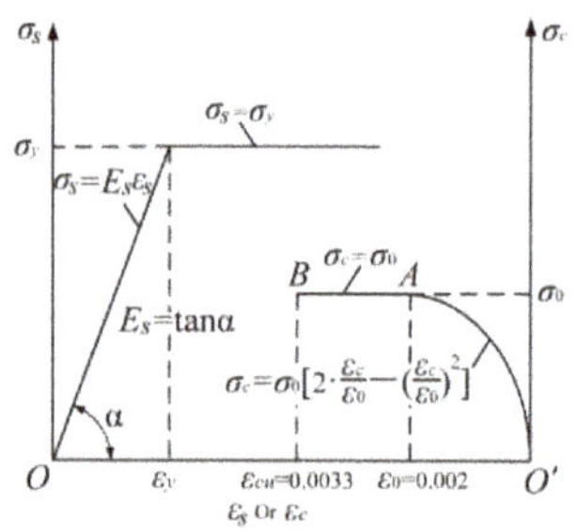

Figure 17. Typical stress–strain constitutive relationships of reinforcement and concrete.

Since the reinforcement N1 is still in an elastic state at this stage, the simplified elastoplastic stress–strain expression can be used for its stress–strain relationship, as shown in Equation (22) and Figure 17.

$$\begin{aligned} \sigma_s &= E_s\varepsilon_s,\ \varepsilon_s\le\varepsilon_y \\ \sigma_s &= \sigma_y,\ \varepsilon_s\ge\varepsilon_y \end{aligned} \tag{22}$$

where ε_y and σ_y are the yield strain and yield strength of steel bars, respectively.

The stress and strain state characteristics of the compressed concrete and tensile steel bars of the pile at this stage are shown in Equation (23). Combining Equations (12) and (23), it can be seen that the strain ($\varepsilon_s=\varepsilon_{tu}$) or stress ($\sigma_s=\varepsilon_{tu}E_s$) of the tensile steel bars can be used as the boundary sign between Stage I and Stage II.

$$\begin{aligned} \varepsilon_c\le\varepsilon_0,\ \varepsilon_s\ge\varepsilon_{tu} \\ \sigma_c\le\sigma_0,\ \sigma_s\ge\varepsilon_{tu}E_s \end{aligned} \tag{23}$$

The calculation method of the bending moment of the pile in Stage II is introduced as follows. Firstly, based on the plane section assumption, the concrete strain at $x_c(\varepsilon_{xc})$ and the concrete strain with a distance y from the neutral axis (ε_{cy}) are determined as follows:

$$\varepsilon_{xc}=\varepsilon'_s\frac{x_c}{x_c-a'_s} \tag{24}$$

$$\varepsilon_{cy} = \varepsilon'_s \frac{y}{x_c - a'_s} \tag{25}$$

Then, the solution expression for the resultant force of the compressive stress of the pile concrete (F_c) is established by integrating the section stress of compressive concrete, as shown in Equation (26). Substituting Equations (21) and (25) into Equation (26), F_c is obtained as:

$$F_c = \int_0^{x_c} \sigma_c b dy \tag{26}$$

$$F_c = b\sigma_0 x_c \left[\frac{\varepsilon'_s}{\varepsilon_0 (1 - \frac{a'_s}{x_c})} - \frac{\varepsilon'_s}{3\varepsilon_0{}^2 (1 - \frac{a'_s}{x_c})^2} \right] \tag{27}$$

Since $a'_s << x_c$ in stage II, $\left(1 - \frac{a_s'}{x_c}\right) \approx 1$ in Equation (27). Making $\eta = \frac{\varepsilon'_s}{\varepsilon_0}$ and simplifying Equation (27) to get F_c:

$$F_c = \sigma_0 b x_c \left(\eta - \frac{1}{3}\eta^2 \right) \tag{28}$$

Secondly, the bending moment of F_c to the center of tensile steel bars is obtained as follows:

$$M_c = \int_0^{x_c} \sigma_c b (h_0 - x_c + y) dy = F_c \left(h_0 - \frac{1 - 0.25\eta}{3 - \eta} x_c \right) \tag{29}$$

Thirdly, the bending moment of the compressive steel bars to the center of the tension steel bars is obtained as follows:

$$M_s = E_s \varepsilon'_s A'_s (h_0 - a'_s) \tag{30}$$

Finally, the bending moment of the section of the test pile in Stage II (M_2) is obtained by combining Equations (29) and (30), as shown in Equation (31).

$$M_2 = M_c + M_s = \sigma_0 b x_c \left(\eta - \frac{1}{3}\eta^2 \right) \left(h_0 - \frac{1 - 0.25\eta}{3 - \eta} x_c \right) + E_s \varepsilon'_s A'_s (h_0 - a'_s) \tag{31}$$

(3) Stage III

When the thrust load behind the cantilever anti-slide pile is large enough, the tensile steel bars gradually yield, and the strains of the steel bars increase rapidly. The concrete in the tension area will no longer bear the tension, and the cracks on the pile gradually rise and expand. Part of the concrete in the compression area enters the stable plastic stage and tends to fail, as shown in Figure 18. When the strain reaches the ultimate compressive strain, the compressive concrete is crushed, and the test pile can no longer bear the external load, which indicates that the test pile has reached the failure state. The stress–strain state characteristics of the steel bars and concrete in this stage are shown in Equation (32), and it can be seen that the strain ($\varepsilon_s = \varepsilon_y$) or stress ($\sigma_s = \sigma_y$) of the tensile steel bars can be used as the boundary sign between Stage II and Stage III.

$$\begin{array}{l} \varepsilon_{xc} \geq \varepsilon_0,\ \varepsilon_s \geq \varepsilon_y \\ \sigma_c \geq \sigma_0,\ \sigma_s \geq \sigma_y \end{array} \tag{32}$$

In the process of calculating the resultant force of compressive stress of the test pile concrete (F_c), the stress distribution curve of compressive concrete should be divided into two sections. The first and second sections are respectively the curve and the straight line (Figure 18), where the straight line section indicates that the stress in the concrete has reached σ_0. Based on the plane section assumption and referring to Equation (25), the height of the boundary point of the concrete stress distribution curve is $\varepsilon_0 x_c / \varepsilon_{xc}$. Based on

Equations (21), (24) and (25), σ_c, ε_{xc} and ε_c can be further obtained respectively. Substituting σ_c, ε_{xc} and ε_c into Equation (26) and integrating in sections, F_c at this stage is:

$$F_c = \int_0^{x_c} \sigma_c b dy = \int_0^{\varepsilon_0 \frac{x_c}{\varepsilon_{xc}}} \sigma_c b dy + \int_{\varepsilon_0 \frac{x_c}{\varepsilon_{xc}}}^{x_c} \sigma_c b dy = \sigma_0 b x_c \left(1 - \frac{1}{3\eta}\right) \tag{33}$$

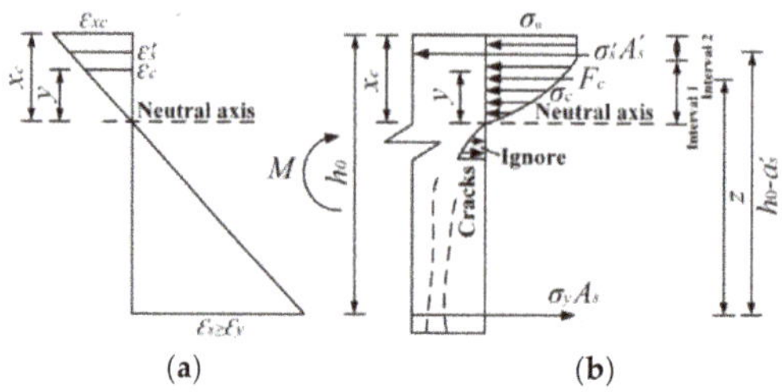

Figure 18. Stress and strain diagrams of the anti-slide pile in normal section stage III. (**a**) Distribution of average strain; (**b**) distribution of normal stress.

The bending moment of F_c to the center of the tensile steel bars in this stage is obtained as follows:

$$M_c = \int_0^{x_c} \sigma_c b(h_0 - x_c + y) dy = F_c\left(h_0 - \frac{6\eta^2 - 4\eta + 1}{12\eta^2 - 4\eta} x_c\right) \tag{34}$$

The calculation of Ms in this stage is the same as Equation (30), so the bending moment of the test pile in Stage III is:

$$M_3 = M_c + M_s = \sigma_0 b x_c \left(1 - \frac{1}{3\eta}\right)\left(h_0 - \frac{6\eta^2 - 4\eta + 1}{12\eta^2 - 4\eta} x_c\right) + E_s \varepsilon'_s A'_s (h_0 - a'_s) \tag{35}$$

In summary, the specific calculation steps of internal force or bending moment in each stage during the whole process of stress and deformation of the cantilever anti-slide pile are represented in Figure 19.

3.3. *Bending Moment Distribution of Test Pile*

According to the optimized calculation method for the bending moment of the cantilever anti-slide pile proposed in this paper, the bending moment distribution of the test pile in the loading process of the trapezoidal load was obtained, as shown in Figure 20. It is obvious that the pile bending moment increased gradually with the trapezoidal load, increasing first and then decreased gradually along the pile length. The maximum bending moment was located at the sliding surface, and the farther away from the sliding surface, the faster the attenuation. At the same time, the evolution process of the pile bending moment mainly included three stages. When the resultant force of the trapezoidal thrust load (F_T) was less than the pile cracking load, the maximum bending moment gradually increased, and increased to 0.86 kN·m. When F_T reached 10.2 kN, subsequently, the test pile entered the working stage with cracks, and the pile bending moment increased rapidly. When F_T increased to 36.6 kN, the maximum bending moment was about 4.16 kN·m. When the steel bars yielded, the external trapezoidal load behind the test pile remained basically unchanged, the test pile tended to be damaged, and the maximum bending moment increased slowly to 4.49 kN·m when the test pile was destroyed.

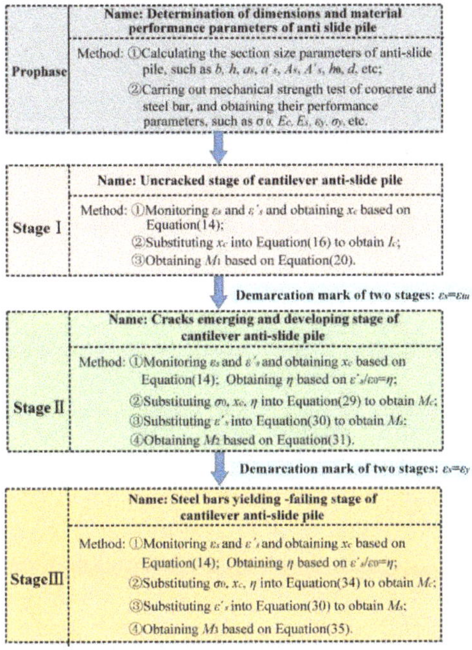

Figure 19. Calculation flow chart of bending moment of anti-slide pile at each stage.

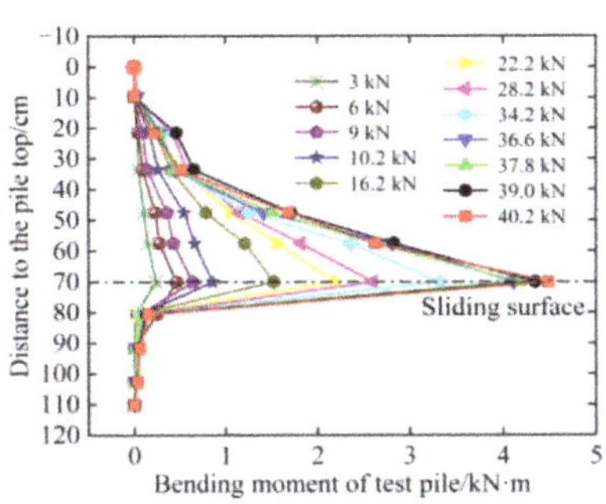

Figure 20. Bending moment distributions along test pile.

Equation (8) in Section 3.2.1 was used to calculate the pile bending moment at the sliding surface, and further compared with the optimized calculation method established in Section 3.2.2, and the comparison results are represented in Figure 21. It can be found that the two are basically the same when the test pile worked in the elastic stage (Stage I). The maximum bending moment at the sliding surface obtained by using the optimized calculation method established in Section 3.2.2 (Equation (20)) is slightly larger. The reason was that the increasing effect of tensile steel bars and compressive steel bars on the inertia moment of pile section was considered in Equation (20), resulting in the calculation result of the inertia moment of pile section larger than that of Equation (8). When the test pile worked in the stage of crack emerging and developing (Stage II), the gap between the two grew gradually with the trapezoidal load. The reason was that the anti-slide pile in stage II was regarded as an elastoplastic body in the optimized calculation method built in this paper (Equation (31)), and the elastoplastic characteristics of compressive concrete were considered. The pile compressive concrete was treated as a linear elastomer during the whole loading process in the traditional method (Equation (8)), resulting in a larger calculation value of bending moment. When the values of F_T were 16.2 kN and 36.6 kN, respectively, the bending moments calculated by the traditional method (Equation

(8)) were 2.87 times and 5.24 times the bending moment obtained by the optimized calculation method established in Section 3.2.2 (Equation (31)). Moreover, the theoretical ultimate flexural bearing capacity of the test pile(M_u) was 3.93 kN·m according to the flexural strength formula of reinforced concrete beam, which is not much different from the maximum bending moment obtained by the optimized calculation method established in this paper (Equation (35)). The bending moment at the sliding surface of the test pile calculated by the traditional method was as high as 35.37 kN·m, which is obviously inconsistent with the fact.

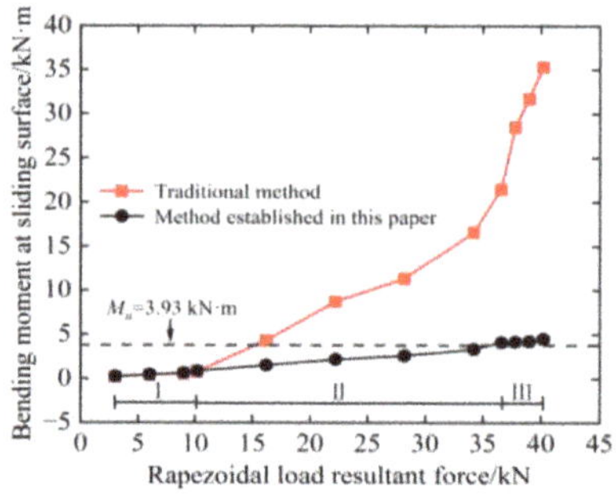

Figure 21. Comparison of the maximum bending moments of the test pile based on the traditional method and the optimized calculation method.

The above results show that when the cantilever anti-slide piles work in the elastic stage or there are no cracks in the pile during the service life, the errors of the bending moment calculated by the traditional method or the optimized calculation method established in Section 3.2.2 are small, and both methods can be used. However, when the cantilever anti-slide pile cracks in service, the traditional method will produce large errors and mislead the evaluation of the service status of the cantilever anti-slide pile in service and the structural design, resulting in a waste of economy and resources.

4. Numerical Analysis and Discussion

4.1. Establishment of the Numerical Model

Three-dimensional numerical nonlinear analysis and parameter analysis of model anti-slide pile were carried out by using ABAQUS numerical simulation software. The model size, external trapezoidal load, and main parameters of the model pile are the same as those in Section 2. The FE model was constructed by embedded element technology, and the damage plasticity model (CDP model) was adopted for the pile concrete, which are defined according to the plasticity and damage parameters. The double broken line model of isotropic plasticity was applied for the constitutive relation of pile steel bars. The boundary conditions of the numerical model mainly set the anchored section as a fixed constraint and the loaded section as a free boundary. The C3D8R element and the T3D2 element were adopted for the pile concrete and steel bars, respectively, and the numerical model was divided into 17,164 elements and 20,134 nodes, as shown in Figure 22.

4.2. Model Calibration

The displacement distribution, the trapezoidal thrust load-pile top lateral displacement curve, and the bending moment distribution of the test pile were applied to verify the rationality of the numerical model developed in Section 4.1. The calibrated trapezoidal load was the representative load in the whole loading process of the cantilever anti-slide piles (Stage I~Stage III), which were 6 kN, 9 kN, 16.2 kN, 34.2 kN, 37.8 kN, and 40.2 kN, respectively. Figure 23a,b, respectively, show the pile lateral displacement distribution and the evolution process of the pile top lateral displacement under external trapezoidal load. It is clear that the lateral displacement distribution and the trapezoidal thrust load-pile top lateral displacement curve of the model pile calculated by ABAQUS are in good agreement with the experimental values. It can be concluded from Figure 23b that the crack distribution and failure form of the model pile are basically consistent with those of the test pile. The comparison results of bending moment distribution under trapezoidal load are shown in Figure 23c. It is clear that the bending moment distributions under different levels of trapezoidal thrust load were unimodal, and the evolution process was consistent with the test pile. The above comparison results comprehensively verify the rationality of the numerical model of the cantilever anti-slide pile built in Section 4.1.

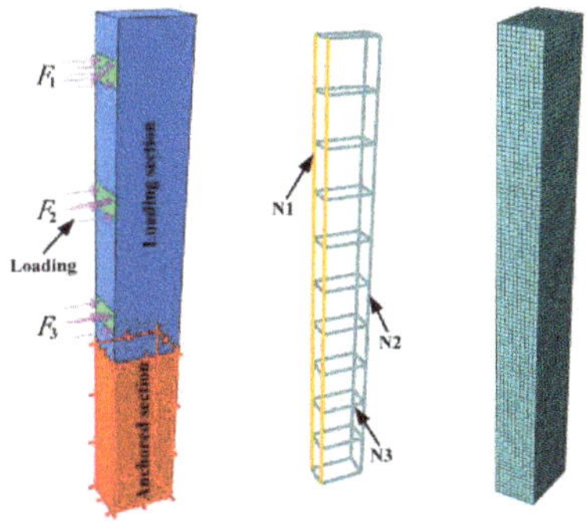

Figure 22. Three-dimensional numerical model and loading diagram of model pile.

4.3. Parametric Study

Based on the above FE model, the critical parameters in the preliminary design and service process of the cantilever anti-slide pile were analyzed, including the material strength and anchored length.

4.3.1. Effect of Concrete Strength and Steel Bars' Strength

The concrete strength and steel bar strength of the cantilever anti-slide pile were in a time-varying evolution states during the service period. Figures 24 and 25, respectively, show the distributions of pile displacement and bending moment for different concrete strengths (40, 35, 30, 25, and 20 MPa) and reinforcement yield strengths (The yield strengths of reinforcement N1 (f_{yN1}) were 460, 432, 402, 371, and 340 MPa, and the yield strengths of reinforcement N2 (f_{yN2}) were 438, 411, 381, 352, and 323 MPa). It can be obtained that the maximal lateral displacements at the pile top of the cantilever anti-slide pile with different concrete strengths under the same thrust load were 55.20 mm, 58.52 mm, 61.04 mm, 67.54 mm, and 86.02 mm, respectively, and the ultimate bending moments when the pile was destroyed are 4.689, 4.58, 4.524, 4.431, and 4.333 kN·m, respectively, indicating that the pile top lateral displacement was negatively related to the concrete strength, and the smaller the concrete strength, the greater the lateral displacement at pile top. However, the reduction of the ultimate bending capacity of the pile is not obvious. The pile top lateral displacements of cantilever anti-slide pile with different steel bars' strengths under the same thrust load were 52.48, 61.04, 72.009, 99.02, and 121.42 mm, respectively, and the ultimate bending moments when the pile was destroyed were 4.812, 4.524, 4.287, 4.112, and 3.705 kN·m, respectively, indicating that the pile top lateral displacement was negatively correlated with the steel bars' strength, and the maximum bending capacity of the pile decreased obviously with the gradual deterioration of steel bars' strength. It can be concluded that the lateral displacement at pile top can be reduced by improving the concrete or steel bars' strength during the preliminary design. However, increasing the concrete strength makes no difference to the pile bearing capacity, while increasing the reinforcement strength has obvious effect. In other words, the pile displacement gradually increases with the deterioration of the material strength during the service life, and the deterioration of the steel bar strength has a stronger deterioration effect on the maximum bearing capacity of the pile than concrete strength.

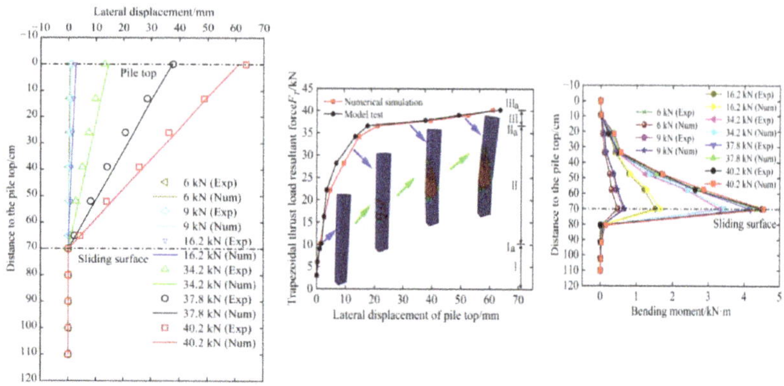

(a) (b) (c)

Figure 23. Comparison of the experimental and numerical simulation results. (**a**) Displacement; (**b**) thrust load-pile top displacement; (**c**) distribution of bending moment.

4.3.2. Effect of Anchor Ratio

The anchor ratio (AR) is another crucial parameter for the structural design of a cantilever anti-slide pile. Figure 26 shows the distributions of the pile displacement and bending moment under five anchor ratios (1/5, 1/4, 1/3, 4/11, 2/5, 1/2). It can be concluded that the pile displacement and bending moment under different anchor ratios changed little. The reason is that the anchored section was set as a fixed constraint in the ABAQUS numerical simulation to simulate the anchored section, indicating that when the strength of the rock formation embedded in the pile anchored section is large enough, blindly increasing the length of the pile anchored section cannot significantly improve the bearing capacity, which is in agreement with the results by [26]. In summary, the length of pile anchored section should be designed reasonably by combining economic factors and safety factors in this case.

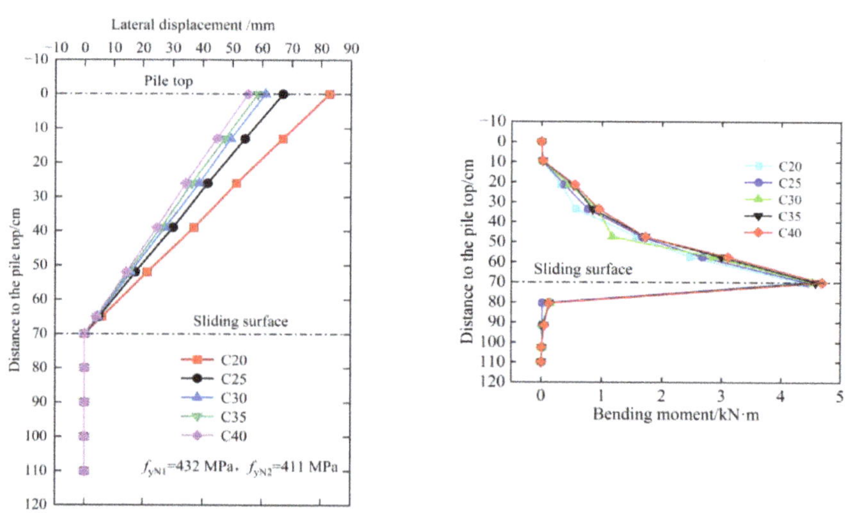

(a) (b)

Figure 24. Variations of the pile displacement and bending moment with respect to concrete strength. (**a**) Pile displacement; (**b**) pile bending moment.

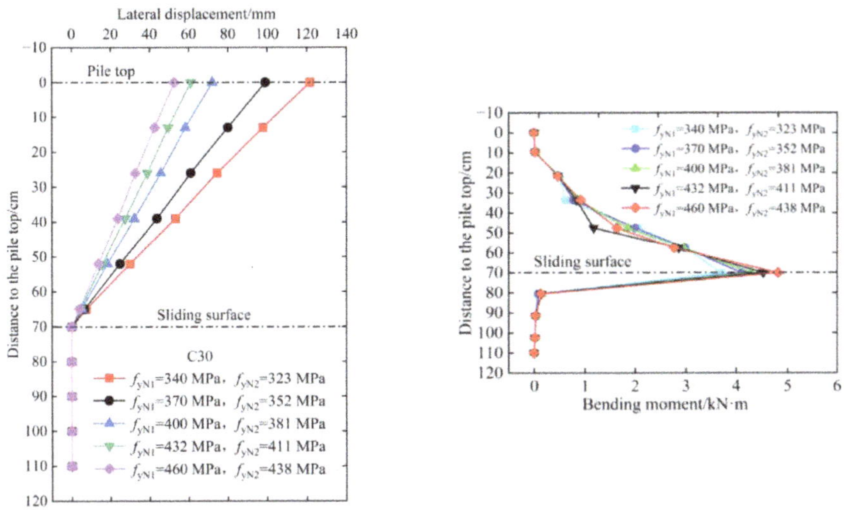

Figure 25. Variations of the pile displacement and bending moment with respect to steel bars' strength. (**a**) Pile displacement; (**b**) pile bending moment.

4.3.3. Discussion

The lateral displacement and bending moment of the cantilever anti-slide pile under the trapezoidal load (TL) and uniform load (UL) were further compared, as shown in Figure 27. When the resultant force of the trapezoidal load (F_T) reached 6 kN, the lateral displacements at the pile top under the trapezoidal load and uniform load were 0.47 mm and 0.9 mm, respectively, and the maximum bending moments at the sliding surface were 0.78 kN·m and 0.97 kN·m, respectively. When F_l reached 28.2 kN, the lateral displacements at pile top under the trapezoidal load and uniform load were 9.3 mm and 49.15 mm, respectively, and the maximum bending moments were 3.42 kN·m and 4.21 kN·m, respectively. This shows that the lateral displacement and bending moment of the anti-slide pile under the uniform load and trapezoidal load have little difference when the external thrust load is small. As the external thrust load increases, the lateral displacement and bending moment of the anti-slide pile under the uniform load are gradually larger than those under the trapezoidal load, especially the lateral displacement, and the gaps under the two loads gradually increase with the external thrust load. In summary, the distribution form of the external thrust load behind the cantilever anti-slide pile will significantly affect its stress and deformation. In the process of the structural design of the cantilever anti-slide pile, the distribution form of the external thrust load should be judged as accurately as possible to make it more consistent with the actual stress distribution, so as to make the structural design of the anti-slide pile more reasonable and economical.

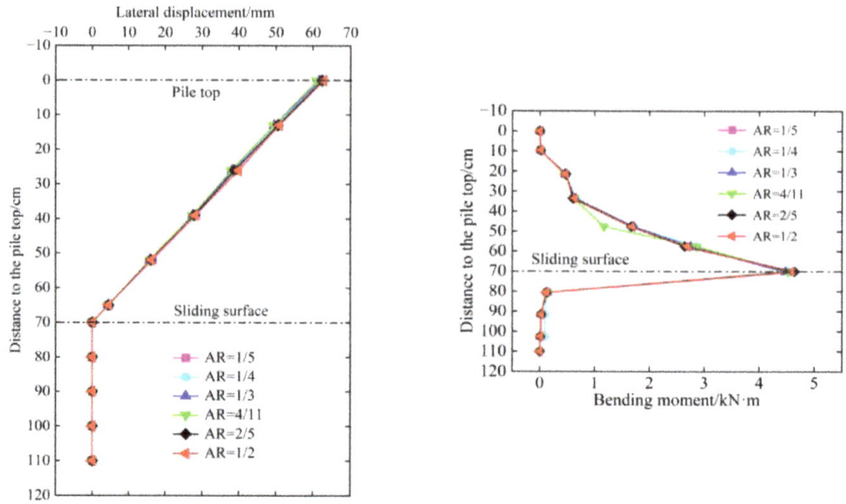

(**a**) (**b**)

Figure 26. Variations of the pile displacement and bending moment with respect to anchor ratio. (**a**) Pile displacement; (**b**) pile bending moment.

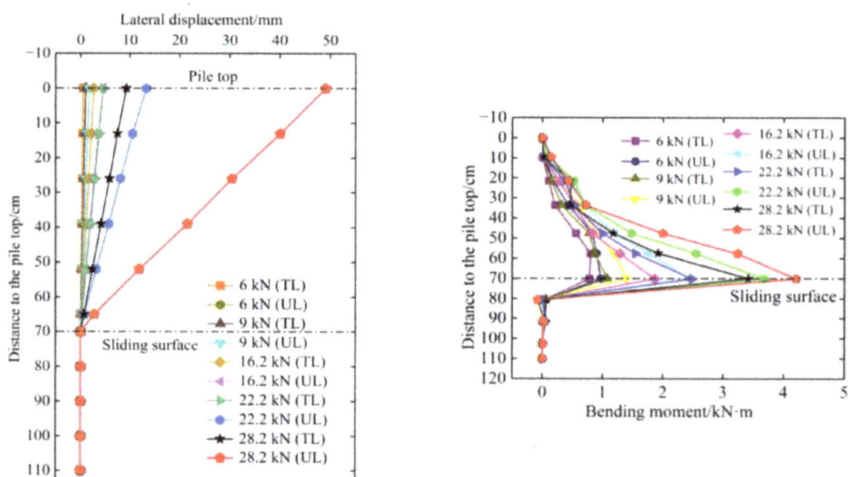

(**a**) (**b**)

Figure 27. Displacement and bending moment comparisons for the anti-slide pile under trapezoidal load and uniform load. (**a**) Displacement; (**b**) bending moment.

5. Conclusions

In this paper, the evolution processes of the internal force and deformation of the cantilever anti-slide pile under trapezoidal thrust load were investigated through experiments and FE methods, and a new method for calculating the internal force of the cantilever anti-slide pile is proposed. The major conclusions are drawn as follows.

1. The whole development processes of the stress and deformation of the cantilever anti-slide pile under trapezoidal load mainly include three stages: the uncracked stage, the cracks emerging and developing stage, and the steel bar yielding–failing stage. The cracking load and yield load of the pile account for 25.37% and 91.04% of the failure load, and the cracks of the pile are concentrated in the range of 0.09 times the pile length above and below the sliding surface.
2. The traditional calculation method for the bending moment of an anti-slide pile is unreasonable because the contribution of reinforcement is ignored in the calculation of the section inertia moment, and the calculation result is small when the anti-slide pile works in the uncracked stage. In the crack emerging and developing stage, when the resultant force of the external trapezoidal load reaches 16.2 kN and 36.6 kN, respectively, the bending moments calculated by the traditional method are 2.87 times and 5.24 times that of the optimized calculation method of this study, respectively. The traditional calculation method is no longer applicable. The optimized calculation method for the bending moment of cantilever an anti-slide pile established in this paper is highly feasible when considering the elastoplastic characteristics of reinforcement materials and concrete materials.
3. The pile displacement and bending moment are negatively and positively related to the strength of the pile material, respectively. When the strength of the pile concrete deteriorates from 40 MPa to 20 MPa, the maximum displacement of the pile increases by 55.8% and the bending moment of the pile deteriorates by 7.59%. When the strength of the pile reinforcement deteriorates from 460 MPa to 340 MPa, the maximum displacement of the pile increases by 131.4% and the bending moment of the pile deteriorates by 23%. When the strength of the rock stratum embedded in the pile anchored section is large enough, increasing the length of anchored section cannot significantly improve its bearing capacity. The displacement and bending moment of the anti-slide pile under the uniform load are greater than those under the trapezoidal load. When the thrust loads are 6 kN and 28.2 kN, the maximum displacement of the pile under uniform load increases by 91.5% and 428.5%, respectively, compared with the trapezoidal load, and the maximum bending moment at the sliding surface increases by 24.4% and 23.1%.

Author Contributions: Conceptualization, methodology, data curation, and formal analysis, H.J.; writing—original draft, Q.R.; investigation, S.X. All authors have read and agreed to the published version of the manuscript.

Funding: This work was supported by the National Natural Science Foundation of China (No. U20A20314, 41472262 and 52104184), the Key Projects of Chongqing Natural Science Foundation (No. cstc2020jcyj-zdxmX0012), the Innovation Research Group of Universities in Chongqing (No. CXQT19021), the First Batch of Innovative Leading Talents in Chongqing (No. CQYC201903026), the Chongqing Graduate Research Innovation Project (No. CYB21210) and the Special Support of Chongqing Postdoctoral Research Project (No. 2021XM3038).

Institutional Review Board Statement: Not applicable.

Informed Consent Statement: Not applicable.

Data Availability Statement: No applicable.

Conflicts of Interest: The authors declare no conflict of interest.

References

1. He, C.C.; Hu, X.L.; Liu, D.Z.; Xu, C.; Wu, S.S.; Wang, X.; Zhang, H. Model tests of the evolutionary process and failure mechanism of a pile-reinforced landslide under two different reservoir conditions. *Eng. Geol.* **2020**, *277*, 105811. [CrossRef]
2. Zhang, Y.M.; Hu, X.L.; Tannant, D.D.; Zhang, G.C.; Tan, F.L. Field monitoring and deformation characteristics of a landslide with piles in the Three Gorges Reservoir area. *Landslides* **2018**, *15*, 581–592. [CrossRef]
3. Zhou, Y.; Wang, Z.K.; Yang, X.H. Monitoring analysis of anti-slide piles for Jiangdingya landslide in Zhouqu. *J. Eng. Geol.* **2022**, *30*, 197–204. [CrossRef]
4. Zhang, H.; Xing, H.F.; Liu, L.L.; Luo, Y. Field test and numerical analysis on deformation response of H-type antislide pile: Case study of Longjiayan landslide, China. *Nat. Hazards Rev.* **2021**, *22*, 05021015. [CrossRef]
5. Zhang, L.; Shi, B.; Zhang, D.; Han, H.M.; Wei, G.Q. Evaluation and analysis of working state of anti-slide pile with BOTDR technology. *J. Eng. Geol.* **2019**, *27*, 1464–1472. [CrossRef]
6. Wang, C.T.; Wang, H.; Qin, W.N.; Tian, H.N. Experimental and numerical studies on the behavior and retaining mechanism of anchored stabilizing piles in landslides. *Bull. Eng. Geol. Environ.* **2021**, *80*, 7507–7524. [CrossRef]
7. Hu, X.L.; Zhou, C.; Xu, C.; Liu, D.G.; Wu, S.S.; Li, L.X. Model tests of the response of landslide-stabilizing piles to piles with different stiffness. *Landslides* **2019**, *16*, 2187–2200. [CrossRef]
8. Zhang, J.W.; Wang, X.J.; Wang, H.; Qin, H.Y. Model test and numerical simulation of single pile response under combined loading in slope. *Appl. Sci.* **2020**, *10*, 6140. [CrossRef]
9. Wei, S.W.; Sui, Y.Y.; Yang, J.M. Model tests on anti-sliding mechanism of circular and rectangular cross section anti-sliding piles. *Rock. Soil. Mech.* **2019**, *40*, 951–961. [CrossRef]
10. Xie, Z.W. Study on Strain Characteristics and Internal Force Nonlinearity of Anti-Slide Pile. Master's Thesis, China University of Geosciences, Wuhan, China, 2019.
11. Cao, Y. Experimental Study on the Mechanical Behavior of Double-Row Anti-Slide Piles Installed in Parallel. Master's Thesis, Chang'an University, Xi'an, China, 2019.
12. Li, Z.; Zhu, Z.G.; Zhang, J.; Cao, S.L.; Li, Y.Y. In-situ model test of a cantilever and full-buried single-row double-pile stabilizing system on a loess slope. *China J. Highw. Transp.* **2020**, *33*, 14–23. [CrossRef]
13. Li, Y.; Nan, Y.L.; He, H.C.; Zhang, P. Research on model test of double row anti slide piles in loess. *J. Safety Environ.* **2022**, *22*, 1314–1321. [CrossRef]
14. *GB/T 38509-2020*; Code for the Design of Landslide Stabilization. National Standardization Management Committee of the People's Republic of China: Beijing, China, 2020.
15. Liu, X.Y.; Cai, G.J.; Liu, L.L.; Zhou, Z.J. Investigation of internal force of ant-slide pile on landslides considering the actual distribution of soil resistance acting on anti-slide piles. *Nat. Hazards* **2020**, *102*, 1369–1392. [CrossRef]
16. *GB 175-2020*; Common Portland Cement. China Architecture Press: Beijing, China, 2020.
17. *JGJ 55-2011*; Specification for Mix Proportion Design of Ordinary Concrete. State Administration of Market Supervision and Administration of the People's Republic of China: Beijing, China, 2011.
18. He, C.C.; Hu, X.L.; Dwayne, D.; Tannant, T.F.; Zhang, Y.M.; Zhang, H. Response of a landslide to reservoir impoundment in model tests. *Eng. Geol.* **2018**, *247*, 84–93. [CrossRef]
19. Zhou, C.; Hu, X.L.; Zheng, W.B.; Xu, C.; Wang, Q. Displacement characteristic of landslides reinforced with flexible piles: Field and physical model test. *J. Mt. Sci.* **2020**, *17*, 787–800. [CrossRef]
20. Chen, J.G.; Guo, X.P.; Tian, D.; Yu, S.B. Effect of anti-slide piles anchorage lengths on slip surface and anti-slide capacity of homogeneous slope. *J. Tongji Univ.* **2022**, *50*, 42–49. [CrossRef]
21. *GB/T 50152-2012*; Standard for Test Methods of Concrete Structures. China Architecture Press: Beijing, China, 2012.
22. Xin, J.Z.; Jiang, Y.; Zhou, J.T.; Peng, L.L.; Liu, S.Y.; Tang, Q.Z. Bridge deformation prediction based on SHM data using improved VMD and conditional KDE. *Eng. Struct.* **2022**, *261*, 114285. [CrossRef]
23. Jiang, Y.; Hui, Y.; Wang, Y.; Peng, L.L.; Huang, G.Q.; Liu, S.Y. A novel eigenvalue-based iterative simulation method for multi-dimensional homogeneous non-Gaussian stochastic vector fields. *Struct. Saf.* **2023**, *100*, 102290. [CrossRef]
24. Liu, D.Z.; Hu, X.L.; Zhou, C.; Xu, C.; He, C.C.; Zhang, H.; Wang, Q. Deformation mechanisms and evolution of a pile-reinforced landslide under long-term reservoir operation. *Eng. Geol.* **2020**, *275*, 105747. [CrossRef]
25. *GB 50010-2019*; Code for Design of Concrete Structures. China Architecture & Building Press: Beijing, China, 2019.
26. Qiu, Y.H. Stability Analysis and Anti-Slide Pile Treatment Study of Gaojia Fangji Landslide. Master's Thesis, Southwest Jiaotong University, Chengdu, China, 2021.

buildings

MDPI

Article

A Rapid Identification Technique of Moving Loads Based on MobileNetV2 and Transfer Learning

Yilun Qin [1], Qizhi Tang [1], Jingzhou Xin [1,*], Changxi Yang [1], Zixiang Zhang [2] and Xianyi Yang [1]

1 State Key Laboratory of Mountain Bridge and Tunnel Engineering, Chongqing Jiaotong University, Chongqing 400074, China
2 CCCC Second Highway Consultants Co., Ltd., Wuhan 430058, China
* Correspondence: xinjz@cqjtu.edu.cn

Abstract: Rapid and accurate identification of moving load is crucial for bridge operation management and early warning of overload events. However, it is hard to obtain them rapidly via traditional machine learning methods, due to their massive model parameters and complex network structure. To this end, this paper proposes a novel method to perform moving loads identification using MobileNetV2 and transfer learning. Specifically, the dynamic responses of a vehicle–bridge interaction system are firstly transformed into a two-dimensional time-frequency image by continuous wavelet transform to construct the database. Secondly, a pre-trained MobileNetV2 model based on ImageNet is transferred to the moving load identification task by transfer learning strategy for describing the mapping relationship between structural response and these specified moving loads. Then, load identification can be performed through inputting bridge responses into the established relationship. Finally, the effectiveness of the method is verified by numerical simulation. The results show that it can accurately identify the vehicle weight, vehicle speed information, and presents excellent strong robustness. In addition, MobileNetV2 has faster identification speed and requires less computational resources than several traditional deep convolutional neural network models in moving load identification, which can provide a novel idea for the rapid identification of moving loads.

Keywords: bridge engineering; moving loads identification; MobileNetV2; transfer learning

Citation: Qin, Y.; Tang, Q.; Xin, J.; Yang, C.; Zhang, Z.; Yang, X. A Rapid Identification Technique of Moving Loads Based on MobileNetV2 and Transfer Learning. *Buildings* **2023**, *13*, 572. https://doi.org/10.3390/buildings13020572

Academic Editor: Jorge de Brito

Received: 4 January 2023
Revised: 13 February 2023
Accepted: 17 February 2023
Published: 20 February 2023

1. Introduction

Moving load is the main external load acting on the bridge [1,2]. Due to the inefficient supervision of overweight vehicles, bridge collapse accidents caused by overweight vehicles have occurred occasionally in recent years [3–5]. Therefore, the accurate and rapid recognition of moving loads is beneficial to the early warning and control of the overweight vehicle, thereby ensuring the safe operation of the bridge [6,7]. Traditional moving loads identification method primarily rely on a bridge weigh-in-motion (WIM) system. However, WIM may harm the road surface, and the sensor is prone to be damaged under long-term moving load, which increases the operation and maintenance costs [8]. Therefore, it is urgent to indirectly identify the moving load using the dynamic response by a more efficient and economic method, i.e., moving load identification (MLI) methods.

MLI methods can roughly be classified into two categories, i.e., intelligent optimization methods and machine learning methods. Among them, intelligent optimization methods compute the optimal solution of the loss function to obtain the load parameters with the smallest loss function [9]. For example, Wang et al. [10] applied simulated annealing algorithm to identify multi-axis moving train loads, and the experimental results demonstrated that the proposed method exhibits excellent robustness and accuracy. Pan et al. [11] proposed a moving loads identification method based on the firefly algorithm, in which, vehicle load information can be accurately identified with only a small number of sensors. Liu et al. [12] recognized the constant component of the moving load with the help of particle swarm algorithm and used the hybrid measurement response to further improve the identification accuracy. Ali R. Vosoughi et al. [13] applied a genetic algorithm for moving load identification by defining a root mean square error function between the measured and calculated responses, and the results showed that the accuracy and efficiency of this method higher than the Newmark's method. Although the intelligent optimization methods can effectively obtain the moving load information from the bridge response, the optimization process often requires searching a huge solution space, which leads to computational inefficiency and is not conducive to the rapid identification of moving loads [14].

With the rapid development of artificial intelligence, machine learning (especially deep learning) has shown great advantages in feature extraction, target detection, and recognition [15], etc., and is also widely applied in moving load identification. For instance, Yang et al. [16] applied a neural network to acquire the information of moving load through structural dynamic strain and discussed the influence of activation function on identification accuracy. Zhou et al. [17] developed a moving load identification algorithm, which converted the bridge acceleration response into a two-dimensional map as the network input. Chen et al. [18] reconstructed and located impact load based on deep convolution recurrent neural network and feature learning, which avoided the infeasibility and ill-posedness of nonlinear structure when identifying random impact loads. Zhang [19] applied a long short-term memory neural network to obtain the information of moving loads through the dynamic responses of the bridge, and the results revealed that the information of the moving load can be recognized synchronously with great accuracy. The above literature confirms the great potential of machine learning methods in the accurate and efficient identification of moving loads. However, these machine learning methods often encounter a heavy computational burden, due to the large model parameters and complex network structure, which leads to an inefficient identification process.

Fortunately, lightweight convolutional neural network has a faster identification speed. Compared with traditional deep convolutional neural network models, separable convolution is used in lightweight convolutional neural network model, which greatly reduces the model parameters without sacrificing the accuracy of the model. As a lightweight convolutional neural network model with superior performance [20], the MobileNetV2 model has not yet been used in moving load identification. Therefore, this paper proposes a moving load identification method based on MobileNetV2 and transfer learning, which has faster identification speed and requires less computing resource. Concretely, the continuous wavelet transform (CWT) is first applied to convert the dynamic responses of vehicle-bridge interaction (VBI) system into images to construct the data set for the moving load identification task. Secondly, a pre-trained MobileNetV2 model is applied to the load identification task through transfer learning strategy to enhance the efficiency of the model. Then, the information of moving loads can be acquired through inputting responses of bridge into the completely trained model. Finally, the feasibility of the method is demonstrated in the numerical modeling case.

The major contributions of this paper in comparison with the published literature are summarized in the following.

(1) MobileNetV2 has been introduced into moving load identification to improve the identification efficiency. Case study shows that the MobileNetV2 has faster identification speed and requires less computational resources than traditional deep convolutional neural network models in moving load identification.
(2) The influence of several types of dynamic response on moving loads identification is discussed. The results demonstrate that the displacement response may be the most suitable input for vehicle load identification, while acceleration response may be more suitable for vehicle speed identification, which provides a guideline for the accurate identification of moving loads.

This paper is organized as follows. In Section 2, the theoretical background involved in this paper is introduced. In Section 3, the process of this method is described. The case study of identification task is conducted in Section 4. In Section 5, the performance of this method is discussed and analyzed. In Section 6, several conclusions are described.

2. Theoretical Background

In this section, firstly, the VBI dynamic model is described. Then, the popular deep convolutional neural network (DCNN) models and lightweight convolutional neural network models are introduced. Finally, the basic concept of transfer learning is introduced.

2.1. The VBI Dynamic Model

The VBI dynamic model is a complex dynamic system, composed of moving vehicles and bridges. In this subsection, the Newmark-β method is used to solve the dynamic equation of the VBI model, thereby obtaining the dynamic response of the model.

2.1.1. Road Surface Roughness

PSD function is used to transformed road surface roughness from spatial frequency domain to the circular frequency domain [21], as follows:

$$\begin{cases} S_{rr}(\Omega) = S_{rr}(\Omega_0)(\Omega/\Omega_0)^{-2} & (\Omega \leq \Omega_0) \\ S_{rr}(\Omega) = S_{rr}(\Omega_0)(\Omega/\Omega_0)^{-1.5} & (\Omega > \Omega_0) \end{cases} \tag{1}$$

Road surface roughness can be calculated by the inverse Fourier transform of the road surface roughness spectrum, as follows [22]:

$$r(x) = \sum_{i=1}^{N} \sqrt{\Delta n} \cdot 2^k \cdot 10^{-3} \cdot \left(\frac{n_0}{i \cdot \Delta n}\right) \cos(2\pi i \cdot \Delta n x + \phi_i) \tag{2}$$

where $r(x)$ is a variable about bridge length L, $\Delta n = 1/L$; N is the number of data points, and k is a constant integer increasing from 3 to 9. ϕ_i is the random phase angle distributed uniformly between 0 and 2π.

2.1.2. The Vehicle Model

In this section, a vehicle model is established, which involves mass, spring, and damper [23]. The influence of road roughness on vehicle-bridge interaction vibration are considered, as shown in Figure 1. These assumptions are used in the VBI dynamic model: (i) the vehicle is traveling on the bridge with a constant speed v; (ii) the wheel is always in contact with the beam by point contact; (iii) the displacement of the wheel and the beam at the contact point is consistent.

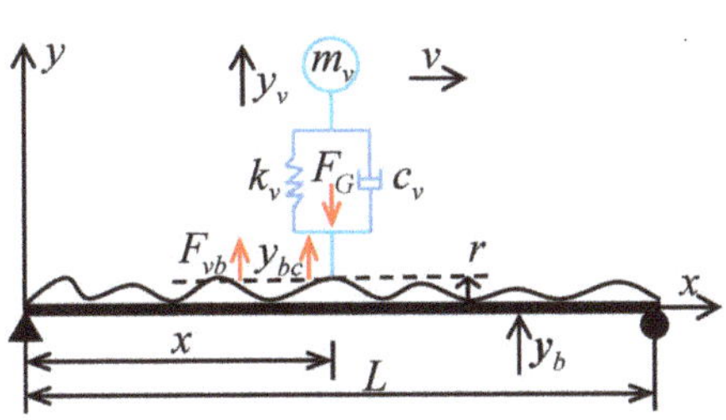

Figure 1. The VBI system.

According to the Newton's second law, the vibration equation of the vehicle can be written as:

$$m_v \ddot{y}_v = F_{vb} - F_G \tag{3}$$

where m_v and y_v are the vehicle weight and vehicle displacement, respectively; F_G and F_{vb} are the vehicle gravity and the interaction force, respectively. F_{vb} can be calculated by:

$$F_{vb} = -k_v(y_v - y_{bc} - r) - c_v(\dot{y}_v - \dot{y}_{bc} - \dot{r}) \tag{4}$$

where k_v, c_v and y_{bc} are spring stiffness, the damping, the bridge deflection, respectively. y_{bc} can be calculated by:

$$y_{bc} = \boldsymbol{N}_b \cdot \boldsymbol{y}_b \tag{5}$$

where $\boldsymbol{y}_b$ denotes the global displacement vector of the bridge, $\boldsymbol{N}_b$ denotes the bridge shape function.

Combining the Equations (3)–(5), we can obtain:

$$m_v\ddot{y}_v - \boldsymbol{C}_{vb} \cdot \dot{\boldsymbol{y}}_b + c_v\dot{y}_v - \boldsymbol{K}_{vb} \cdot \boldsymbol{y}_b + k_v y_v = F_{vr} - F_G \tag{6}$$

where $\boldsymbol{C}_{vb} = c_v \cdot \boldsymbol{N}_b$, $\boldsymbol{K}_{vb} = c_v \cdot \dot{\boldsymbol{N}}_b + k_v \cdot \boldsymbol{N}_b$, $F_{vr} = c_v\dot{r} + k_v r$ are the vehicle additional damping, stiffness, and load terms, respectively.

2.1.3. The Bridge Model

Under vehicle load, bridge dynamic equation can be expressed as:

$$\boldsymbol{M}_b\ddot{\boldsymbol{y}}_b + \boldsymbol{C}_b\dot{\boldsymbol{y}}_b + \boldsymbol{K}_b\boldsymbol{y}_b = -\boldsymbol{F}_{bv} \tag{7}$$

where $\boldsymbol{M}_b$ is bridge mass matrices, $\boldsymbol{C}_b$ is bridge damping matrices, and $\boldsymbol{K}_b$ is bridge stiffness matrices; $\boldsymbol{F}_{bv}$ is the equivalent nodal force of F_{bv}. It has the following relationship:

$$\boldsymbol{F}_{bv} = \boldsymbol{N}_b^T \cdot F_{bv} = \boldsymbol{N}_b^T \cdot F_{vb} \tag{8}$$

Substituting Equations (8) and (4) into Equation (7), we can obtain:

$$\boldsymbol{M}_b\ddot{\boldsymbol{y}}_b + (\boldsymbol{C}_b + \boldsymbol{C}_{bb})\dot{\boldsymbol{y}}_b - \boldsymbol{C}_{bv} \cdot \dot{y}_v + (\boldsymbol{K}_b + \boldsymbol{K}_{bb} + \boldsymbol{K}_{bc})\boldsymbol{y}_b - \boldsymbol{K}_{bv} \cdot y_v = -\boldsymbol{K}_b^T \cdot F_{vr} \tag{9}$$

where $\boldsymbol{C}_{bb} = \boldsymbol{N}_b^T \cdot c_v \cdot \boldsymbol{N}_b$, $\boldsymbol{C}_{bv} = \boldsymbol{N}_b^T \cdot c_v$, $\boldsymbol{K}_{bb} = \boldsymbol{N}_b^T \cdot k_v \cdot \boldsymbol{N}_b$, $\boldsymbol{K}_{bc} = \boldsymbol{N}_b^T \cdot c_v \cdot \dot{\boldsymbol{N}}_b$ and $\boldsymbol{K}_{bv} = \boldsymbol{N}_b^T \cdot k_v$ are the bridge additional damping and stiffness, respectively.

2.1.4. The VBI Dynamic Model

The VBI dynamic equation can be obtained by combining Equations (6) and (9) in the matrix form:

$$\begin{bmatrix} \boldsymbol{M}_b & \boldsymbol{0} \\ \boldsymbol{0} & m_v \end{bmatrix}\begin{bmatrix} \ddot{\boldsymbol{y}}_b \\ \ddot{y}_v \end{bmatrix} + \begin{bmatrix} \boldsymbol{C}_b + \boldsymbol{C}_{bb} & -\boldsymbol{C}_{bv} \\ -\boldsymbol{C}_{vb} & c_v \end{bmatrix}\begin{bmatrix} \dot{\boldsymbol{y}}_b \\ \dot{y}_v \end{bmatrix} + \begin{bmatrix} \boldsymbol{K}_b + \boldsymbol{K}_{bb} + \boldsymbol{K}_{bc} & -\boldsymbol{K}_{bv} \\ -\boldsymbol{K}_{vb} & k_v \end{bmatrix}\begin{bmatrix} \boldsymbol{y}_b \\ y_v \end{bmatrix} = \begin{bmatrix} -\boldsymbol{N}_b^T \cdot F_{vr} \\ F_{vr} - F_G \end{bmatrix} \tag{10}$$

To improve the computational efficiency, the model synthesis method is used to reduce the computational degrees of freedom of the bridge, and the vibration equations of the VBI dynamic model are rewritten as follows:

$$\begin{bmatrix} \boldsymbol{I} & 0 \\ 0 & m_v \end{bmatrix}\begin{bmatrix} \ddot{q}_b \\ \ddot{y}_v \end{bmatrix} + \begin{bmatrix} \overline{\boldsymbol{C}}_b + \boldsymbol{\Phi}_b^T \boldsymbol{C}_{bb}\boldsymbol{\Phi}_b & -\boldsymbol{\Phi}_b^T \boldsymbol{C}_{bv} \\ -\boldsymbol{C}_{vb}\boldsymbol{\Phi}_b & c_v \end{bmatrix}\begin{bmatrix} \dot{q}_b \\ \dot{y}_v \end{bmatrix} + \begin{bmatrix} \overline{\boldsymbol{K}}_b + \boldsymbol{\Phi}_b^T(\boldsymbol{K}_{bb} + \boldsymbol{K}_{bc})\boldsymbol{\Phi}_b & -\boldsymbol{\Phi}_b^T \boldsymbol{K}_{bv} \\ -\boldsymbol{K}_{vb}\boldsymbol{\Phi}_b & k_v \end{bmatrix}\begin{bmatrix} q_b \\ y_v \end{bmatrix} = \boldsymbol{F}_R + \boldsymbol{F}_G \tag{11}$$

where $\boldsymbol{\Phi}_b$ is the modal shape matrix of the bridge.

$$\boldsymbol{F}_R = \begin{bmatrix} -\boldsymbol{\Phi}_b^T \boldsymbol{N}_b^T \cdot F_{vr} \\ F_{vr} \end{bmatrix}, \ \boldsymbol{F}_G = \begin{bmatrix} 0 \\ F_G \end{bmatrix}$$

Let $\boldsymbol{M} = \begin{bmatrix} \boldsymbol{M}_b & 0 \\ 0 & m_v \end{bmatrix}$, $\boldsymbol{C} = \begin{bmatrix} \boldsymbol{C}_b + \boldsymbol{C}_{bb} & -\boldsymbol{C}_{bv} \\ -\boldsymbol{C}_{vb} & \boldsymbol{C}_v \end{bmatrix}$, $\boldsymbol{K} = \begin{bmatrix} \boldsymbol{K}_b + \boldsymbol{K}_{bb} + \boldsymbol{K}_{bv} & -\boldsymbol{K}_{bv} \\ -\boldsymbol{K}_{vb} & k_v \end{bmatrix}$, $\boldsymbol{F}_{eq} = \boldsymbol{F}_G + \boldsymbol{F}_R$.

According to the Newmark-β method, one can obtain:

$$K_{eq} y_i = F_{eq} \tag{12}$$

where

$$F_{eq} = F + M[\frac{1}{\beta\Delta t^2} y_i + \frac{1}{\beta\Delta t} \dot{y}_i + (\frac{1}{2\beta} - 1)\ddot{y}_i] + C[\frac{\gamma}{\beta\Delta t} y_i + \frac{\gamma}{\beta} - 1)\dot{y}_i + (\frac{\gamma}{\beta} - 2)\ddot{y}_i]$$
$$K_{eq} = K + \frac{1}{\beta\Delta t^2} M + \frac{\gamma}{\beta\Delta t} C$$

According to the Newmark-β method, when the initial displacement, velocity, and acceleration at the initial time are given, the response of the system at any time can be determined based on Equation (12), and the complete time series data for dynamic response can be obtained.

2.2. Deep Convolutional Neural Network

2.2.1. Popular Models of DCNN

In 1998, Lecun et al. [24] constructed the first convolutional neural network model (LeNet-5), which has excellent identification performance in handwritten font identification tasks. In 2012, Krizhevsky and Hinton [25] proposed AlexNet, which won first place in that year's ImageNet Visual Recognition Challenge. Since then, various artificial intelligence applications based on DCNN have been merged, and DCNN has developed rapidly. Some new models have been proposed, which can be divided into the branchless model, such as VGG, and modular stacked models, such as GoogleNet, ResNet, and DenseNet.

VGGNet [26] constructs a deeper network structure based on AlexNet network to improve the learning ability of image features. Meanwhile, VGGNet stacks smaller convolution kernels to reduce network parameters and iterations. VGGNets are still widely used in image feature extraction due to their excellent performance. In the same year, Google launched GoogleNet [27], which used far less network parameters than VGGNet. Therefore, it takes up less memory and computing resources in computing. It also used a modular network structure containing convolution kernel parallel merging. After years of optimization and improvement, several versions have been derived.

ResNet [28] adopts the same modular stack structure as VGGNet and introduces a novel residual structure to greatly improve the fitting ability and overcome the degradation problem of deep neural network. Subsequently, DenseNet [29] introduces a dense block structure to reuse the features of each layer of feature map, thereby improving the transmission of features in the network, improving the identification efficiency of the network, and reducing the number of network parameters.

Although the deep convolutional network model has excellent performance, its computational efficiency is low, due to the complex network structure, which makes DCNNs difficult to widely applied in practical engineering.

2.2.2. Lightweight Convolutional Neural Networks

Lightweight convolutional neural networks aim to reduce computational storage and increase recognition speed. Lightweight convolutional neural networks mainly include ShuffleNet series and MobileNet series. Howard et al. [30] proposed MobileNetV1, which uses a straight network structure and replace the deep separable convolution instead of traditional convolutional layers. By this improvement, the model parameters can be greatly reduced while ensuring the computational accuracy of the network. Then, Sandler et al. proposed MobileNetV2, which adopts the deep separable convolution instead of the traditional standard convolution and adds the inverted residuals block and linear bottlenecks structure. Therefore, MobileNetV2 reduces the number of model parameters while ensuring accuracy. The convolution process of MobileNetV2 is shown in Figure 2. Due to the excellent performance and lightweight size, MobileNet series models are often used in various recognition fields [31–34]. Because the moving load identification needs to respond rapidly to the vehicle information in driving, this paper introduces MobileNetv2 into the moving load identification.

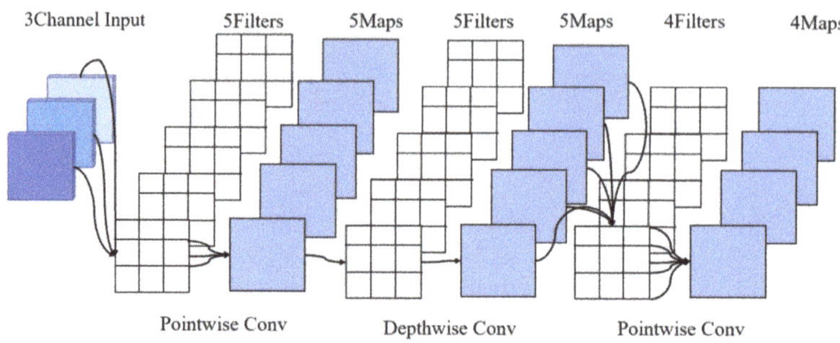

Figure 2. Convolution process of MobileNetV2.

2.3. *Transfer Learning*

The main concept of transfer learning is to utilize data from similar fields to solve the problem of data shortage in target fields. Its goal is to utilize the knowledge learned from an original environment to help learn tasks in a new environment. Depending on the requirements for transfer into the target domain, it can be classified as the following forms [35]: (1) instance transfer; (2) feature representation transfer; (3) parameter transfer; (4) relational knowledge transfer.

The strong transferability of neural network model greatly improves the applicability of transfer learning in the field of deep learning. Compared with the general transfer learning method, the transfer learning strategy in DCNNs transfers the shallow feature extraction ability (i.e., texture, edge feature, etc.) in the source domain to the target domain. Therefore, fine-tuning is commonly used in DCNNs for transfer learning. Specifically, it is to freeze the front several layers network of the pre-trained model, retrain the remaining layers, and replace the task classifier to match the new learning task when the new dataset has a small amount of data, and it is significantly different from the training dataset used by the pre-trained model. An example of transfer learning is shown in Figure 3.

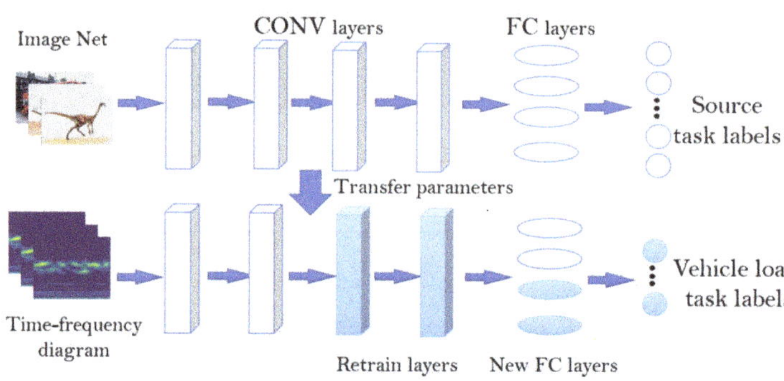

Figure 3. Illustration of transfer learning strategy process.

3. The Proposed Method

This paper proposes a moving load identification method based on MobileNetV2 and transfer learning, which identify the moving load information from responses of bridge, respectively. The training of DCNNs needs to optimize a large number of parameters and construct sufficient samples, and it will take a lot of time to train the model from scratch. Therefore, this paper adopts transfer learning strategy. The implementation process of this method is shown in Figure 4, including the following steps:

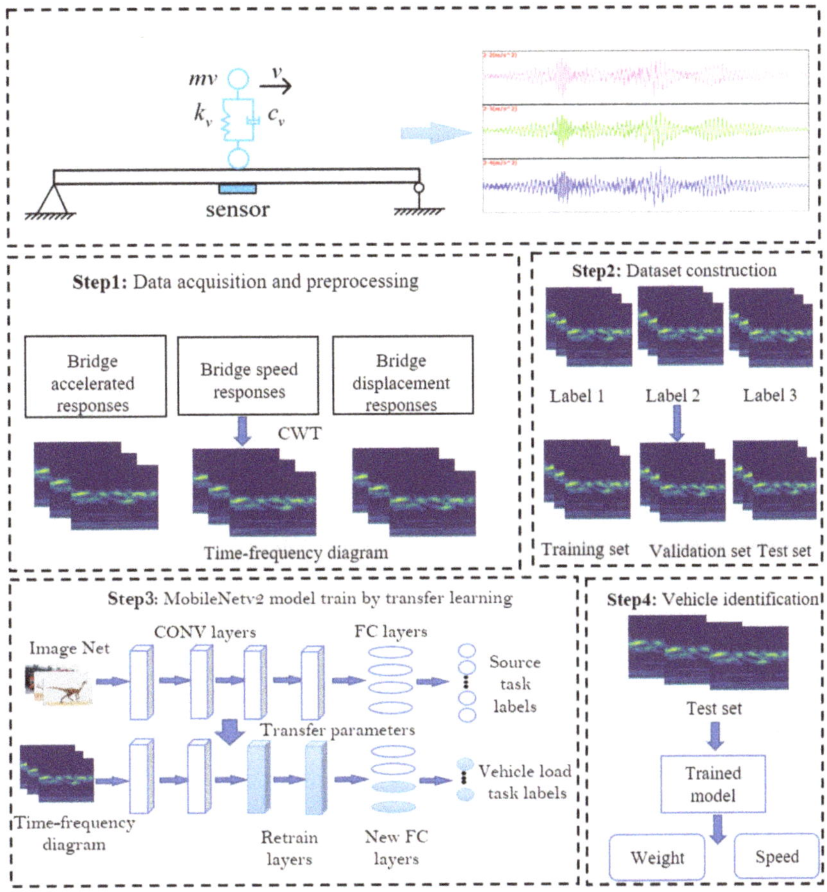

Figure 4. Overview of the proposed method.

(1) Responses acquisition and pre-processing. By solving the VBI dynamic equation, the responses of bridge corresponding to different vehicle parameters are obtained. In order to meet the input requirements of DCNN, the CWT is applied to transform the response into a time-frequency map. The formula of CWT is as follows:

$$WT(a,b) = \frac{1}{\sqrt{a}}\int_{-\infty}^{\infty} x(t)\cdot\psi\left(\frac{t-b}{a}\right)dt \tag{13}$$

where a is the scaling factor which can control the expansion of wavelet, b denotes the shifting factor that identifies its location, and ψ denotes the mother wavelet. In this paper, Complex Morlet wavelet is used as the mother wavelet ψ because it has good resolution in both time and frequency domains [36]. Both the scaling factor a and the shifting factor b are set to 3.

(2) Dataset construction. The size of normalized image samples is adjusted to the input size of the MobileNetV2 model. On this basis, all image samples are labelled with the corresponding VBI system parameters, thus forming the sample library of displacement, velocity, and acceleration responses for the moving load identification task. Then, the samples are divided into the training set, validation set and test set with the ratio as 8:1:1. The network is trained by the training set, the network is verified by the validation set, and the performance of the network is evaluated by the test set.

(3) Model training. A pre-trained MobileNetV2 model is used to moving load identification tasks. On this basis, the hyperparameters (batch size, learning rate, epoch, etc.) of model is adjusted to optimize the network performance. Meanwhile, the bottlenecks of the MobileNetV2 model are retrained, and the original 1000-class target classifiers are replaced with the 5-class target classifiers required for moving load identification in this task.
(4) Moving load identification. Test set samples are input to the trained MobileNetV2 model to obtained information of moving loads and to evaluate the identification performance of the network.

4. Case Study

In order to verify the feasibility of the proposed method, the VBI system with single degree of freedom is taken as the object in this paper. According to the VBI dynamic equation, a sufficient sample database is constructed to perform moving load identification tasks.

4.1. The Numerical Model

In this paper, a 30 m concrete simply supported, single-span bridge is established to verify the method, as shown in Figure 5. The main beam is simulated by beam188 element [37–39] with concrete specification of C50. The elastic modulus is 3.40×10^4 MPa, and Poisson ratio is 0.2, mass density is 2600 kg/m^3. The single-wheeled vehicle model is simulated by spring element. The spring stiffness of vehicle is set to $k_v = 190$ kN/m and the damping of the vehicle is $c_v = 5$ kN·m/s. The information for the vehicle is designed according to the required load conditions.

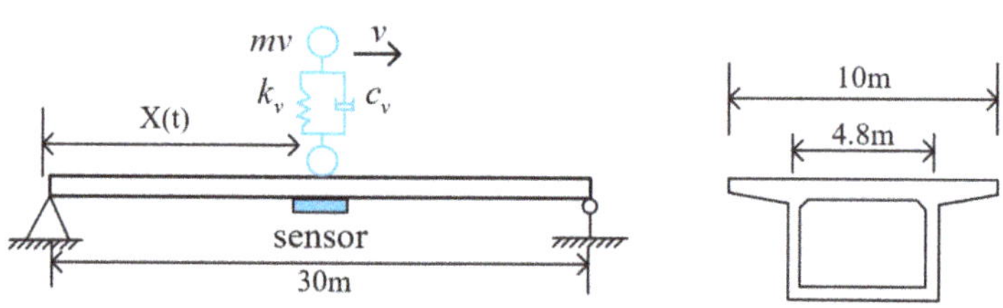

Figure 5. Vehicle-bridge system with single degree of freedom.

4.2. Vehicle Weight Identification

Acquisition of vehicle weight information is critical to the safe operation of bridges. In this section, vehicle weight is classified into 5 categories from light to heavy (i.e., A, B, C, D, E), and the vehicle weight classification table is shown in Table 1.

Table 1. Classification of vehicle weight.

Class	Vehicle Weight (kN)	Mean (kN)
A	0~10	5
B	10~20	15
C	20~30	25
D	30~40	35
E	40~50	45

A total of 500 weight samples are randomly generated, including 100 samples per grade. Then, the bridge responses of each weight sample are calculated when both the speed and road roughness class are fixed. By the aforementioned preprocessing procedure, the response of the VBI dynamic system is converted into images, which were divided into training set, validation set and test set for model training, selection, and evaluation, respectively. Figures 6 and 7 show the displacement sample and acceleration sample images corresponding to five randomly selected vehicle weights, respectively. Visible discrepancy can be observed from those images, which suggests the possibility of accurate identification results. Then, those samples are used as the input of the network model for model training and vehicle weight identification evaluation.

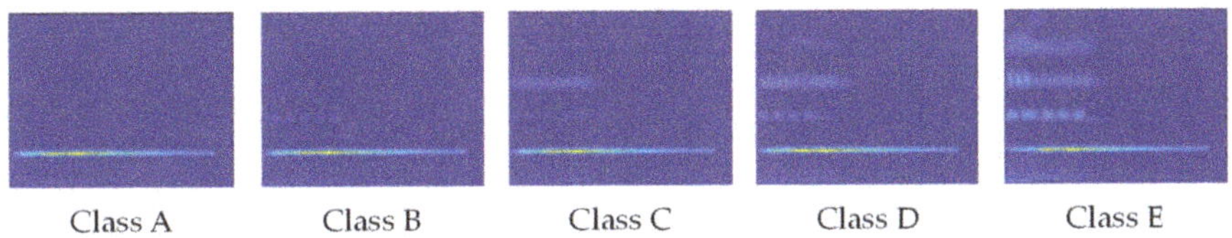

Figure 6. Displacement sample images of five randomly selected vehicle weights.

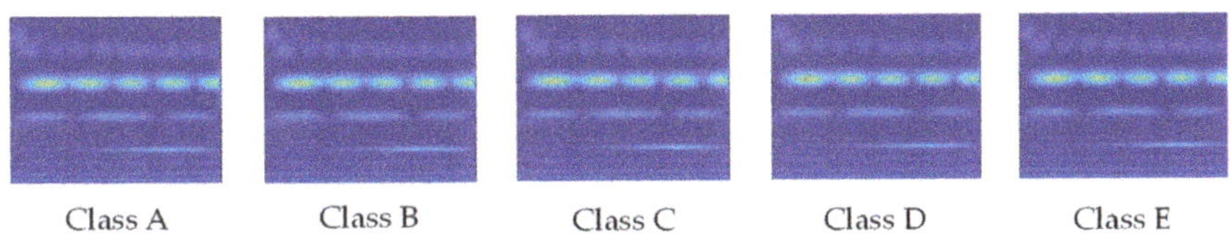

Figure 7. Acceleration sample images of five randomly selected vehicle weights.

The trained MobileNetV2 model is used to identify vehicle weight information. Because the data characteristics of the target domain are quite different from those of the source domain, the MobileNetV2 model is transferred in the form of fine-tuning. The first 12th Bottlenecks of the MobileNetV2 model are frozen, the rest of the bottlenecks are retrained, and the original 1000-class target classifier are replaced with the 5-class target classifier required for moving load identification in this paper.

In this paper, the stochastic gradient descent algorithm is used to train the model. The learning rate is set to 0.0001. According to the scale of sample data, batch size is set to 64 and epoch is set to 50. Based on Pytorch 2.0, the MobileNetV2 model is trained under the operating system of Windows 10 and CPU of AMD Ryzen 53550H @ 2.10 GHz.

The accuracy and loss curves of the model training to convergence process are shown in Figure 8, and the accuracy and loss at the end of training are shown in Table 2. The vehicle weight is identified by using the test set samples. The identification results and confusion matrix are shown in Table 3 and Figure 8. After careful analysis of Figures 8 and 9 and Tables 2 and 3, the following conclusions can be drawn:

(1) The vehicle weight information can be accurately identified from the response samples. For example, it can be seen from Figure 8 that, when the epoch is within 10, the accuracy curve and the loss curve of the training set and verification set input by all the sample change faster. When the epoch is within 10 and 50, the accuracy curve and the loss curve change slowly and gradually tend to be stable. From Table 2, we can see that after the model training completed, the training set accuracy of each input is above 98%, in which the accuracy of the training set of both velocity sample input and acceleration response sample input reached 99%, the accuracy of the validation set of both displacement response sample input and velocity response sample input reached 98%, and the acceleration response sample input also reached 98.83%. It shows that the MobileNetV2 model trained by transfer learning can converge after less iterations. At the same time, the confusion matrix of the test results shows that the proposed method misclassifies only a small number of samples.

(2) The displacement response sample has the best identification effect on vehicle weight identification tasks. From the test results in Table 3, the identification accuracy using the displacement response is the highest, reaching 100%; the lowest identification accuracy of acceleration response is 96.08%. As can be seen from the confusion matrix in Figure 9, the displacement response profile sample input accurately classifies all test samples, and both acceleration and velocity sample inputs misclassify only a

small number of samples. The above analysis shows that the information of vehicle weight is most efficiently identified from displacement responses.

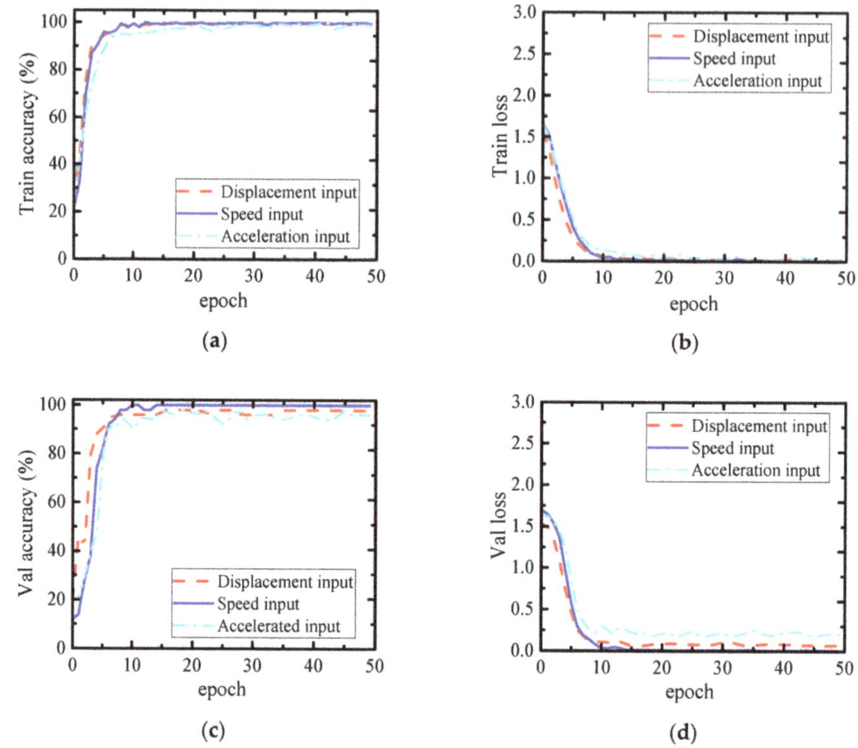

Figure 8. Results of vehicle weights identification: (**a**) Training accuracy curves of response samples; (**b**) Training loss curves of the response samples; (**c**) Validation accuracy curves of the response samples; (**d**) Validation loss curves of response samples.

Table 2. Train results of vehicle weights identification after 50 epochs.

Input	Accuracy (%)		Loss	
	Training Set	**Validation Set**	**Training Set**	**Validation Set**
Displacement	98.83	98.00	0.0239	0.0714
Velocity	99.62	100.00	0.0105	0.0104
Acceleration	99.43	96.08	0.0275	0.2069

Table 3. Identification results of vehicle weights.

Input	Identification Accuracy (%)
Displacement	100.00
Velocity	98.00
Acceleration	96.08

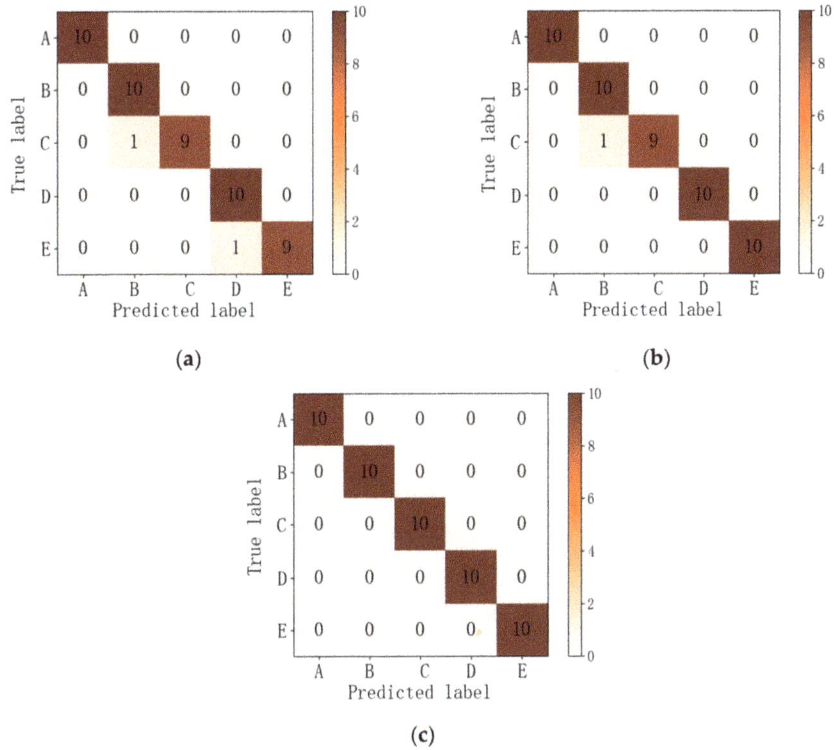

Figure 9. Confusion matrix for vehicle weight identification results: (**a**) Test results of acceleration samples; (**b**) Test results of speed sample; (**c**) Test results of displacement sample.

4.3. Vehicle Speed Identification

Acquisition of vehicle speed information is vital for control of vehicle overspeed. Therefore, this paper carries out a vehicle speed identification task to verify the performance of the proposed method in vehicle speed identification. Similar to vehicle weight identification, speed is divided into five grades from slow to fast (A, B, C, D, E). The speed classification table is shown in Table 4.

Table 4. Classification of vehicle speed.

Class	Vehicle Speed (m/s)	Mean (m/s)
A	0~5	2.5
B	5~10	7.5
C	10~15	12.5
D	15~20	17.5
E	20~25	22.5

Similarly, 500 speed samples are randomly generated, including 100 samples per grade. Then, based on the VBI dynamic equations, the responses of each speed sample are obtained in turn under the condition that the vehicle weight is class B, and the unevenness of the road surface is class B. Thus, the sample images are constructed according to the aforementioned strategy to generate a sample library for the vehicle speed identification task. Then, the samples are used as input to the network model for model training and evaluation of vehicle speed identification, respectively.

The accuracy and loss curves of the type training to convergence process are shown in Figure 10. Accuracy and loss at the end of training are shown in Table 5. The identification results and the confusion matrix are shown in Table 6 and Figure 11. After a detailed analysis of Figures 10 and 11 and Tables 5 and 6, the following conclusions can be drawn:

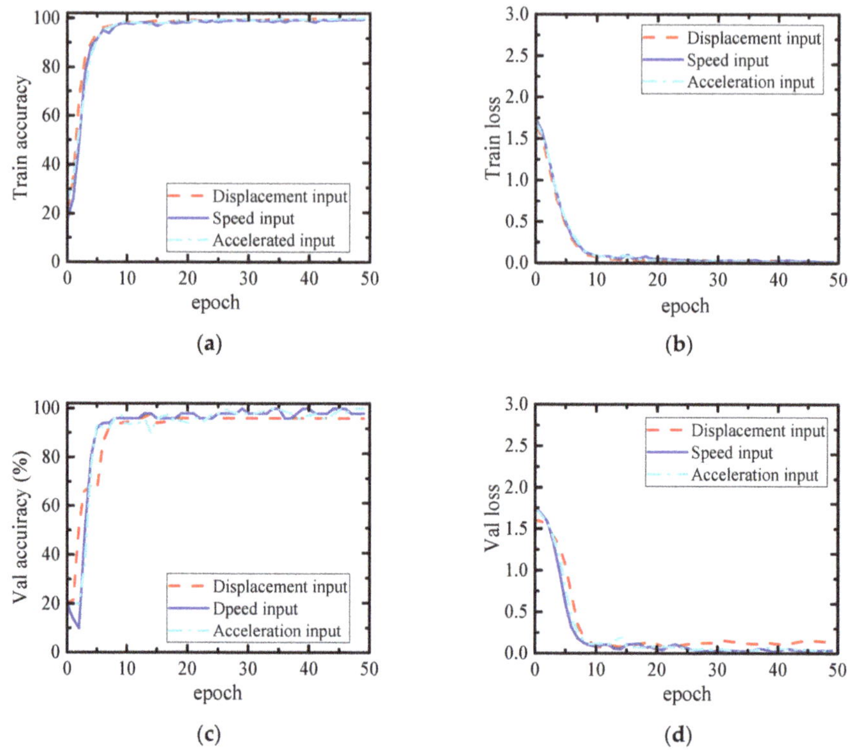

Figure 10. Results of vehicle speed identification: (**a**) Training accuracy curves of response samples; (**b**) Training loss curves of the response samples; (**c**) Validation accuracy curves of the response samples; (**d**) Validation loss curves of response samples.

Table 5. Train results of vehicle speed identification after 50 epochs.

Input	Accuracy (%)		Loss	
	Training Set	**Validation Set**	**Training Set**	**Validation Set**
Displacement	99.43	96.00	0.0298	0.0236
Velocity	99.61	98.00	0.0662	0.0163
Acceleration	100.00	100.00	0.0148	0.0078

Table 6. Identification results for the vehicle weight.

Input	Identification Accuracy (%)
Displacement	96.00
Velocity	98.00
Acceleration	100.00

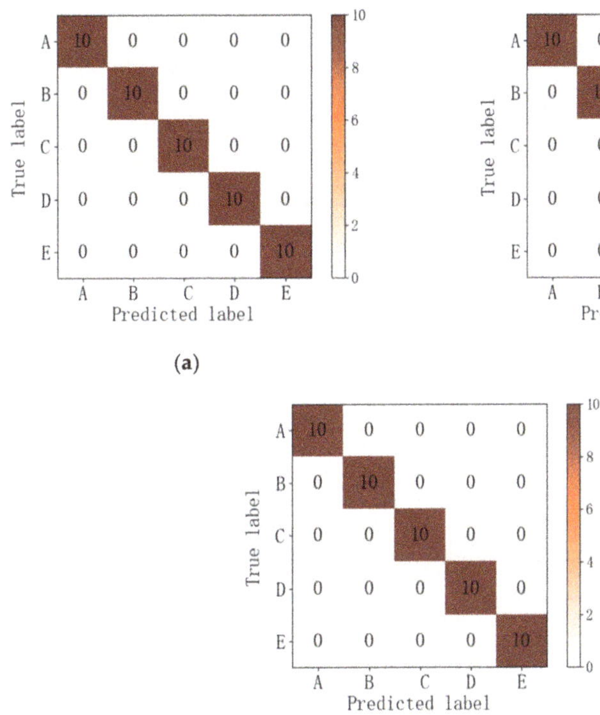

Figure 11. Confusion matrix for vehicle speed identification results: (**a**) Test results of acceleration samples; (**b**) Test results of speed sample; (**c**) Test results of displacement sample.

(1) The vehicle speed information can be accurately identified from response samples. For example, as can be seen in Figure 10 and Table 5 that the accuracy and loss of the training set, validation set of the proposed method are gradually stabilized after 10 epochs. The accuracy of each input training set reaches 99% after training, the loss of both the training and validation sets are within 0.03. At the same time, it can be seen from the test results and confusion matrix that this method misclassifies only a small number of samples. Those analysis demonstrate that the method has excellent performance in vehicle weight identification tasks.

(2) The acceleration response sample has the best identification effect on vehicle speed identification tasks. From the test results in Table 6, the highest identification accuracy of 100% was achieved using acceleration response as input. It can be seen from the confusion matrix of Figure 11 that all test samples are accurately classified by using acceleration response as input, and the information of vehicle speed is most efficiently identified from acceleration responses.

5. Discussion and Analysis

5.1. Comparison of Popular Network Models

To further validate the efficiency of the proposed method, the test results of MobileNetV2 model are compared with those of AlexNet, VGG16, and ResNet. Specifically, the above network models are transferred to identification tasks in the same fine-tuning form as the MobileNetV2 model. The acceleration samples are used as input on vehicle speed identification task, and the displacement samples are used as input on vehicle weight identification task. To compare the computational efficiency of the models, the training time complete 50 epochs, the weight file of model generated after identification of 50 samples in the test set, and the identification time for each model are compared.

The training results for vehicle weights identification task are shown in Figure 12, After 50 epochs, the results of the identification of the test set are shown in Tables 7 and 8. As can be seen from Figure 12, the MobileNetV2 model and each of the other models have converged after 50 epochs. As can be seen from Table 7, the final accuracy of the training set is above 98% for MobileNetV2 and other models. For the final accuracy of the validation set, VGG16 model has the highest accuracy of 100%, AlexNet has the lowest accuracy of 94%, and MobileNetV2 and ResNet both reach 98%. For training time, MobileNetV2 takes 53 min to train, which is only 10% of VGG16 and 55% of AlexNet and ResNet. In terms of identification speed, MobileNetV2 took 11.04 s to identify 50 test samples, which was only 27.8% of VGG16, 44.3% of AlexNet, and 58.8% of ResNet.

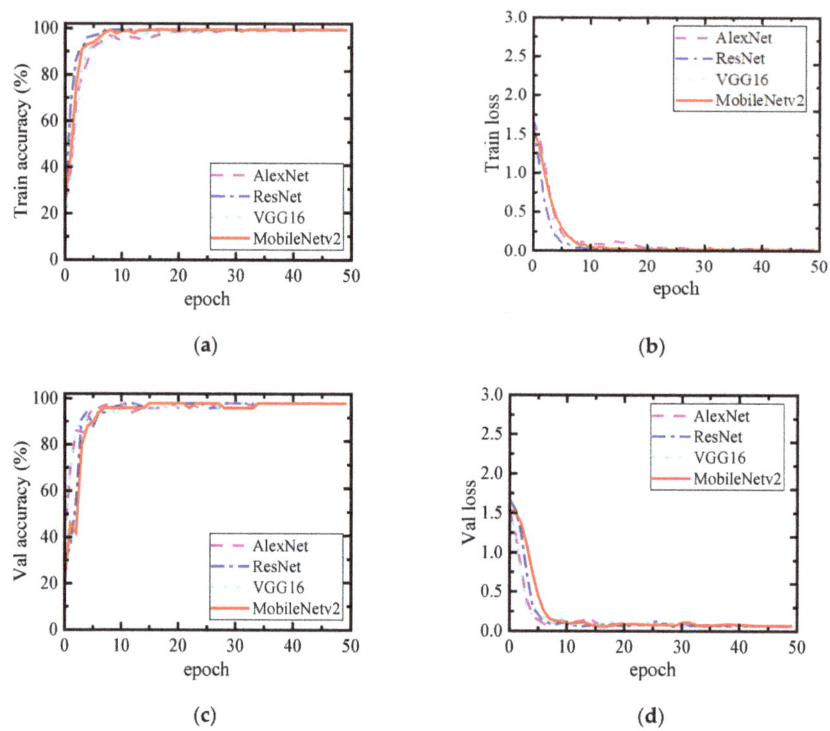

Figure 12. Results of vehicle weights identification of different models: (**a**) Training accuracy curves of response samples; (**b**) Training loss curves of the response samples; (**c**) Validation accuracy curves of the response samples; (**d**) Validation loss curves of response samples.

Table 7. Train results of vehicle weight identification of different models after 50 epochs.

Model	Accuracy (%)		Loss		Total Training Time (Min)
	Training Set	Validation Set	Training Set	Validation Set	
AlexNet	98.63	94.00	0.0580	0.0837	98
ResNet	99.80	98.00	0.0143	0.0437	95
VGG16	99.80	100.00	0.0137	0.0142	508
MobileNetv2	98.83	98.00	0.0275	0.0714	**53**

Table 8. Identification results of vehicle weight for different models.

Model	Accuracy (%)	Weight File (MB)	Identification Time (s)
AlexNet	98.00	684.31	24.91
ResNet	98.00	134.26	18.78
VGG16	98.00	1573.64	39.76
MobileNetv2	100.00	27.12	11.04

Training results of vehicle speed identification task are shown in Figure 13. The identification results of the test set after 50 epochs are shown in Tables 9 and 10. From Figure 13, it can be seen that the MobileNetV2 model and each of the other models have converged after 50 epochs, On the accuracy curves of the training and validation sets, ResNet and MobileNetV2 rise faster and fluctuate less, while AlexNet and VGG16 rise slower and fluctuate more. The accuracy of both the training and validation sets after training is 100% for MobileNetV2 and 98% for all the other three models. Based on the identification results in Table 10, MobileNetV2 and VGG16 have the highest accuracy of 100%, while AlexNet has the lowest accuracy of 94%. The weight file size of MobileNetV2 is only 1.7% of VGG16, 3.9% of AlexNet, and 20% of ResNet, which has obvious advantages. In terms of identification speed, MobileNetV2 takes 9.58S to identify 50 test samples, which is only 24.8% of VGG16, 49.7% of AlexNet, and 50% of ResNet.

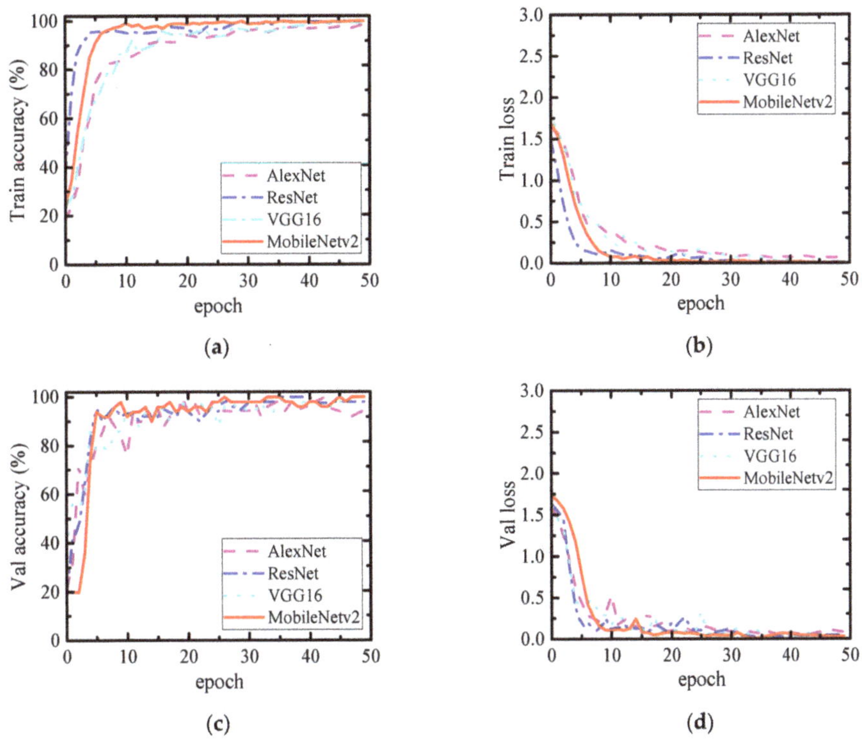

Figure 13. Results of vehicle speed identification of different models: (**a**) Training accuracy curves of response samples; (**b**) Training loss curves of the response samples; (**c**) Validation accuracy curves of the response samples; (**d**) Validation loss curves of response samples.

Table 9. Train results for vehicle speed identification of different models after 50 epochs.

Model	Accuracy (%)		Loss		Total Time (Min)
	Training Set	Validation Set	Training Set	Validation Set	
AlexNet	99.23	98.00	0.0251	0.0711	99
ResNet	98.43	98.00	0.0219	0.0751	95
VGG16	98.43	98.00	0.0241	0.0680	508
MobileNetv2	100.00	100.00	0.0148	0.0078	**53**

Table 10. Identification results of vehicle speed for different models.

Model	Accuracy	Weight File (MB)	Identification Time (S)
AlexNet	94.00	684.31	20.40
ResNet	98.00	134.26	19.15
VGG16	100.00	1573.64	38.59
MobileNetv2	100.00	27.12	9.58

In summary, MobileNetV2 outperforms the other three models in both vehicle weight identification and vehicle speed identification tasks, in terms of identification accuracy. At the same time, MobileNetV2 is superior to other models in terms of training time and memory resource occupancy on the premise of ensuring accuracy. Additionally, MobileNetV2 has a major advantage in the speed of identification from the test set. It can be seen that this method has the best performance in terms of identification speed and accuracy, and the short training time and identification time make this method more suitable for practical applications.

5.2. Robustness Analysis

Anti-noise ability is a key basis for judging the practicability of the method. Therefore, this section evaluates its robustness by considering the measurement noise. In this paper, to simulate the actual test environment, the following noise levels of white noise were added to the sample data. In accordance with the implementation steps of the proposed method, the training and testing of the models for the two identification tasks were carried out under different noise levels, with the same transfer process and parameter settings as in the absence of noise; acceleration is used to identify vehicle speed, and displacement is used to identify vehicle weight.

The training results of the vehicle weight identification task are shown in Figure 14. As can be seen from Figure 14, the accuracy and loss curves of the training set do not fluctuate much under different noise levels and the accuracy and loss curves of the validation set fluctuate in the first 20 epochs under 10 dB noise condition, but become smooth at 30 to 50 epochs. The models have converged after 50 epochs for different noise conditions.

The training results for the vehicle speed identification task are shown in Figure 15. It can be seen that the training set and validation set curves of vehicle speed identification and vehicle weight identification at different noise levels have the same pattern. The accuracy and loss curves of the training set do not fluctuate much. The accuracy and loss curves of the validation set have small fluctuations in the first 20 epochs under 10 dB noise conditions, but become smooth in 30 to 50 epochs.

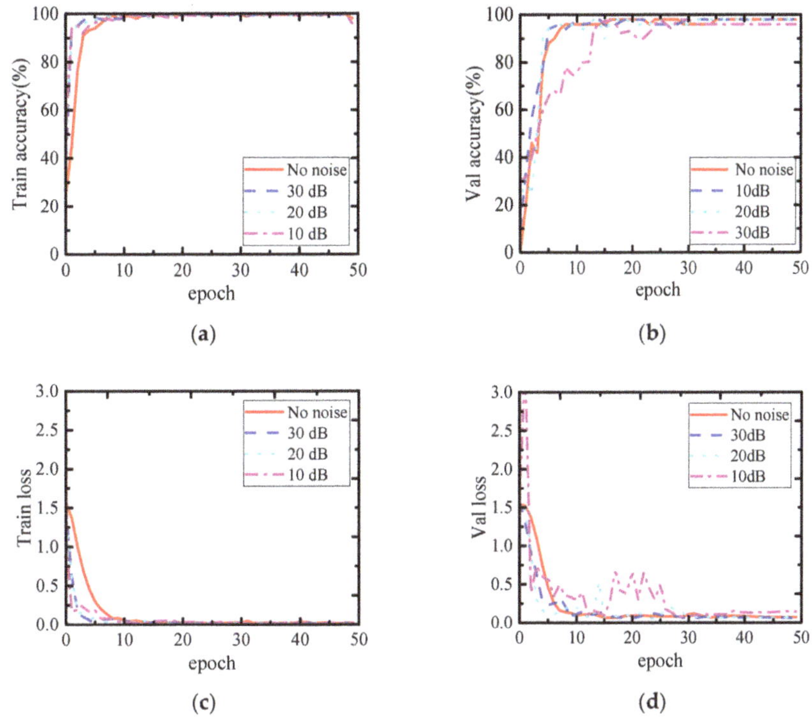

Figure 14. Results of vehicle weights identification under different noise levels: (**a**) Training accuracy curves of response samples; (**b**) Validation accuracy curves of the response samples; (**c**) Training loss curves of the response samples; (**d**) Validation loss curves of response samples.

From the test results in Table 11, the identification accuracy of the test set of the vehicle weight identification task decreases and stabilizes at 98%. With the noise level increases, identification accuracy of test set in vehicle speed identification decreased slightly. The accuracy at 30 dB and 10 dB was 98%, and the accuracy of 20 dB was still 100%. It shows that there is no obvious downward trend in the accuracy of model test, and there is only a slight fluctuation. It can be concluded that the proposed method presents excellent, strong robustness.

Table 11. Model Identification results under different noise levels.

Identification Task	No Noise	30 dB	20 dB	10 dB
vehicle weight	100%	98%	98%	98%
vehicle speed	100%	98%	100%	98%

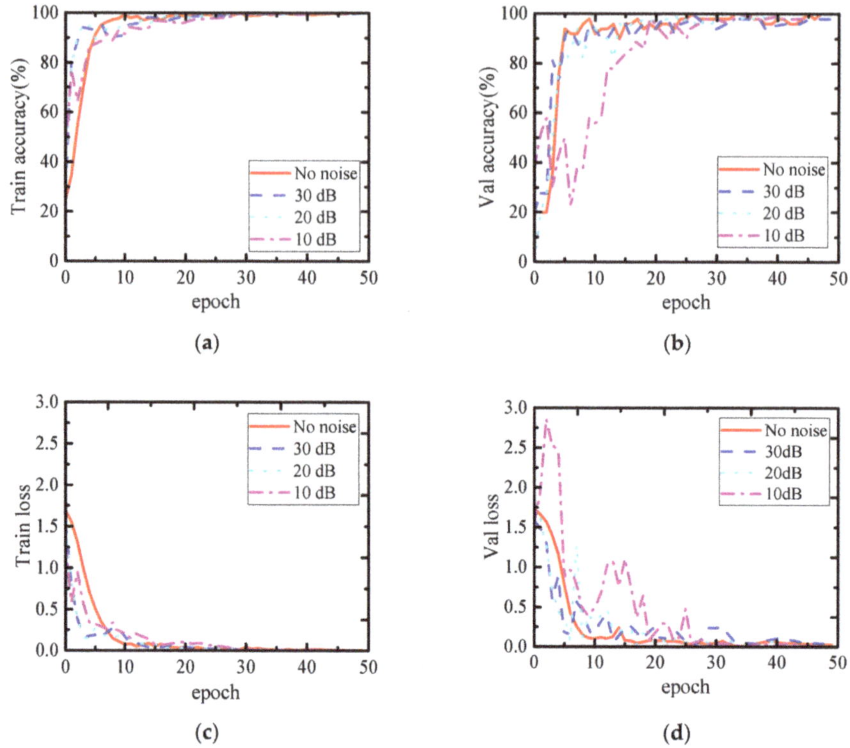

Figure 15. Results of vehicle speed identification under different noise levels: (**a**) Training accuracy curves of response samples; (**b**) Validation accuracy curves of the response samples; (**c**) Training loss curves of the response samples; (**d**) Validation loss curves of response samples.

6. Conclusions

This paper presents a moving load identification method based on MobileNetV2 and transfer learning. The performance of the proposed method is verified by the numerical simulation and parameter analysis. The main conclusions were obtained as follows:

(1) Vehicle load information can be accurately separated from the responses of the established VBI dynamic model by the proposed method. The displacement response is the most effective model input on vehicle weight identification task, with an identification accuracy of 100%. The acceleration response is the most effective input of model for vehicle speed identification task, with 100% identification accuracy.

(2) The proposed method has higher identification efficiency. Lightweight convolutional neural networks and transfer learning strategies can improve identification efficiency. Compared to VGG16, AlexNet, and ResNet, the training time of MobileNetV2 is reduced by more than 50%, and the identification speed of MobileNetV2 is increased by more than 42.2% on vehicle weight identification and more than 50% on vehicle speed identification. In addition, the storage occupancy of MobileNetV2 is only 1.7% of VGG16, 3.9% of AlexNet, and 20% of ResNet.

(3) The proposed method has excellent robustness. The robustness analysis shows that the method can still maintain excellent identification ability when encountering a higher noise level. At 10 dB noise level, the identification accuracy of vehicle speed and vehicle weight still reached 98%.

Although this method has achieved satisfactory identification results on the established numerical scenarios, it cannot directly address cases where vehicles are distributed in multiple lanes and drive in opposite direction. In addition, for vehicles with multiple axles, it is difficult to identify the axle weight of each wheel. Therefore, the future work will focus on further improving the practicality of the proposed method.

Author Contributions: Conceptualization, Y.Q.; methodology, Y.Q.; software, Y.Q., Q.T. and J.X.; validation, Y.Q., Q.T. and J.X.; formal analysis, Q.T. and J.X.; writing—original draft preparation, Y.Q.; writing—review and editing, Y.Q., J.X., C.Y., Z.Z. and X.Y.; visualization, Q.T. and J.X.; supervision, Q.T. and C.Y.; funding acquisition, J.X. All authors have read and agreed to the published version of the manuscript.

Funding: This research was funded by the Natural Science Foundation of China (grant Nos. 52278292, 51978111), Chongqing Technology Innovation and Application Development Special Key Project (grant No. CSTB 2022TIAD-KPX0205), Chongqing Transportation Science and Technology Project (2022-01), China Postdoctoral Science Foundation (grant No. 2022MD713699), Special Funding of Chongqing Postdoctoral Research Project (grant No. 2021XM1016), and Chongqing Zhongxian Science and Technology Plan Project (grant No. zxkyxm202202).

Data Availability Statement: Data available on request, due to restrictions.

Conflicts of Interest: The authors declare no conflict of interest.

Nomenclature

L	Bridge length
N	Number of data points
ϕ_i	Random phase angle
F_G	Vehicle gravity
F_{vb}	Interaction force
k_v	Spring stiffness
c_v	Damping
m_v	Vehicle weight
y_v	Vehicle displacement
y_{bc}	Bridge deflection
$\boldsymbol{y}_b$	Bridge global displacement vector
$\boldsymbol{N}_b$	Bridge shape function
$\boldsymbol{M}_b$	Bridge mass matrices
$\boldsymbol{C}_b$	Bridge damping matrices
$\boldsymbol{K}_b$	bridge stiffness matrices
$\boldsymbol{F}_{bv}$	Quivalent nodal force
$\boldsymbol{C}_{vb}$	Vehicle additional damping
$\boldsymbol{K}_{vb}$	Vehicle additional stiffness
F_{vr}	Load term of the vehicle
$\boldsymbol{C}_{bb}$	Bridge additional damping
$\boldsymbol{C}_{bv}$	Bridge additional damping
$\boldsymbol{K}_{bb}$	Bridge additional stiffness
$\boldsymbol{K}_{bc}$	Bridge additional stiffness
$\boldsymbol{K}_{bv}$	Bridge additional stiffness

References

1. Tang, Q.Z.; Zhou, J.T.; Xin, J.Z. Fast identification of random loads using the transmissibility of power spectral density and improved adaptive multiplicative regularization. *J. Sound Vib.* **2022**, *534*, 117033. [CrossRef]
2. Xin, J.Z.; Jiang, Y.; Zhou, J.T. Bridge deformation prediction based on SHM data using improved VMD and conditional KDE. *Eng. Struct.* **2022**, *261*, 114285. [CrossRef]
3. Zhang, H.; Li, H.X.; Zhou, J.T. A multi-dimensional evaluation of wire breakage in bridge cable based on self-magnetic flux leakage signals. *J. Magn. Magn. Mater.* **2023**, *566*, 170321. [CrossRef]
4. Jiang, Y.; Hui, Y.; Wang, Y. A novel eigenvalue-based iterative simulation method for multi-dimensional homogeneous non-Gaussian stochastic vector fields. *Struct. Saf.* **2023**, *100*, 102290. [CrossRef]
5. Li, S.J.; Xin, J.Z.; Jiang, Y. Temperature-induced deflection separation based on bridge deflection data using the TVFEMD-PE-KLD method. *J. Civ. Struct. Health* **2023**. [CrossRef]

6. Zhu, X.Q.; Law. S.S, Recent developments in inverse problems of vehicle–bridge interaction dynamics. *J. Civ. Struct. Health* **2016**, *6*, 107–128. [CrossRef]
7. Zhang, H.; Zhou, Y.; Quan, L. Identification of a moving mass on a beam bridge using piezoelectric sensor arrays. *J. Sound Vib.* **2021**, *491*, 115754. [CrossRef]
8. Yu, Y.; Cai, C.S.; Deng, L. State-of-the-art review on bridge weigh-in motion technology. *Adv. Struct. Eng.* **2016**, *19*, 1514–1530. [CrossRef]
9. Lee, S. An advanced coupled genetic algorithm for identifying unknown moving loads on bridge decks. *Math. Probl. Eng.* **2014**, *2014*, 462341. [CrossRef]
10. Wang, Y.; Qu, W. Experimental study on moving train loads identification from bridge responses. *Adv. Mater. Res.* **2010**, *143–144*, 32–37. [CrossRef]
11. Pan. C.D.; Yu, L. Moving force identification based on firefly algorithm. *AMR* **2014**, *919–921*, 329–333. [CrossRef]
12. Liu, H.; Yu, L. Moving force identification based on particle swarm optimization. In Proceedings of the 2016 12th International Conference on Natural Computation, Fuzzy Systems and Knowledge Discovery (ICNC-FSKD), Changsha, China, 13–15 August 2016; pp. 825–829. [CrossRef]
13. Vosoughi, A.R.; Anjabin, N. Dynamic moving load identification of laminated composite beams using a hybrid FE-TMDQ-GAs method. *Inverse Probl. Sci. Eng.* **2017**, *25*, 1639–1652. [CrossRef]
14. Liu, R.; Dobriban, E.; Hou, Z. Dynamic load identification for mechanical systems: A review. *Arch. Comput. Methods Eng.* **2022**, *29*, 831–863. [CrossRef]
15. Wang, C.; Ansari, F.; Wu, B.; Li, S.; Morgese, M.; Zhou, J. LSTM approach for condition assessment of suspension bridges based on time-series deflection and temperature data. *Adv. Struct. Eng.* **2022**, *25*, 3450–3463. [CrossRef]
16. Yang, H.; Yan, W.; He, H. Parameters identification of moving load using ANN and dynamic strain. *Shock Vib.* **2016**, *2016*, 8249851. [CrossRef]
17. Zhou, Y.; Pei, Y.; Zhou, S. Novel methodology for identifying the weight of moving vehicles on bridges using structural response pattern extraction and deep learning algorithms. *Measurement* **2021**, *168*, 108384. [CrossRef]
18. Chen, T.; Guo, L.; Duan, A. A feature learning-based method for impact load reconstruction and localization of the plate-rib assembled structure. *Struct. Health Monit.* **2022**, *21*, 1590–1607. [CrossRef]
19. Zhang, H.; Zhou, Y. AI-based modeling and data-driven identification of moving load on continuous beams. *Fundam. Res.* **2022**, *in press*. [CrossRef]
20. Sandler, M.; Howard, A.; Zhu, M.; Zhmoginov, A.; Chen, L.C. MobileNetV2: Inverted residuals and linear bottlenecks. In Proceedings of the 2018 IEEE/CVF Conference on Computer Vision and Pattern Recognition, Salt Lake City, UT, USA, 18–23 June 2018; pp. 4510–4520. [CrossRef]
21. Pourzeynali, S.; Zhu, X.; Zadeh, A.G. Comprehensive study of moving load identification on bridge structures using the explicit form of Newmark-β method: Numerical and Experimental Studies. *Remote Sens.* **2021**, *13*, 2291. [CrossRef]
22. Agostinacchio, M.; Ciampa, D.; Olita, S. The vibrations induced by surface irregularities in road pavements-a Matlab approach. *Eur. Transp. Res. Rev.* **2013**, *6*, 267–275. [CrossRef]
23. Tang, Q.Z.; Zhou, J.T.; Xin, J.Z. Novel identification technique of moving loads using the random response power spectral density and deep transfer learning. *Measurement* **2022**, *195*, 111120. [CrossRef]
24. Lecun, Y.; Bottou, L.; Bengio, Y. Gradient-based learning applied to document recognition. *Proc. IEEE* **1998**, *86*, 2278–2324. [CrossRef]
25. Krizhevsky, A.; Sutskever, I.; Hinton, G. Imagenet classification with deep convolutional neural networks. *Commun. ACM* **2017**, *60*, 84–90. Available online: https://dl.acm.org/doi/10.1145/3065386 (accessed on 3 January 2023). [CrossRef]
26. Simonyan, K.; Zisserman, A. Very deep convolutional networks for large-scale image recognition. *arXiv* **2014**, arXiv:1409.1556. [CrossRef]
27. Szegedy, C.; Liu, W.; Jia, Y. Going deeper with convolutions. In Proceedings of the 2015 IEEE Conference on Computer Vision and Pattern Recognition (CVPR), Boston, MA, USA, 7–12 June 2015; pp. 1–9. [CrossRef]
28. He, K.; Zhang, X.; Ren, S.; Sun, J. Deep residual learning for image recognition. In Proceedings of the 2016 IEEE Conference on Computer Vision and Pattern Recognition (CVPR), Las Vegas, NV, USA, 27–30 June 2016; pp. 770–778. [CrossRef]
29. Huang, G.; Liu, Z.; Laurens, V.; Weinberger, K.Q. Densely connected convolutional networks. In Proceedings of the 2017 IEEE Conference on Computer Vision and Pattern Recognition (CVPR), Honolulu, HI, USA, 21–26 July 2017; pp. 2261–2269. [CrossRef]
30. Howard, A.; Zhu, M.; Chen, B. MobileNets: Efficient convolutional neural networks for mobile vision applications. *arXiv* **2017**, arXiv:1704.04861. [CrossRef]
31. Zhuo, D.; Cao, H. Damage identification of bolt connections base on wavelet time-frequency diagrams and lightweight convolution neural networks. *Eng. Mech.* **2021**, *38*, 228–238. [CrossRef]
32. Li, Y.T.; Huang, H.S.; Xie, Q.S. Research on a surface defect detection algorithm based on MobileNet-SSD. *Appl. Sci.* **2018**, *8*, 1678. [CrossRef]
33. Pan, H.; Pang, Z.; Wang, Y.; Chen, L. A New Image recognition and classification method combining transfer learning algorithm and MobileNet model for welding defects. *IEEE Access* **2020**, *8*, 119951–119960. [CrossRef]
34. Srinivasu, P.N.; Sivasai, J.G.; Ijaz, M.F. Classification of skin disease using deep learning neural networks with MobileNet V2 and LSTM. *Sensors* **2021**, *21*, 2852. [CrossRef]

35. Pan, S.J.; Yang, Q. A survey on transfer learning. *IEEE Trans. Knowl. Data Eng.* **2010**, *10*, 1345–1359. [CrossRef]
36. Liang, P.F.; Deng, C.; Wu, J. Intelligent fault diagnosis of rotating achinery via wavelet transform, generative adversarial nets and convolutional neural network. *Measurement* **2020**, *159*, 107768. [CrossRef]
37. Yang, J.; Chen, R.; Zhang, Z. Experimental study on the ultimate bearing capacity of damaged RC arches strengthened with ultra-high performance concrete. *Eng. Struct.* **2023**, *279*, 115611. [CrossRef]
38. Tan, H.; Qian, D.; Xu, Y.; Yuan, M.; Zhao, H. Analysis of vertical temperature gradients and their effects on hybrid girder cable-stayed bridges. *Sustainability* **2023**, *15*, 1053. [CrossRef]
39. Zeng, Y.; He, H.; Qu, Y.; Sun, X.; Tan, H.; Zhou, J. Numerical simulation of fatigue cracking of diaphragm notch in orthotropic steel deck model. *Materials* **2023**, *16*, 67. [CrossRef] [PubMed]

buildings

Article

Prediction of the Debonding Failure of Beams Strengthened with FRP through Machine Learning Models

Tianyu Hu [1,2], Hong Zhang [1,2,*] and Jianting Zhou [1,2]

[1] State Key Laboratory of Mountain Bridge and Tunnel Engineering, Chongqing Jiaotong University, Chongqing 400074, China
[2] School of Civil Engineering, Chongqing Jiaotong University, Chongqing 400074, China
* Correspondence: hongzhang@cqjtu.edu.cn

Abstract: Plate end (PE) debonding and intermediate crack (IC) debonding are the two main failure modes of beams strengthened with fiber-reinforced polymer (FRP) in flexure. Therefore, it is essential to clarify the force state of the structure when debonding occurs in strengthened beams. This paper collected 229 beams with debonding failure as the database, of which 128 were PE debonding and 101 were IC debonding. Correlation and grey correlation analysis were used to establish the indicator systems for predicting PE and IC debonding and to identify the critical indicators among them. Five machine learning models, linear regression, ridge regression, decision trees, random forests, and back propagation (BP) neural networks, were used to build the two debonding prediction models. Optimization of the best prediction among the five machine learning models took place using the Dung Beetle Optimizer (DBO) algorithm, which has competitive performance with state-of-the-art optimization approaches in terms of convergence rate, solution accuracy, and stability. Finally, the optimal prediction model was compared with the models suggested by codes, and it was found that the established model can well predict PE and IC debonding.

Keywords: plate end debonding; intermediate crack debonding; fiber-reinforced polymer; machine learning; dung beetle optimizer (DBO)

Citation: Hu, T.; Zhang, H.; Zhou, J. Prediction of the Debonding Failure of Beams Strengthened with FRP through Machine Learning Models. *Buildings* **2023**, *13*, 608. https://doi.org/10.3390/buildings13030608

Academic Editor: Krishanu Roy

Received: 29 January 2023
Revised: 21 February 2023
Accepted: 24 February 2023
Published: 25 February 2023

1. Introduction

In recent years, deformation and strengthening of structures have been widely studied [1–4]; fiber-reinforced polymers (FRP) have been widely used to strengthen structures [5]. However, due to the linear elasticity of the material, debonding failure occurs in FRP-strengthened reinforced concrete (RC) beams before the FRP has reached its full performance, which largely limits its further application. [6–10]. PE and IC debonding are the two failure modes for FRP-strengthened RC beams (Figure 1). Generally, IC debonding occurs in beams with a relatively large shear-span-to-depth ratio, while for beams with a relatively small shear-span-to-depth ratio, since the bending moment is minor at this time, the beams are mainly subjected to shear strength, so PE debonding occurs [6]. Nevertheless, due to the complex mechanism of debonding failure of FRP-strengthened RC beams in flexure, it is not realistic to judge the failure modes in strengthened beams merely by shear-span-to-depth ratio [11,12].

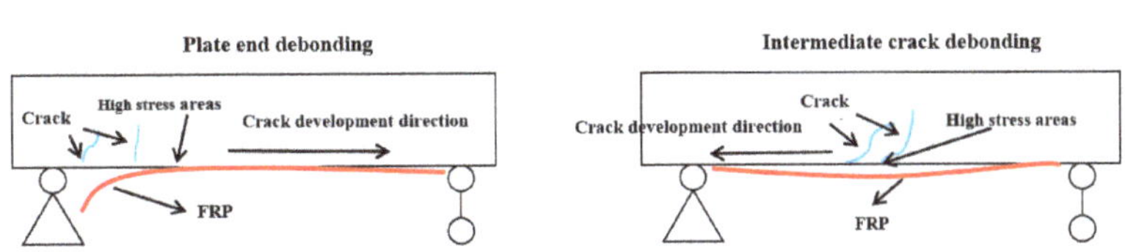

Figure 1. Debonding modes.

To clarify the failure mechanism of FRP-strengthened RC beams, researchers have conducted a lot of experiments and proposed some models. For PE debonding, in 1992, Oehlers proposed a model for strength based on the forces of shear and flexural moments which act on the end of the plate [11]. In 1997, Jansze proposed a plate-end debonding strength model for steel-plated beams. The model focuses on PE debonding of the strengthened beams at the onset of shear cracking [13]. Ahmed and Van Germert, in 1999, modified Jansze's model to take into account the differences between FRP and steel properties and the effect of shear reinforcement [14]. In 2001, fib Bulletin adopted Blaschko's beam-based model for the shear strength of concrete [15]. In 2002, Smith and Teng proposed a concrete shear strength model for PE debonding [16]. Yao and Teng, in 2007, conducted tests on FRP-strengthened RC beams and modified the expressions proposed by Oehlers [17]. TR55 in 2012, ACI and AS in 2017 recommended an upper limit on the shear force in the plate end region to avoid PE debonding [18–20]. El-Sayed proposed a model for predicting PE debonding based on the beam's concrete shear strength, which considers the parameters that affect the opening of shear cracks in 2021 [21]. For IC debonding, in 2001, fib and JSCE proposed models which limit the tensile stress during debonding [15,22]. CECS 2003, CNR 2004, and TR55 2012 present FRP ultimate debonding strain models which are based on tests of shear at the interface between FRP and concrete. [18,23,24]. In 2013, Kim and Harries proposed a Monte Carlo prediction model for the effective strain of FRP based on the statistical approach [25]. Bilotta et al., in 2013, proposed standard and design values for the maximum tensile strain of FRP when IC debonding occurs based on flexural tests of RC beams strengthened with FRP [26]. In 2016, Lopez-Gonzalez characterized the FRP–concrete interface of the strengthened beams and proposed an interface fracture energy model [27]. ACI modified the model proposed by Teng in 2003 based on the maximum tensile strain of RC beams strengthened with FRP and proposed a model for calculating the allowable debonding strain of FRP in 2017 [20]. Li and Wu investigated the mechanism of IC debonding of FRP-strengthened beams using a finite element analysis model based on the smeared-crack method in 2018. They established a model for IC debonding based on the finite element analysis simulation results [28].

These models for predicting PE and IC debonding have facilitated the study of the debonding failure of FRP-strengthened RC beams. Still, they have significant coefficients of variation between the calculated and experimental values [21,29]. Therefore, it is necessary to conduct a deeper study on debonding failure to establish more accurate models for predicting PE and IC debonding. In recent years, machine learning has been widely used in structural engineering [30–32]. In this paper, a data-driven model was adopted, which reduces the cost of experiments and is scientific in the selection of indicators; 229 beams with debonding failure were collected as a database, of which 128 were PE debonding and 101 were IC debonding. Correlation analysis and grey correlation analysis were used to establish the indicator systems for predicting PE and IC debonding and to identify the important indicators among them. Five machine learning models, linear regression, ridge regression, decision trees, random forests, and BP neural networks, were used to build the two debonding prediction models. Optimization of the best prediction among the five machine learning models took place using the Dung Beetle Optimizer (DBO). Finally, the optimal prediction model was compared with the models suggested by codes.

2. Construct the Indicator System

2.1. Primary Indicator System

Based on the relevant codes and experimental studies [6–8,10–12,25–32], in the primary indicator system, according to the literature [18–24], as shown in Figure 2, independent indicators and mixed indicators systems are constructed, respectively. Among them, the mixed indicator is a linear combination of independent indicators. The combination principle is based on the codes and related literature [16–24]. In contrast, the independent indicator system is the physical composition of the study object itself, including the geometric and material properties of the reinforced material itself and the geometric and material properties of the reinforced material.

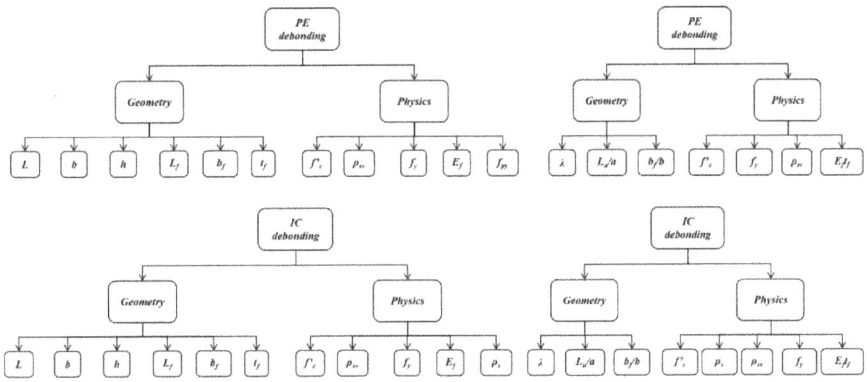

PS: $\lambda = a/h$, a = distance from the point of concentrated load to the edge of the support, L_u(anchorage length) = $L_f/2+a-L/2$

Figure 2. Primary indicator system.

2.2. Data Collection

The data collection criteria are as follows:

(a) Debonding failure occurred in all 229 strengthened beams, including IC debonding and PE debonding, as shown in Table 1.

(b) The FRP sheets were not prestressed. The ends of the strengthened beams were not anchored.

(c) The geometric and material properties of the strengthened beams, FRP, and reinforcements are clear.

Table 1. Data sources and failure modes.

Reference	Beam	Number	Failure Mode
[33]	C D G I M	5	PE
[34]	B2 B3 B4 B6	4	PE
[35]	A4 A5 B3	3	PE
[36]	1Bu 2Bu 3Bu 1Cu 2Cu 3Cu	6	PE
[37]	B1U,1.0 B3U,1.0 B4U,1.0 B1U,2.3	4	PE
[38]	VR5 VR6 VR7 VR8	4	PE
[39]	P2 P3 P4 P5	4	PE
[40]	2	1	PE
[41]	A3 A8 C2	3	PE
[42]	AF.2 AF.2.1 AF.3 AF.4 BF.2-1 BF.3-1 CF.2-1 CF.3-1 CF.4-1 DF2 DF3 DF4 EF.1-1 EF.3-1 EF.4-1	15	PE PE
[43]	BF2 BF3 BF4 BF5	4	PE
[44]	F5 F6 F7 F10	4	PE
[45]	1N4 1N6	2	PE
[46]	Beam3 Beam4 Beam5	3	PE
[47]	A950 A1100 A1150	3	PE
[48]	A4 A6 A8 B5	4	PE
[49]	1A 1B 2A 2B 3A 3B	6	PE
[50]	A-S1 B-S1	2	PE
[51]	A1-1 A2-1 A3-1 A4-1 A5-1 A1-11 A2-11 A3-11 A4-11 A5-11	10	PE PE
[52]	1T6LN 2T6LN 2T4LN	3	PE
[53]	A0 B0	2	PE

Table 1. *Cont.*

Reference	Beam	Number	Failure Mode
[54]	Bb1 Bb3	2	PE
[55]	E1a E1b E2a E2b E3a E3b	11	PE
	E4a E4b E5a E5b E3b2		PE
[56]	RB1	1	PE
[57]	B3-12D-2L15 B4-12D-3L15	2	PE
[17]	CS-B CS-L3-B CS-W100-B CP-B	4	PE
[58]	A3 B2 B3	3	PE
[59]	B85P B70P B25P	3	PE
[60]	C2	1	PE
[10]	S-0.5-35-360	1	PE
[61]	S(AT1)C B(AT1)C	2	PE
[62]	BC1-1 BC1-2 BC2-1 BC3-1 BC4-1 BC4-2	6	PE
[55]	B	1	IC
[6]	SM2 SM3 MM2 MM3	4	IC
[63]	A1.1 A3.1	2	IC
[64]	S-A R-A	2	IC
[65]	H-75-2 C3	2	IC
[66]	CB3-2S CB4-2S CB5-3S CB6-3S	8	IC
	CB7-1S CB8-1SB CB9-1SB CB10-2SB		IC
[67]	A11 A12 A21 A31 A32	9	IC
[68]	A41 A42 A51 A52		IC
[51]	BP BF	2	IC
	A3 A4 A5 A6 B3 B4 B5 B6	8	IC
[54]	A10 A20 B10 B20	4	IC
[69]	S1a S1b S2a S2b S3a S3b	6	IC
[70]	B11 B12 B21 B22 B31 B32	6	IC
	Beam 2 Beam 6	2	IC
[71]	S2PF7M S3PS1M S3PS2M	3	IC
[72]	B5 B6 B7 B8	4	IC
[73]	A2	1	IC
	NFCB1 NFCBW2	2	IC
[74]	EBR	1	IC
[58]	1 2 4 3 9	10	IC
[75]	10 11 12 5 6		IC
	B-04/0.5S B-08/S1 BF-04/0.5S BF-06/S	4	IC
[76]	S-0 S-0.4 S-0.6	3	IC
[77]	W-CSP1-0 W-CSP1-0E W-CSP1-4(1) W-CSP2-3-0	17	IC
[78]	W-CSP2-3-4(1) W-CSP2-3-4(2) W-CSP2-3-4(3)		
	W-CSP6-9-0 W-CSP6-9-4(1) W-CSP6-9-4(2) W-CSP6-9-4(3)		
	P-CSP1(1) P-CSP2-3(1) P-CSP2-3(2) P-CSP6-9(1) P-CSP6-9(2)		

The range of the parameters is large, which includes the general situation in practical applications.

2.3. Data Analysis and Indicator System Establishment

There are many independent indicators affecting PE and IC debonding of the strengthened beams and it is unclear whether there is a relationship between each independent indicator and the debonding failure, so correlation and grey correlation analysis were used to identify the redundant indicators.

2.3.1. Correlation Analysis

The correlation between the indicators is expressed by Pearson correlation coefficient, with a correlation coefficient close to 1 for a positive correlation and close to −1 for a negative correlation [79]. The formula for its calculation is given in Equation (1).

$$r_{xy} = \frac{Cov(X,Y)}{S_x S_y} \tag{1}$$

where $Cov(x,y) = \frac{\sum_{1}^{n}(X_i - X).(Y_i - Y)}{n-1}$, $S_x = \sqrt{\frac{\sum_{1}^{n}(X_i - \overline{X})^2}{n-1}}$ and $Cov(x,y)$ denotes the covariance of x and y, S_x denotes the standard deviation of variable x, and S_y denotes the standard deviation of variable y

The correlation coefficients (expressed as P) between independent indicators were calculated as shown in Figure 3. In it, the absolute value of the correlation coefficient of the indicator is compared with 0.5, and the correlation between indicators is strong if it is greater than 0.5 and the correlation is weak if it is less than 0.5.

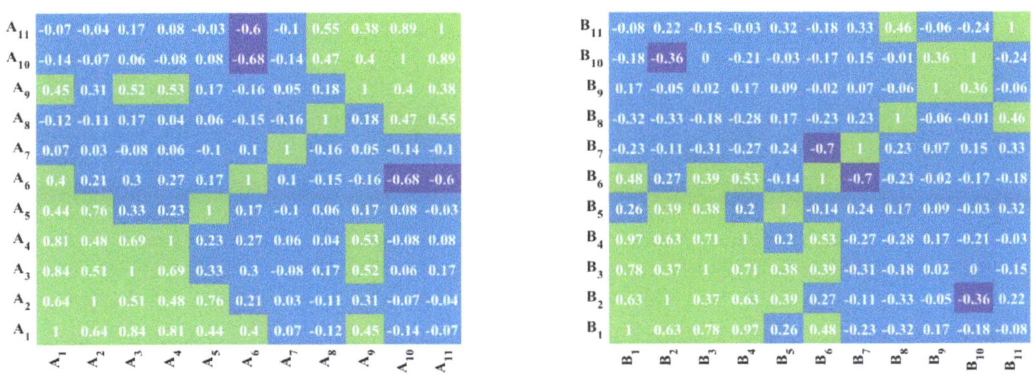

Figure 3. p-value of independent indicators of PE and IC failure.

In Figure 3, A_1 to A_{11} are L, b, h, L_f, b_f, t_f, f'_c, ρ_{sv}, f_y, E_f, f_{py}; B_1 to B_{11} are L, b, h, L_f, b_f, t_f, f'_c, ρ_{sv}, f_y, E_f, ρ_s. It can be seen from Figure 3 that there is a significant correlation between A_1 and A_2, A_3, A_4, a significant correlation between A_2 and A_3, A_5, a significant correlation between A_3 and A_2, A_4, A_9, a significant correlation between A_4 and A_1, A_3, A_9, a significant correlation between A_5 and A_2, a significant correlation between A_6 and A_{10}, A_{11}, a significant correlation between A_8 and A_{11}, a significant correlation between A_9 and A_3, A_4, a significant correlation between A_{10} and A_6, and a significant correlation between A_{11} and A_6. Additionally, there is a significant correlation between B_1 and B_2, B_3, and B_4, a significant correlation between B_2 and B_1 and B_4, a significant correlation between B_3 and B_1 and B_4, a significant correlation between B_4 and B_1, B_2, B_3, and B_6, and a significant correlation between B_6 and B_7. In summary, the correlations among the independent indicators of PE failure are complicated, and most have significant correlations; the relationships among some indicators of IC failure are complicated and cannot be eliminated; therefore, mixed indicators are considered in establishing the indicator system. The correlation analysis results of the mixed indicators are shown in Figure 4.

In Figure 4, a_1 to a_7 are λ, L_u/a, b_f/b, f'_c, f_y, ρ_{sv}, $E_f t_f$; b_1 to b_8 are λ, L_u/a, b_f/b, f'_c, ρ_s, ρ_{sv}, f_y, $E_f t_f$. It can be seen from Figure 4 that the mixed indicators of PE failure indicators are weakly or lowly correlated (because the P-value between these indicators are less than 0.5), and no indicators need to be removed; the mixed indicators of IC failure are all weakly or lowly correlated with each other also, and as with PE failure, there is no need to eliminate any indicator.

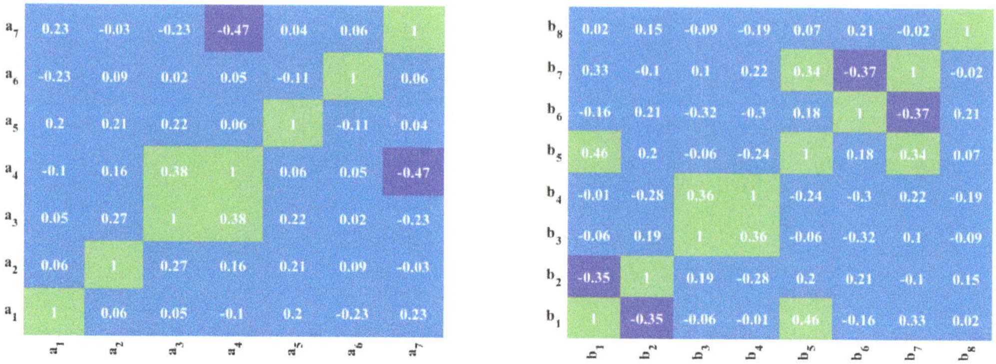

Figure 4. p-value of mixed indicators of PE and IC failure.

2.3.2. Grey Correlation Analysis

Grey correlation analysis is used to determine the degree of influence of each factor on the system [80]. The basic steps are as follows.

(a) Determine the reference series reflecting the characteristics of the system behavior and the comparison series affecting the system behavior.
(b) Dimensionless processing of the reference and comparison series.
(c) Find the grey correlation coefficient between the reference series and the comparison series, noted here as ξ, which is shown in Equation (2).

$$\xi(x_0(k), x_i(k)) = \frac{a + \rho b}{|x_0(k) - x_i(k)| + \rho b}, \forall i, k \tag{2}$$

where a denotes the minimum difference of the data in the subsequence and b denotes the maximum difference of the data in the subsequence; ρ is the resolution factor and generally taken as 0.5.

(d) Search for correlation. Because the correlation coefficient is the value of the degree of correlation between the comparison series and the reference series at each point, it has more than one number, and the information is too scattered to facilitate a holistic comparison. Therefore, it is necessary to pool the correlation coefficients at each point into one value; that is, to find its average value, as a quantitative representation of the degree of correlation between the comparison series and the reference series, the correlation degree is noted as r_i, and its formula is as follows in Equation (3).

$$r_i = \frac{1}{N}\sum_{k=1}^{N} \xi_i(k) \tag{3}$$

The grey correlation of each indicator of PE and IC with failure is shown in Figure 5.

From Figure 5, it can be seen that the degree of the influence of the parameters on PE debonding are FRP stiffness (Eftf), concrete strength (f′c), the ratio of sheet width to beam width (bf/b), stirrup reinforcement ratio (ρsv), tensile strength of tensile reinforcement (fy), shear span ratio (λ), and location of FRP cut-off point (Lu/a); the output parameter is the shear strength ($V_{db,end}$). As for IC debonding, the degree of influence of the parameters on it are concrete strength (f′c), FRP stiffness (Eftf), shear span ratio (λ), stirrup reinforcement ratio (ρsv), the ratio of sheet width to beam width (bf/b), tensile strength of tensile reinforcement (fy), tensile reinforcement ratio (ρs), and location of FRP cut-off point (Lu/a); the output parameter is the debonding strain of FRP (ε_{fd}). It can be obtained that FRP stiffness (Eftf) and concrete strength (f′c) have a large effect on both PE and IC debonding and this conclusion is roughly the same as the models suggested by codes [15,18–20,22–24].

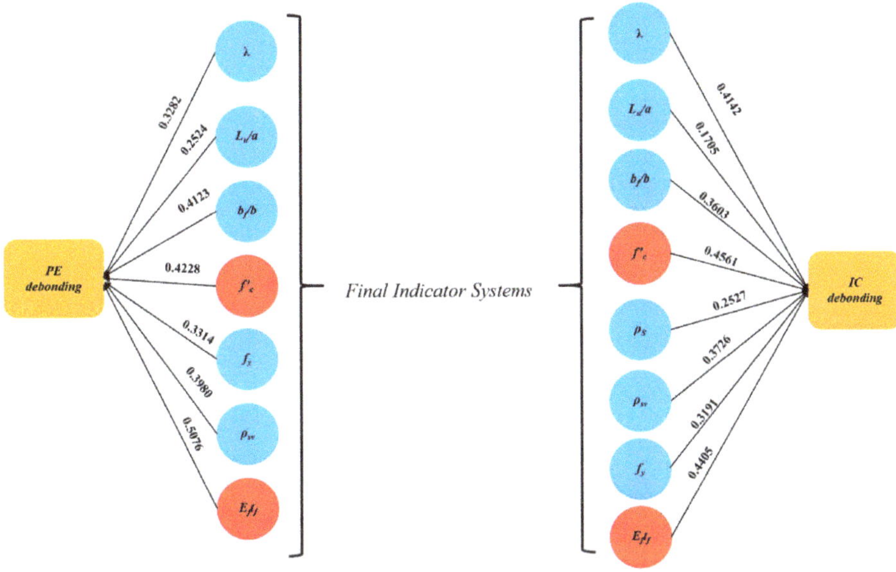

Figure 5. Indicator systems of the debonding failure.

3. Construction of the Machine Learning Models

3.1. Machine Learning Algorithms

3.1.1. Linear Regression

Linear regression (LR) can be classified as univariate linear regression, and multivariate linear regression, which establishes a functional relationship between the dependent and independent variables and is a supervised machine learning algorithm [81]. This paper uses multivariate linear analysis because of the large number of independent variables. Its expression is given in Equation (4).

$$Y = \beta 0 + \sum_{j=1}^{m} Xj\beta j \quad (4)$$

where Y denotes the dependent variable, which in this paper represents the shear force at the end of the beam when PE debonding occurs and the strain of the sheet at the middle of the beam when IC debonding occurs; β_0 is the regression constant; m denotes the number of independent variables; X_j denotes the independent variable, which in this paper is the parameter for predicting the debonding; and β_j is the regression coefficient. The root mean square error is minimized by the gradient descent method, and the best combination of regression coefficients is subsequently obtained, which in turn leads to the best-fit line.

3.1.2. Ridge Regression

The ridge regression (RR) is a modified linear regression method. By giving up the unbiased nature of linear regression, the regression method obtains more realistic and reliable regression coefficients at the cost of losing some information and accuracy, and the fit to the pathological data is stronger than that of the ordinary linear regression method [82].

3.1.3. Decision Tree

The decision tree (DT) is a classic machine learning algorithm. It mainly consists of nodes and directed edges. There are two types of nodes: the internal node and the leaf node. Internal nodes represent features or attributes, and leaf nodes represent classes or values. When the regression is performed with a decision tree, each feature of the sample is tested from the root node, and the sample is assigned to its child nodes according to the test results; at this time, each child node corresponds to one of the values taken for the feature. The samples are tested and assigned in this way recursively until they reach the leaf nodes [83].

3.1.4. Random Forest

The random forest (RF) is an algorithm that integrates multiple trees through the concept of integrated learning. Its basic unit is a decision tree, and its essence goes with an important branch of machine learning–integration learning [83]. As shown in Figure 6, from an intuitive point of view, each decision tree is a decision maker. N trees will have N decision. The random forest integrates all the decision and designates the consequence with the most decision as the final output.

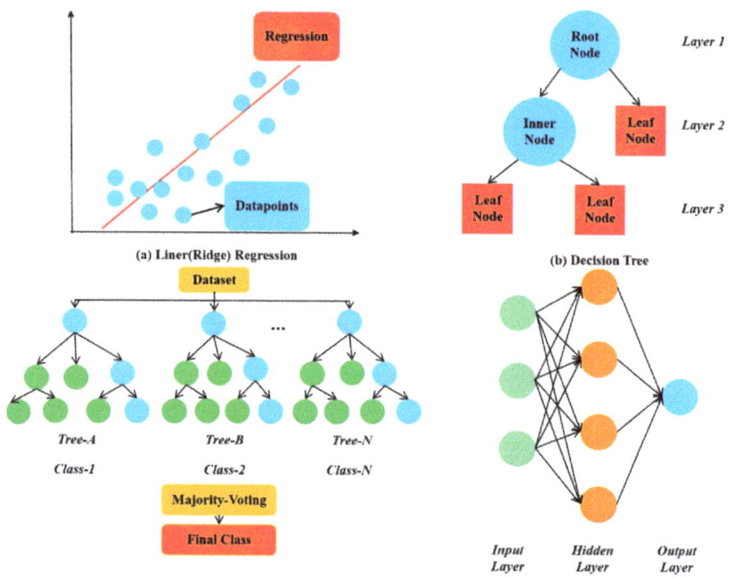

Figure 6. Architecture of the machine learning models.

3.1.5. BP Neural Network

Back propagation (BP) neural network is a multilayer forward neural network trained according to the error back propagation algorithm. Its gradient descent method is used to adjust the weights and thresholds of neurons in each layer to reduce the error of the network output. BP neural networks can achieve arbitrary nonlinear mapping of input and output, and have the characteristics of self-learning and simple structure [84].

The general architecture of the above machine learning algorithms is shown in Figure 6.

3.2. Model Construction and Evaluation

The data used for model construction were 128 sets of PE debonding and 101 sets of IC debonding, as listed in Table 1. The model's input parameters are identified in Section 2 of this paper, shown in Figure 5. For PE debonding, the input parameters are FRP stiffness ($E_f t_f$), concrete strength (f'_c), the ratio of sheet width to beam width (b_f/b), stirrup reinforcement ratio (ρ_{sv}), tensile strength of tensile reinforcement (f_y), shear span ratio (λ), and location of FRP cut-off point (L_u/a). For IC debonding, the input parameters are concrete strength (f'_c), FRP stiffness ($E_f t_f$), shear span ratio (λ), stirrup reinforcement ratio (ρ_{sv}), the ratio of sheet width to beam width (b_f/b), tensile strength of tensile reinforcement (f_y), tensile reinforcement ratio (ρ_s), location of FRP cut-off point (L_u/a). The training set, validation set, and test set of the model are 60%, 20%, and 20%, respectively.

In the paper, *MAE* and R^2 are used to evaluate the performance of the model, where *MAE* indicates the mean of the absolute error between the predicted value of the model and the actual value of the sample, and R^2 indicates the degree of fit between the predicted value of the model and the actual value of the sample, and their equation are shown in Equation (5).

$$MAE = \frac{1}{n}\sum_{i=1}^{n}|y_i - p_i|$$
$$R^2 = 1 - \frac{\sum_{i=1}^{n}(y_i-p_i)^2}{\sum_{i=1}^{n}(y_i-\overline{y}_i)^2} \quad (5)$$

where y_i is the true value of the sample, p_i is the predicted value of the model, and $\overline{y}_\mathrm{i}$ is the mean value of the sample.

The *MAE* and R^2 of the training set, the validation set, and the testing set of the machine learning model are shown in Figure 7; the ratio of the training, the validation, and the testing sets is 70%, 15%, and 15%. The validation set is the samples left during the model training, which can be used to adjust the hyperparameters of the model and evaluate the ability of the model. The testing set is used to evaluate the performance of the final model.

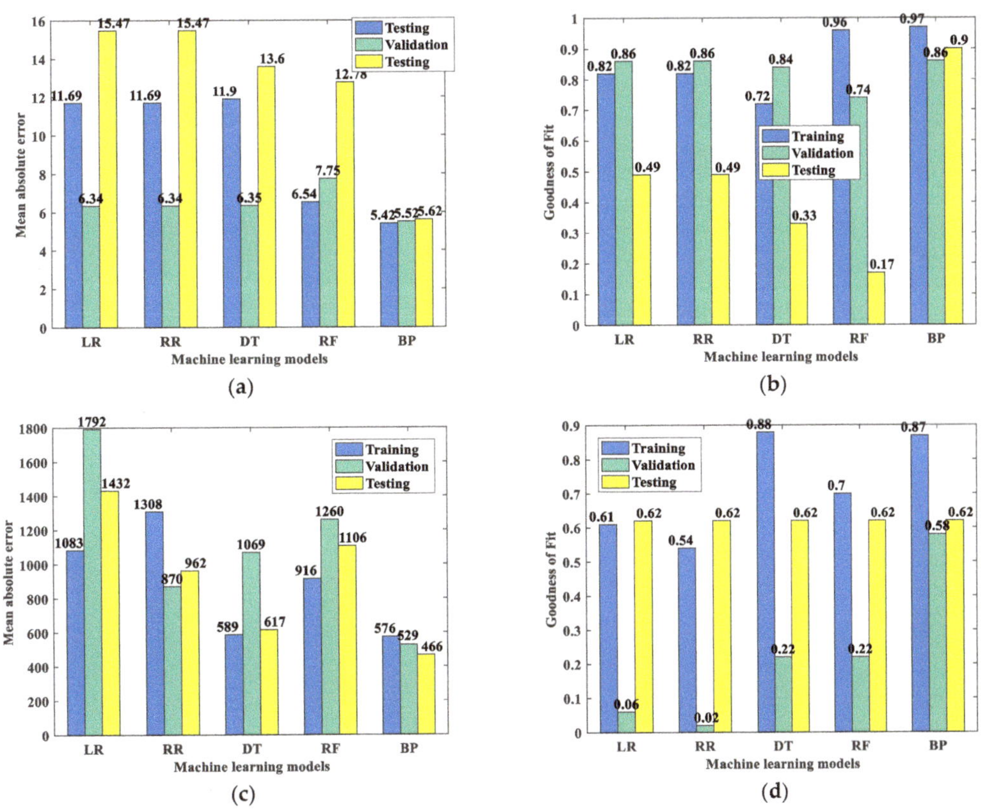

Figure 7. (**a**) *MAE* for PE debonding prediction model; (**b**) R^2 for PE debonding prediction model; (**c**) *MAE* for IC debonding prediction model; (**d**) R^2 for IC debonding prediction model.

From Figure 7a,b, it can be seen that for PE debonding, the *MAE* and R^2 of LR, RR, DT, and RF in the training set, the validation set, and the testing set differ significantly, indicating that the generalization ability of these models is poor, while the *MAE* of BP neural network in the training set, the validation set, and the testing set are 5.42, 5.52, and 5.62, respectively, which are the most minor and most average among all models. Meanwhile, the R^2 of the training set, the validation set, and the BP neural network testing set are 0.97, 0.86, and 0.90, respectively, which are also the highest among all the models. From Figure 7c,d, it can be seen that for IC debonding, the *MAE* and R^2 of LR, RR, DT, and RF in the training set, the validation set, and the testing set differ greatly, indicating that these models have poor generalization ability, while the *MAE* of BP neural network in the training set, the validation set, and the testing set are only 576, 529, and 466, respectively, which are the lowest and the most average among all models. In addition, the R^2 of the BP neural network in the training, validation, and testing sets reach 0.87, 0.58, and 0.62, respectively. Although the R^2 of the training set is lower than that of the DT, the R^2 of its validation set is higher than that of the DT. In summary, the BP neural network outperforms other machine learning models regarding prediction accuracy and debonding failure's generalization ability.

3.3. Model Optimization

For the prediction of debonding failure, it is known from Section 3.2 that the BP neural network model with the best performance still has the problem of poor generalization ability, so the BP neural network model for debonding failure prediction is considered for optimization. To overcome the shortcomings of the BP algorithm in the generation of weights and thresholds, which are easy to fall into the local optimum, and slow convergence speed, the Dung Beetle Optimizer algorithm (DBO) is used to optimize it. The Dung Beetle Optimizer algorithm is a novel swarm intelligence optimization algorithm proposed in November 2022, mainly inspired by the ball rolling, dancing, foraging, stealing, and reproduction behaviors of dung beetles and the algorithm presents substantially competitive performance with state-of-the-art optimization approaches in terms of the convergence rate, solution accuracy, and stability [85]. The coefficients of variation of the predicted and actual values of the 50 runs of DBO-BP and BP for PE and IC debonding are shown in Figure 8.

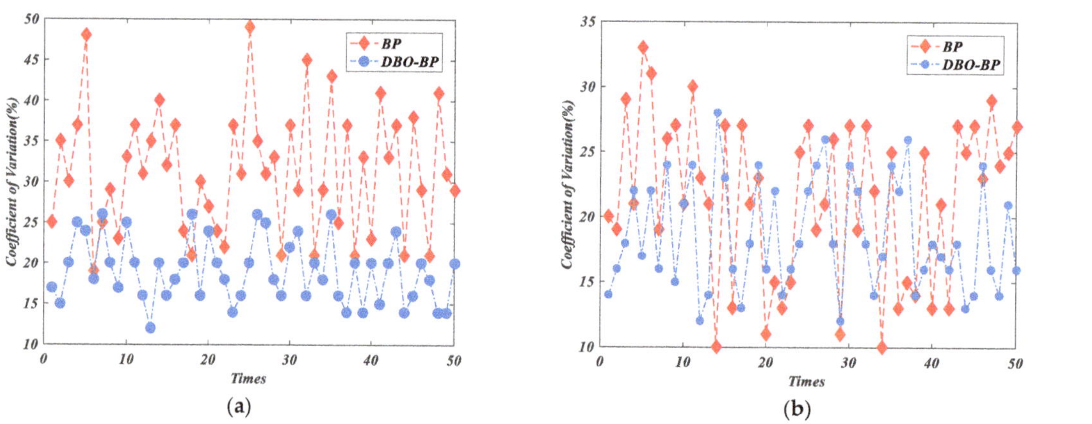

Figure 8. Comparison of model performance before and after optimization. (**a**) PE debonding; (**b**) IC debonding.

It can be seen from Figure 8 that the overall coefficient of variation of the prediction of debonding failure by the BP neural network model optimized by the DBO algorithm is smaller than that of the BP neural network model. The accuracy of the optimized model is significantly better than that of the traditional BP neural network.

4. Evaluation of Models Recommended by Codes

The models recommended by codes for PE and IC debonding are shown in Table 2. In all, 128 sets of PE debonding and 101 sets of IC debonding samples collected in 'Table 1′ are used to analyze the two types of models separately. The computed values of the models are compared with the actual values of the samples in Figure 9. The evaluation of the model is based on two indicators: coefficient of variation and conservativeness. In general, the smaller the coefficient of variation, the smaller the dispersion between the two sets of data, indicating a better predictive value of the model. The model is conservative if more than ninety percent of the predicted values are less than the test values. The performance evaluation of the codes' models is shown in Table 3.

Table 2. Models recommended by codes.

Codes	PE Debonding	IC Debonding
ACI440.2R	$V_{db,end} < 0.67V_c$	$\varepsilon_{fd} = 0.41\sqrt{\frac{f_c'}{E_f t_f}} \leq 0.9\varepsilon_{fu}$
fib	$V_{db,end} < 0.15 f_{ck}^{1/3} bd$	$\varepsilon_{fd} = \alpha c_1 k_c k_b \sqrt{\frac{f_{ct}}{E_f t_f}}$
TR55	$V_{db,end} < 0.67V_{rd}$	$\varepsilon_{fd} = 0.5k_b\sqrt{\frac{f_{ct}}{E_f t_f}}$
AS 5100.8	$V_{db,end} < 0.67V_u$	
CNR		$\varepsilon_{fd} = 0.484\sqrt{\frac{k_b\sqrt{f_c' f_{ct}}}{E_f t_f}}$
JSCE		$\varepsilon_{fd} \leq \sqrt{\frac{2G_f}{E_f t_f}}$
CECS		$\varepsilon_{fd} = k_b f_{ct}\left(\frac{1}{\sqrt{E_f t_f}} - \frac{0.2}{L_d}\right)$

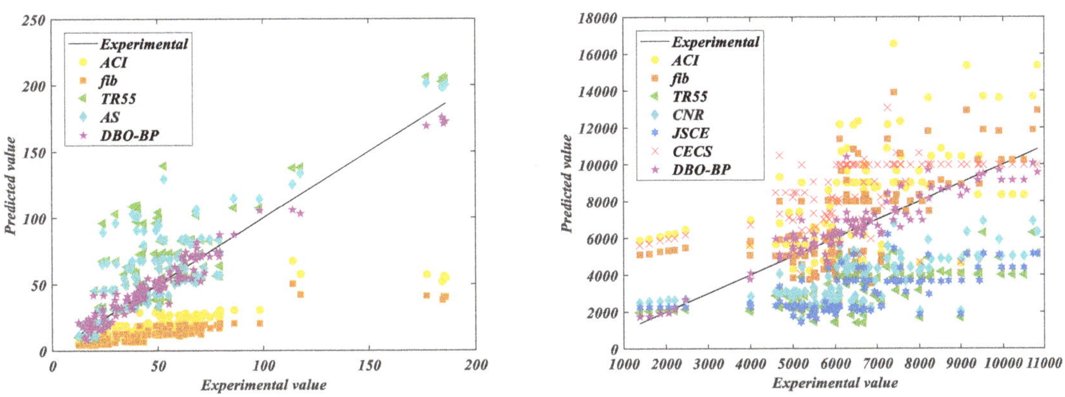

(a) (b)

Figure 9. Evaluation of the models. (**a**) PE debonding; (**b**) IC debonding.

As can be seen from Figure 9 and Table 3, for PE debonding, the models suggested by ACI and *fib* are a bit conservative, and their calculated values are lower than the experimental values. The coefficient of variation between the calculated values of the models suggested by TR55 and AS and the experimental values exceeds 40%, which is difficult to apply in practice. For IC debonding, the models suggested by TR55, CNR, and JSCE are a little conservative, while the models suggested by ACI and CECS have the risk of overestimating the FRP strains in case of the IC debonding, and the coefficients of variation between the calculated values of all the models suggested by the codes and experimental values are above 38%, which is difficult to apply in practice. For PE and IC debonding, the DBO-BP model has the lowest coefficient of variation between predicted values and experimental values among all models, which are 19% and 10%, and its predicted values are more stable, which shows a relatively balanced proportion of conservative values in the predicted values of the models.

Table 3. Model Evaluation.

	PE Debonding		IC Debonding	
Models	Coefficient of Variation	Conservative Value	Coefficient of Variation	Conservative Value
ACI440.2R	26.6%	100%	46.8%	29%
fib	25.1%	100%	45.4%	40%
TR55	48.3%	39%	38.6%	96%
AS 5100.8	44.5%	38%		
CNR			40.1%	94%
JSCE			44.7%	94%
CECS			38.8%	14%
DBO-BP	19%	48%	10%	42%

5. Parametric Study

The importance of the parameters is calculated based on the connection weights and excitation functions between the input layer and the hidden layer, and between the hidden layer and the output layer of the DBO-BP neural network model. In this study, the connection functions of the input layer and the hidden layer are "hyperbolic tangent function" and "linear function", which are expressed in Equations (6) and (7).

$$y_i = f \cdot \left(\sum_i w_{ij} x_i + \phi_j \right) = \frac{2}{1 + e^{-2(\sum_i w_{ij} x_i + \phi_j)}} - 1 \tag{6}$$

$$y_i = f \cdot \left(\sum_i w_{ij} x_i + \phi_j \right) = \sum_i w_{ij} x_i + \phi_j \tag{7}$$

where x_i denotes the value of the ith input metric, w_{ij} is the connection weight of the ith metric to the jth neuron, and Φ_j is the bias of the jth neuron.

The interlayer connection weights and biases for PE and IC debonding are shown in Tables 4 and 5.

The importance results of the indicators calculated according to Tables 4 and 5 are shown in Figure 10.

Table 4. Layer-to-layer connection weights for the predictive model of PE debonding.

P		P													
		H													O
		h_1	h_2	h_3	h_4	h_5	h_6	h_7	h_8	h_9	h_{10}	h_{11}	h_{12}	h_{13}	V
I	Φ	−0.9	−0.5	−0.3	−0.4	0.3	0.5	−0.6	1.2	−0.5	−0.4	−0.9	−0.7	−0.7	
	a_1	1.0	−0.4	−0.2	−0.9	−0.1	−0.3	−0.1	−0.2	−0.5	−0.1	0.8	0.6	0.2	
	a_2	0.8	0.4	−1.1	0.3	0.9	−1.5	0.7	0.1	0.4	0.5	−0.6	0.4	−0.1	
	a_3	−1.2	0.0	−0.3	−1.1	−0.1	0.3	−0.8	0.6	0.6	−0.1	−0.4	−0.4	−0.1	
	a_4	−0.8	0.4	−1.2	1.2	−1.1	1.7	0.0	−1.1	0.1	−0.2	−0.8	−0.7	0.4	
	a_5	0.0	−0.1	0.9	−1.6	0.7	−1.0	−0.6	0.9	0.0	−0.3	−1.2	1.3	0.4	
	a_6	0.0	1.5	−0.8	−0.2	0.5	0.7	0.8	0.6	0.2	−0.1	−1.4	0.1	−0.8	
	a_7	0.0	0.8	0.8	0.6	0.0	−0.7	−0.1	0.5	0.2	0.3	−1.0	−0.9	−0.4	

Table 4. *Cont.*

P		*P*												
		H												O
		h_1	h_2	h_3	h_4	h_5	h_6	h_7	h_8	h_9	h_{10}	h_{11}	h_{12}	h_{13}	V
H	Φ														0.4
	h_1														1.0
	h_2														−1.1
	h_3														0.9
	h_4														−1.5
	h_5														−0.7
	h_6														0.9
	h_7														1.3
	h_8														0.4
	h_9														0.5
	h_{10}														0.3
	h_{11}														−1.5
	h_{12}														−1.2
	h_{13}														1.3

Table 5. Layer-to-layer connection weights for the predictive model of IC debonding.

		P					
		H					O
		h_1	h_2	h_3	h_4	h_5	ε
I	Φ	−0.6	0.4	−0.3	1.1	−0.1	
	b_1	0.9	−1.2	−0.4	−0.2	−0.8	
	b_2	−0.4	−0.6	0.7	−0.6	1.0	
	b_3	−0.1	0.3	0.1	0.6	0.7	
	b_4	−0.4	0.8	1.2	−0.1	−0.8	
	b_5	−0.6	0.1	0.2	0.0	0.8	
	b_6	−1.6	−0.8	−0.2	1.1	1.0	
	b_7	0.3	0.4	0.0	0.4	−0.8	
	b_8	−0.6	−1.0	−0.2	−0.9	0.6	
H	Φ						0.7
	h_1						1.5
	h_2						0.7
	h_3						−0.6
	h_4						−0.8
	h_5						1.5

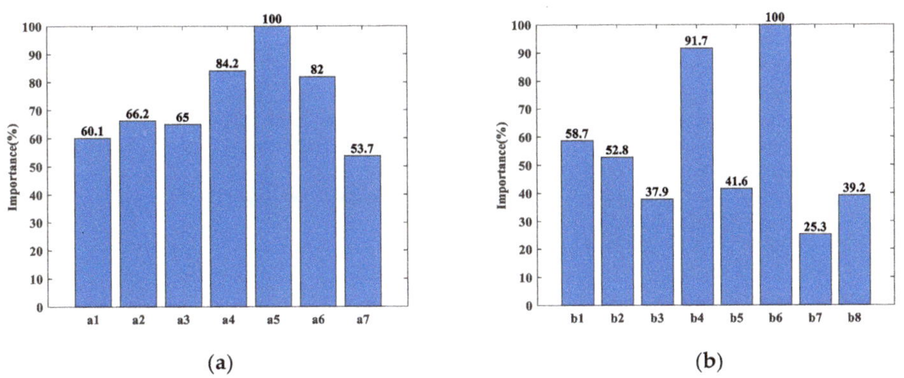

Figure 10. Importance of parameters. (**a**) PE debonding; (**b**) IC debonding.

From Figure 10, it can be seen that for PE debonding, the degree of influence of each parameter on the output result is a5, a4, a6, a2, a3, a1, a7; for IC debonding, the degree of influence of each indicator on the output is b6, b4, b1, b2, b5, b8, b3, b7.

6. Discussion and Conclusions

Researchers have proposed different prediction models for PE and IC debonding and these models for predicting PE and IC debonding have facilitated the study of the debonding failure of FRP-strengthened RC beams. Still, they have significant coefficients of variation between the calculated and experimental values.

This paper used correlation analysis and grey correlation analysis to construct the indicator system for debonding failure. PE and IC debonding prediction models for FRP-strengthened RC beams were developed by several machine learning algorithms. Optimization of the model took place using the Dung Beetle Optimizer and comparison with the model suggested by codes. We can draw the following conclusions:

(1) The correlation between the indicators in the established mixed indicator system is insignificant, and the correlation with the predicted values is significant. Additionally, FRP stiffness (Eftf) and concrete strength (f'c) have a large effect on both PE and IC debonding.
(2) For the prediction of PE and IC debonding, the BP neural network has the best performance compared to other machine learning algorithms.
(3) The model optimized by the DBO algorithm is significantly accurate and can better predict PE and IC debonding. The models suggested by codes have a significant coefficient of variation and are conservative or overestimate the state of the strengthened beam in case of failure.
(4) The experimental data used in this paper are unevenly distributed, which will affect the accuracy of the model to some extent. More data need to be collected in the future to build a more reliable model. Additionally, the parameters considered in this paper are not complete and more parameters need to be investigated in the future.

Author Contributions: Conceptualization, methodology, software, validation, writing—original draft, writing, T.H.; supervision, funding acquisition, writing, H.Z.; funding acquisition, supervision, J.Z. All authors have read and agreed to the published version of the manuscript.

Funding: This study was supported by the National Natural Science Foundation of China (52278291), and the Chongqing Natural Science Foundation of China (CSTB2022NSCQ-LZX0006, cstc2020jcyj-jqX0006, cstc2022ycjh-bgzxm0086).

Data Availability Statement: Not applicable.

Conflicts of Interest: The authors declare no conflict of interest.

References

1. Li, S.; Xin, J.; Jiang, Y.; Wang, C.; Zhou, J.; Yang, X. Temperature-induced deflection separation based on bridge deflection data using the TVFEMD-PE-KLD method. *J. Civ. Struct. Health Monit.* **2023**, 1–17. [CrossRef]
2. Jiang, Y.; Hui, Y.; Wang, Y.; Peng, L.; Huang, G.; Liu, S. A novel eigenvalue-based iterative simulation method for multi-dimensional homogeneous non-Gaussian stochastic vector fields. *Struct. Saf.* **2023**, *100*, 102290. [CrossRef]
3. Zhang, H.; Li, H.; Zhou, J.; Tong, K.; Xia, R. A multi-dimensional evaluation of wire breakage in bridge cable based on self-magnetic flux leakage signals. *J. Magn. Magn. Mater.* **2023**, *566*, 170321. [CrossRef]
4. Xin, J.; Jiang, Y.; Zhou, J.; Peng, L.; Liu, S.; Tang, Q. Bridge deformation prediction based on SHM data using improved VMD and conditional KDE. *Eng. Struct.* **2022**, *261*, 114285. [CrossRef]
5. Kotynia, R.; Oller, E.; Marí, A.; Kaszubska, M. Efficiency of shear strengthening of RC beams with externally bonded FRP materials—State-of-the-art in the experimental tests. *Compos. Struct.* **2021**, *267*, 113891. [CrossRef]
6. Bakay, R.; Sayed-Ahmed, E.Y.; Shrive, N.G. Interfacial debonding failure for reinforced concrete beams strengthened with carbon-fibre-reinforced polymer strips. *Can. J. Civ. Eng.* **2009**, *36*, 103–121. [CrossRef]
7. Ceroni, F.; Pecce, M.; Matthys, S.; Taerwe, L. Debonding strength and anchorage devices for reinforced concrete elements strengthened with FRP sheets. *Compos. Part B Eng.* **2008**, *39*, 429–441. [CrossRef]

8. Al-Saawani, M.A.; El-Sayed, A.K.; Al-Negheimish, A.I. Assessment of plate-end debonding design provisions for RC beams strengthened with FRP. *Lat. Am. J. Solids Struct.* **2020**, *17*. [CrossRef]
9. Chen, J.; Sun, B.; Bao, Z. A brief discussion of the prestressed carbon fiber board reinforcement material system. In Proceedings of the Twelfth National Academic Exchange Conference on Building Inspection, Appraisal and Reinforcement, Guangzhou, China, 6–7 December 2004.
10. Al-Saawani, M.; El-Sayed, A.K.; Al-Negheimish, A.I. Effect of basic design parameters on IC debonding of CFRP-strengthened shallow RC beams. *J. Reinf. Plast. Compos.* **2015**, *34*, 1526–1539. [CrossRef]
11. Oehlers, D.J. Reinforced Concrete Beams with Plates Glued to their Soffits. *J. Struct. Eng.* **1992**, *118*, 2023–2038. [CrossRef]
12. Zhang, S.-S.; Teng, J.G. End cover separation in RC beams strengthened in flexure with bonded FRP reinforcement: Simplified finite element approach. *Mater. Struct.* **2015**, *49*, 2223–2236. [CrossRef]
13. Jansze, W. Strengthening of RC Members in Bending by Externally Bonded Steel Plates. Ph.D. Thesis, Delft University of Technology, Delft, The Netherlands, 1997.
14. Ahmed, O.; Van Gemert, D. Effect of Longitudinal Carbon Fiber Reinforced Plastic Laminates on Shear Capacity of Reinforced Concrete Beams. *Spec. Publ.* **1999**, *188*, 933–944. [CrossRef]
15. FIB International. *Externally Applied FRP Reinforcement for Concrete Structures*; FIB International: Lausanne, Switzerland, 2001.
16. Smith, S.; Teng, J. FRP-strengthened RC beams. II: Assessment of debonding strength models. *Eng. Struct.* **2002**, *24*, 397–417. [CrossRef]
17. Yao, J.; Teng, J. Plate end debonding in FRP-plated RC beams—I: Experiments. *Eng. Struct.* **2007**, *29*, 2457–2471. [CrossRef]
18. Concrete Society. *Design Guidance for Strengthening Concrete Structures Using Fibre Composite Materials*, 2nd ed.; Concrete Society Technical Rep. 55; Concrete Society: Surrey, UK, 2004.
19. American Concrete Institute Committee. *440 Report. Guide for the Design and Construction of Externally Bonded FRP Systems for Strengthening Concrete Structures*; American Concrete Institute Committee: Farmington Hills, MI, USA, 2017.
20. *AS 5100.8*; Bridge Design, Part 8: Rehabilitation and Strengthening of Existing Bridges. Standards Australia, SA1 Global Limited: Sydney, Austrilia, 2017.
21. El-Sayed, A.K.; Al-Saawani, M.A.; Al-Negheimish, A.I. Empirical shear based model for predicting plate end debonding in FRP strengthened RC beams. *J. Civ. Eng. Manag.* **2021**, *27*, 117–138. [CrossRef]
22. Japan Society of Civil Engineers (JSCE). *Recommendations for Upgrading of Concrete Structures with Use of Continuous Fiber Sheets*; Japan Society of Civil Engineers (JSCE): Tokyo, Japan, 1997.
23. China Association for Engineering Construction Standardization (CECS146). *Technical Specification for Strengthening Concrete Structure with Carbon Fiber Reinforced Polymer Laminate*; China Planning Press: Beijing, China, 2003.
24. *CNR-DT, 200*; Guide for the Design and Construction of Externally Bonded FRP Systems for Strengthening Existing Structures. Italian National Research Council: Rome, Italy, 2004.
25. Kim, Y.J.; Harries, K.A. Statistical Characterization of Reinforced Concrete Beams Strengthened with FRP Sheets. *J. Compos. Constr.* **2013**, *17*, 357–370. [CrossRef]
26. Bilotta, A.; Faella, C.; Martinelli, E.; Nigro, E. Design by testing procedure for intermediate debonding in EBR FRP strengthened RC beams. *Eng. Struct.* **2013**, *46*, 147–154. [CrossRef]
27. López-González, J.C.; Fernández-Gómez, J.; Díaz-Heredia, E.; López-Agüí, J.C.; Villanueva-Llauradó, P. IC debonding failure in RC beams strengthened with FRP: Strain-based versus stress increment-based models. *Eng. Struct.* **2016**, *118*, 108–124. [CrossRef]
28. Li, X.-H.; Wu, G. Finite-Element Analysis and Strength Model for IC Debonding in FRP-Strengthened RC Beams. *J. Compos. Constr.* **2018**, *22*, 04018030. [CrossRef]
29. Zhang, J.; Guo, R.; Li, S.; Zhao, S. Prediction of the IC debonding failure of FRP-strengthened RC beams based on the cohesive zone model. *Lat. Am. J. Solids Struct.* **2020**, *17*. [CrossRef]
30. Fang, Z.; Roy, K.; Chen, B.; Sham, C.-W.; Hajirasouliha, I.; Lim, J.B. Deep learning-based procedure for structural design of cold-formed steel channel sections with edge-stiffened and un-stiffened holes under axial compression. *Thin-Walled Struct.* **2021**, *166*, 108076. [CrossRef]
31. Fang, Z.; Roy, K.; Ingham, J.; Lim, J.B.P. Assessment of end-two-flange web crippling strength of roll-formed aluminium alloy perforated channels by experimental testing, numerical simulation, and deep learning. *Eng. Struct.* **2022**, *268*, 114753. [CrossRef]
32. Fang, Z.; Roy, K.; Xu, J.; Dai, Y.; Paul, B.; Lim, J.B. A novel machine learning method to investigate the web crippling behaviour of perforated roll-formed aluminium alloy unlipped channels under interior-two flange loading. *J. Build. Eng.* **2022**, *51*, 104261. [CrossRef]
33. Ritchie, P.A.; Thomas, D.A.; Lu, L.W.; Conelly, G.M. External reinforcement of concrete beams using fibre reinforced plastics. *ACI Struct. J.* **1991**, *88*, 490–500.
34. Quantrill, R.J.; Hollaway, L.C.; Thorne, A.M. Experimental and analytical investigation of FRP strengthened beam response: Part I. *Mag. Concr. Res.* **1996**, *48*, 331–342. [CrossRef]
35. Arduini, M.; Di Tommaso, A.; Nanni, A. Brittle failure in FRP plate and sheet bonded beams. *ACI Struct. J.* **1997**, *94*, 363–370.
36. Garden, H.N.; Hollaway, L.C.; Thorne, A.M. A preliminary evaluation of carbon fibre reinforced polymer plates for strengthening reinforced concrete members. *Proc. Inst. Civ. Eng. Struct. Build.* **1997**, *123*, 127–142. [CrossRef]
37. Garden, H.N.; Quantrill, R.J.; Hollaway, L.C.; Thorne, A.M.; Parke, G.A.R. An experimental study of the anchorage length of carbon fiber composite plates used to strengthen reinforced concrete beams. *Constr. Build. Mater.* **1998**, *12*, 203–219. [CrossRef]

38. Beber, A.J.; Campos Filho, A.; Campagnolo, J.L. Flexural strengthening of R/C beams with CFRP sheets. In Proceedings of the 8th International Conference in Structural Faults - Repair-99, London, UK, 13–15 July 1999.
39. David, E.; Djelal, C.; Ragneau, E.; Bodin, F.B. Use of FRP to strengthen and repair RC beams: Experimental study and numerical simulations. In Proceedings of the Eighth International Conference on Advanced Composites for Concrete Repair, London, UK, 1999.
40. Hau, K.M. Experiments on Concrete Beams Strengthened by Bonding Fibre Reinforced Plastic Sheets. Master's Thesis, The Hong Kong Polytechnic University, Hong Kong, China, 1999.
41. Tumialan, G.; Serra, P.; Nanni, A.; Belarbi, A. Concrete cover delamination in reinforced concrete beams strengthened with carbon fiber reinforced polymer sheets. In Proceedings of the Fourth International Symposium on Fiber Reinforced Polymer Reinforcement for Reinforced Concrete Structures, Baltimore, MD, USA, 1 January 1999; pp. 725–735.
42. Ahmed, O.A.F. Strengthening of R. C. Beams by Means of Externally bonded CFRP Laminates: Improved Model for Plateend shear. Ph.D. Thesis, Department of Civil Engineering, Catholicke University of Leuven, Leuven, Belgium, 2000.
43. Matthys, S. Structural Behavior and Design of Concrete Beams Strengthened with Externally Bonded FRP Reinforcement. Ph.D. Thesis, Ghent University, Ghent, Belgium, 2000.
44. Fanning, P.J.; Kelly, O. Ultimate Response of RC Beams Strengthened with CFRP Plates. *J. Compos. Constr.* **2001**, *5*, 122–127. [CrossRef]
45. Gao, B.; Leung, W.; Cheung, C.; Kim, J.K.; Leung, C.K.Y. Effects of adhesive properties on strengthening of concrete beams with composite strips. In Proceedings of the International Conference on FRP Composites in Civil Engineering, Hong Kong, China, 12–15 December 2001; pp. 423–432.
46. Maalej, M.; Bian, Y. Interfacial shear stress concentration in FRP-strengthened beams. *Compos. Struct.* **2001**, *54*, 417–426. [CrossRef]
47. Nguyen, D.M.; Chan, T.K.; Cheong, H.K. Brittle Failure and Bond Development Length of CFRP-Concrete Beams. *J. Compos. Constr.* **2001**, *5*, 12–17. [CrossRef]
48. Rahimi, H.; Hutchinson, A. Concrete Beams Strengthened with Externally Bonded FRP Plates. *J. Compos. Constr.* **2001**, *5*, 44–56. [CrossRef]
49. Smith, S.; Teng, J.G. Shear-Bending Interaction in Debonding Failures of FRP-Plated RC Beams. *Adv. Struct. Eng.* **2003**, *6*, 183–199. [CrossRef]
50. Valcuende, M.; Benlloch, J.; Parra, C.J. Ductility of reinforced concrete beams strengthened with CFRP strips and fabric. In Proceedings of the Sixth International Symposium on FRP Reinforcement for Concrete Structures, Singapore, 8–10 July 2003; pp. 337–346.
51. Breña, S.F.; Macri, B.M. Effect of Carbon-Fiber-Reinforced Polymer Laminate Configuration on the Behavior of Strengthened Reinforced Concrete Beams. *J. Compos. Constr.* **2004**, *8*, 229–240. [CrossRef]
52. Gao, B.; Kim, J.K.; Leung, C.K.Y. Taper ended FRP strips bonded to RC beams: Experiments and FEM analysis. In Proceedings of the Second International Conference on FRP in Civil Engineering, Adelaide, Australia, 8–10 December 2004; pp. 399–405.
53. Gao, B.; Kim, J.-K.; Leung, C.K. Experimental study on RC beams with FRP strips bonded with rubber modified resins. *Compos. Sci. Technol.* **2004**, *64*, 2557–2564. [CrossRef]
54. Grace, N.F.; Singh, S.B. Durability evaluation of carbon fiber-reinforced polymer strengthened concrete beams: Experimental study and design. *ACI Struct. J.* **2005**, *102*, 40–48.
55. Pham, H.B.; Al-Mahaidi, R. Prediction algorithms for debonding failure loads of carbon fiber reinforced polymer retrofitted reinforced concrete beams. *J. Compos. Constr.* **2006**, *10*, 48–59. [CrossRef]
56. Benjeddou, O.; Ben Ouezdou, M.; Bedday, A. Damaged RC beams repaired by bonding of CFRP laminates. *Constr. Build. Mater.* **2007**, *21*, 1301–1310. [CrossRef]
57. Esfahani, M.R.; Kianoush, M.R.; Tajari, A.R. Flexural behavior of reinforced concrete beams strengthened by CFRP sheets. *Eng. Struct.* **2007**, *29*, 2428–2444. [CrossRef]
58. Ceroni, F. Experimental performances of RC beams strengthened with FRP materials. *Constr. Build. Mater.* **2010**, *24*, 1547–1559. [CrossRef]
59. Al-Tamimi, A.K.; Hawileh, R.; Abdalla, J.; Rasheed, H.A. Effects of Ratio of CFRP Plate Length to Shear Span and End Anchorage on Flexural Behavior of SCC RC Beams. *J. Compos. Constr.* **2011**, *15*, 908–919. [CrossRef]
60. Sadrmomtazi, A.; Rasmi Atigh, H.; Sobhan, J. Experimental and analytical investigation on bond performance of the interfacial debonding in flexural strengthened RC beams with CFRP sheets at tensile face. *Asian J. Civ. Eng. (BHRC)* **2014**, *15*, 391–410.
61. Hasnat, A.; Islam, M.M.; Amin, A.F.M.S. Enhancing the Debonding Strain Limit for CFRP-Strengthened RC Beams Using U-Clamps: Identification of Design Parameters. *J. Compos. Constr.* **2016**, *20*. [CrossRef]
62. Skuturna, T.; Valivonis, J. Experimental study on the effect of anchorage systems on RC beams strengthened using FRP. *Compos. Part B Eng.* **2016**, *91*, 283–290. [CrossRef]
63. Saadatmanesh, H.; Ehsani, M.R. RC beams strengthened with GFRP plates. I: Experimental study. *J. Struct. Eng.* **1991**, *117*, 3417–3433. [CrossRef]
64. Arduini, M.; Nanni, A. Behavior of Precracked RC Beams Strengthened with Carbon FRP Sheets. *J. Compos. Constr.* **1997**, *1*, 63–70. [CrossRef]
65. Spadea, G.; Swamy, R.N.; Bencardino, F. Strength and Ductility of RC Beams Repaired with Bonded CFRP Laminates. *J. Bridg. Eng.* **2001**, *6*, 349–355. [CrossRef]

66. White, T.W.; Soudki, K.A.; Erki, M.-A. Response of RC Beams Strengthened with CFRP Laminates and Subjected to a High Rate of Loading. *J. Compos. Constr.* **2001**, *5*, 153–162. [CrossRef]
67. Grace, N.F.; Abdel-Sayed, G.; Ragheb, W. Strengthening of concrete beams using innovative ductile fiber-reinforced polymer fabric. *Struct. J.* **2002**, *99*, 692–700.
68. Alagusundaramoorthy, P.; Harik, I.E.; Choo, C.C. Flexural Behavior of R/C Beams Strengthened with Carbon Fiber Reinforced Polymer Sheets or Fabric. *J. Compos. Constr.* **2003**, *7*, 292–301. [CrossRef]
69. Maalej, M.; Leong, K. Effect of beam size and FRP thickness on interfacial shear stress concentration and failure mode of FRP-strengthened beams. *Compos. Sci. Technol.* **2005**, *65*, 1148–1158. [CrossRef]
70. Zhang, A.-H.; Jin, W.-L.; Li, G.-B. Behavior of preloaded RC beams strengthened with CFRP laminates. *J. Zhejiang Univ. A* **2006**, *7*, 436–444. [CrossRef]
71. Lee, S.; Moy, S. A Method for Predicting the Flexural Strength of RC Beams Strengthened with Carbon Fiber Reinforced Polymer. *J. Reinf. Plast. Compos.* **2007**, *26*, 1383–1401. [CrossRef]
72. Rusinowski, P.; Täljsten, B. Intermediate Crack Induced Debonding in Concrete Beams Strengthened with CFRP Plates—An Experimental Study. *Adv. Struct. Eng.* **2009**, *12*, 793–806. [CrossRef]
73. Gunes, O.; Buyukozturk, O.; Karaca, E. A fracture-based model for FRP debonding in strengthened beams. *Eng. Fract. Mech.* **2009**, *76*, 1897–1909. [CrossRef]
74. Pan, J.; Leung, C.K.; Luo, M. Effect of multiple secondary cracks on FRP debonding from the substrate of reinforced concrete beams. *Constr. Build. Mater.* **2010**, *24*, 2507–2516. [CrossRef]
75. Sena-Cruz, J.M.; Barros, J.A.; Coelho, M.R.; Silva, L.F. Efficiency of different techniques in flexural strengthening of RC beams under monotonic and fatigue loading. *Constr. Build. Mater.* **2012**, *29*, 175–182. [CrossRef]
76. Kotynia, R. Debonding failures of RC beams strengthened with externally bonded strips. In Proceedings of the International Symposium on Bond Behaviour of FRP in Structures (BBFS 2005), Hong Kong, China, 7–9 December 2005.
77. Al-Zaid, R.A.; El-Sayed, A.K.; Al-Negheimish, A.I.; Shuraim, A.B.; Alhozaimy, A.M. Strengthening of structurally damaged wide shallow RC beams using externally bonded CFRP plates. *Lat. Am. J. Solids Struct.* **2014**, *11*, 946–965. [CrossRef]
78. Yalim, B.; Kalayci, A.S.; Mirmiran, A. Performance of FRP-Strengthened RC Beams with Different Concrete Surface Profiles. *J. Compos. Constr.* **2008**, *12*, 626–634. [CrossRef]
79. Bi, Y.; Wu, S.; Pei, J.; Wen, Y.; Li, R. Correlation analysis between aging behavior and rheological indices of asphalt binder. *Constr. Build. Mater.* **2020**, *264*, 120176. [CrossRef]
80. Kong, L.; Chen, X.; Du, Y. Evaluation of the effect of aggregate on concrete permeability using grey correlation analysis and ANN. *Comput. Concr.* **2016**, *17*, 613–628. [CrossRef]
81. Huang, C.-H.; Hsieh, S.-H. Predicting BIM labor cost with random forest and simple linear regression. *Autom. Constr.* **2020**, *118*, 103280. [CrossRef]
82. Wang, X.; Liang, Y.; Wang, Q.; Zhang, Z. Empirical models for tool forces prediction of drag-typed picks based on principal component regression and ridge regression methods. *Tunn. Undergr. Space Technol.* **2017**, *62*, 75–95. [CrossRef]
83. Dabiri, H.; Farhangi, V.; Moradi, M.J.; Zadehmohamad, M.; Karakouzian, M. Applications of Decision Tree and Random Forest as Tree-Based Machine Learning Techniques for Analyzing the Ultimate Strain of Spliced and Non-Spliced Reinforcement Bars. *Appl. Sci.* **2022**, *12*, 4851. [CrossRef]
84. Moradi, N.; Tavana, M.H.; Habibi, M.R.; Amiri, M.; Moradi, M.J.; Farhangi, V. Predicting the Compressive Strength of Concrete Containing Binary Supplementary Cementitious Material Using Machine Learning Approach. *Materials* **2022**, *15*, 5336. [CrossRef]
85. Xue, J.; Shen, B. Dung beetle optimizer: A new meta-heuristic algorithm for global optimization. *J. Supercomput.* **2022**, 1–32. [CrossRef]

buildings

MDPI

Article

Prediction Study on the Alignment of a Steel-Concrete Composite Beam Track Cable-Stayed Bridge

Xiaogang Li [1,2,*], Haoran Luo [2], Peng Ding [3], Xiaohu Chen [3] and Shulin Tan [3]

[1] Chongqing Key Laboratory of Energy Engineering Mechanics & Disaster Prevention and Mitigation, Chongqing University of Science & Technology, Chongqing 400074, China
[2] School of Civil Engineering and Architecture, Chongqing University of Science & Technology, Chongqing 400074, China
[3] T. Y. Lin International Engineering Consulting (China) Co., Ltd., Chongqing 401121, China
* Correspondence: 2021058@cqust.edu.cn; Tel.: +86-135-9409-4526

Abstract: Due to the alignment of track bridges directly affecting the safety and comfort of rail traffic operation, the alignment prediction of track bridges needs to be accurate. However, the structure of steel-concrete composite beam (SCCB) cable-stayed bridges is more complex, and the alignment prediction needs to be more accurate. To further improve the accuracy of alignment prediction for large-span SCCB track cable-stayed bridges, a method based on the response surface method (RSM) is proposed. In this paper, the Nanjimen Yangtze River Track Special Bridge was taken as a case for research. Considering the randomness of the influencing factors, the 95% confidence interval was obtained by using Monte Carlo (MC) sampling analysis, and the predicted values were within the confidence interval. The results show that the method integrates the confidence interval under each confidence level by simulating the long-term deformation of different years after bridge completion. The method could accurately predict the alignment of large-span SCCB track cable-stayed bridges, and thereby provide technical support for alignment control and ensure the safe and comfortable operation of rail transit.

Keywords: large-span track bridge; steel-concrete composite beam; cable-stayed bridge; alignment prediction; response surface method; Monte Carlo

Citation: Li, X.; Luo, H.; Ding, P.; Chen, X.; Tan, S. Prediction Study on the Alignment of a Steel-Concrete Composite Beam Track Cable-Stayed Bridge. *Buildings* **2023**, *13*, 882. https://doi.org/10.3390/buildings13040882

Academic Editors: Humberto Varum and Ahmed Senouci

Received: 15 February 2023
Revised: 9 March 2023
Accepted: 21 March 2023
Published: 28 March 2023

1. Introduction

Bridges are key elements of the transportation system. Health monitoring and assessment of large-span bridges, mastering the evolution law of structural state, predicting structural performance changes and early warning of emergencies are effective means to ensure bridge operation safety, and are the priorities and research direction in bridge engineering. As a response parameter of the bridge structure, the bridge alignment can directly reflect the overall state and performance degradation of the structure (i.e., force change, material degradation, stiffness degradation, etc. all could cause the alignment change) and is an important parameter for assessment of the structure state. Structural state monitoring, analysis and assessment with bridge alignment as the key index is one of the important aspects of bridge health monitoring. Bridge alignment prediction is an area of intense research in academia and industry which has important theoretical significance and engineering application value.

SCCB are a new type of structure developed on the basis of steel and concrete. This structure can resist the lifting and opposite sliding at the interface of the steel beam and concrete bridge deck by setting shear connectors (shear nails, bending bars, etc.) between the steel beam and the concrete bridge deck, so as to make it work together as a whole. The SCCB bridge has the advantages of both steel and concrete structures, and is the main development direction of long-span bridges. The research on the alignment prediction of SCCB bridges is of great significance. Due to the influence of large traffic volume, narrow width, and high frequency of operation, the error between the predicted alignment and actual alignment of large-span SCCB track cable-stayed bridges can be large. The alignment state of such bridges directly impacts their safe construction and operation. If the error of alignment prediction is too large, it can cause a great threat to the safety of the whole bridge. Alignment analyses of large-span SCCB track cable-stayed bridges have been carried out in several studies, starting from the response surface method (RSM), and accurate alignment predictions were obtained.

Changes in bridge alignment are closely related to the bridge structure system, material characteristics, construction technology, etc. Researchers have carried out a large number of related studies and achieved fruitful research results. Ahad et al. [1,2] explored the role of bonding between ultra-high-performance concrete and reinforcement through experiments. Xiong [3] studied the influence of different influencing factors on the construction control of SCCB cable-stayed bridges through a geometric control method and identified the parameter sensitivity affecting the alignment of the main beam.

In order to reduce the influence of the construction process, several researchers have proposed different methods to improve the efficiency of alignment analysis for large-span bridges as bridge construction technology continues to mature. Xin et al. [4] proposed a new method based on improved variational mode decomposition (IVMD) and conditional kernel density estimation (CKDE) analysis data to obtain a high reliability bridge alignment prediction. Lu et al. [5] have proposed an agent model which is based on mind evolutionary computation-back propagation (MEC-BP) to improve the efficiency of the finite dimension analysis through the assistance of the model, using the model to study the alignment of a large span of waveform steel webs. Zhou et al. [6] proposed an optimized extreme learning machine algorithm to obtain the optimum extreme learning machine (ELM) data through an MEC search and then added in the ELM for the training so as to obtain a model with an average error of only 0.225 cm to generate a construction alignment prediction of large-span continuous rigid bridges. Chen et al. [7] established the multivariable grey model (MGM) (1,2) on the basis of grey system theory in order to combine the grey prediction model and the characteristics of the arch rib space of the shaped arch. They used the actual work for comparison to prove that the MGM(1,2) model is an accurate and reasonable method to predict the arch rib space of a shaped arch bridge alignment. Li and Zhu et al. [8] evaluated the long-term state of the Qingma Bridge under traffic load by a suspension cable monitoring system. Li and Wei et al. [9] proposed a cable state evaluation method based on machine learning, which showed that the cable tension ratio is only related to the properties of the cable and the transverse position of the vehicle on the deck. The model analysis showed that the tension ratio can be considered an effective feature reflecting the cable state which can be used effectively for cable state evaluation. Considering the inaccuracy of monitoring systems caused by temperature in some specific cases, Li et al. [10] proposed a method to extract bridge deformation under the influence of temperature and effectively analyzed the influence of temperature. Zhang et al. [11] established the dual defects magnetic dipole model with double defects in the cable and evaluated the state of the cable by quantifying the fracture width. Statistical methods are also becoming more widely used in the field of civil engineering. Among them, multiple linear regression is a statistical method that uses mathematical models to describe the relationship between one or more dependent variables. Yang et al. [12] analyzed eight parameters using the multiple linear regression method. In this way, the energy consumption of buildings could be analyzed and predicted. This statistical method is also applicable to the prediction research of bridge engineering. RSM is often used as a statistical method in engineering research. Li et al. [13] revised the model of cable-stayed bridges based on RSM. Using RSM, Ma et al. [14] performed synchronous revision of the structural parameters of the multi-scale finite element model (FEM) for a concrete-filled steel tube bridge. Xin et al. [15] used RSM to predict the long-term deformation of track cable-stayed bridges in a probabilistic sense considering five main structural parameters. Liu et al. [16] combined the RSM and equivalent normalization method (JC method) to analyze the reliability of large-span bridges.

However, the RSM has been applied less to the prediction of the alignment for large-span SCCB track cable-stayed bridges. There is less research on how to accurately predict the alignment for large-span SCCB track cable-stayed bridges. In this study, the RSM was used to conduct the alignment prediction based on environmental factors, actual situation on site, applicable standards and specifications, and other factors. It can supplement the alignment prediction research of entire large-span track bridges, and provide a reference for bridge health monitoring, assessment and reasonable alignment control during the operation period.

2. Experiment Principles

2.1. Design of Experiments

The structural alignment of bridges is influenced by many factors, many of which are random. The randomness of influencing factors should be considered in the prediction of bridge alignment. When selecting the samples of influencing factors, samples that can reflect the spatial characteristics and laws [17–21] should be selected as far as possible to improve the accuracy and efficiency of the experiment and achieve more precise results [22,23].

At present, the most common experimental design methods are uniform and orthogonal. Compared with the orthogonal design method, the uniform design method, which can solve more influencing factors in a shorter test time, is more suitable for large-span track SCCB cable-stayed bridges. This method is characterized by uniform dispersion, neatness, and comparability, making it an application of the pseudo MC method [24]. In this paper, the test sample points were selected with reference to the U_8 (8^8) uniform design table based on the aspects of design complexity, environmental randomness and practical stability.

2.2. Experiment Methods

2.2.1. Response Surface Method

The multivariate linear regression method [25–27] is adopted to convert the nonlinear system into a linear system. The partial least squares method is adopted to optimize and find the matching function. The method of minimum error squared is adopted to fit the model prediction. The principle of the method is as follows.

1. The independent variable matrix $A_{\alpha \times \beta}$ is a standardized treatment to obtain V_0 and its column vectors $V_{01}, \ldots, V_{0\beta}$; the dependent variable matrix $B_{\alpha \times \delta}$ is a standardized treatment to obtain M_0 and its column vectors $M_{01}, \ldots, M_{0\delta}$, where V_0 is expressed as $V_0 = (V_{01}, \ldots, V_{0\beta})_{\alpha \times \beta}$ and M_0 is expressed as $M_0 = (M_{01}, \ldots, M_{0\delta})_{\alpha \times \beta}$.
2. The transpose matrix of V_0 and M_0 are denoted by V_0^T and M_0^T. The characteristic vectors φ_1, φ_2 corresponding to the maximum eigenvalues of $V_0^T M_0 M_0^T V_0$ and $V_1^T M_0 M_0^T V_1$, respectively, are calculated. Then t_1 and t_2 are calculated, where $t_1 = V_0\varphi_1$, $V_1 = V_0 - t_1 G_1'$, $G_1' = \frac{V_0^T t_1}{\|t_1\|^2}$, $t_2 = V_1\varphi_2$, $V_2 = V_1 - t_2 G_2'$, and $G_2' = \frac{V_1^T t_2}{\|t_2\|^2}$. This step is repeated until step m when t_m, and V_{m-1} are obtained.
3. To establish a suitable model, m finite components $t_1, \ldots, t_n$ are selected. After executing to step m, the regression equation can be obtained, which is given as:

$$M_{0k} = \alpha_{k1} V_{01} + \cdots + \alpha_{k\beta} V_{0\beta} + M_k \quad (1)$$

where $\alpha_{k1}, \ldots, \alpha_{k\beta}$ are the regression coefficients.

2.2.2. Accuracy Test

After obtaining the data, the accuracy of the model needs to be tested. The multivariate RSM model was used in this work, so it is reasonable to use an R^2 test to judge the accuracy of the model. which is given as:

$$R^2 = 1 - \frac{\Sigma_{j=1}^{m}\left[f(x)_j - \hat{f}(x)_j\right]^2}{\Sigma_{j=1}^{m}\left[f(x)_j - \overline{f}(x)\right]^2} \tag{2}$$

where $f(x)_j$ is the jth sample point response value, $\hat{f}(x)_j$ is the calculated value of the response surface corresponding to the jth sample point, and $\overline{f}(x)$ is the average value, which is obtained from $\overline{f}(x) = \frac{1}{m}\Sigma_{j=1}^{m} f(x)_j$. The calculated R^2 is between 0 and 1, and the closer the value is to 1, the more accurate it is.

2.2.3. Monte Carlo Sampling Analysis

In order to improve the accuracy of the predicted values represented by the response surface function, sample selection is significant. The MC method was selected for sampling in this study. To begin with, x_i (i = 1, 2, ... , n) are the independent random parameters in the influence model function $f(x)$, and each corresponding independent parameter x_i (i = 1, 2, ... , n) is randomly sampled to obtain a random sample of $x_1, x_2, \ldots, x_n$. Then, x_i is substituted into a specific program for repeated random sampling and analysis. The response values of the M group structure are obtained, which are denoted as $f(x)_1, f(x)_2, \ldots, f(x)_m$. Furthermore, the average value μ and standard deviation σ of $f(x)$ are calculated, which can be expressed as:

$$\mu = \overline{f}(x) = \frac{1}{m}\sum_{j=1}^{m} f(x)_j \tag{3}$$

$$\sigma = \sqrt{\frac{1}{m-1}\sum_{j=1}^{m}\left[f(x)_j - \mu\right]^2} \tag{4}$$

3. Application Analysis

3.1. Project Profile

The Nanjimen Yangtze River Track Special Bridge connects Yuzhong District and Nan'an District and is an important link in the second phase project of Chongqing Rail Transit Line 10. The main bridge is a five-span double-cable plane and high-low tower cable-stayed bridge with a semi-floating system. The span arrangement of the main bridge is 34.5 + 180.5 + 480 + 215.5 + 94.5 m. The main girder is a SCCB, and the portal tower form is used. The bridge type layout, cross-section view of the main beam, the main tower, and the cable are shown in Figures 1–4, respectively.

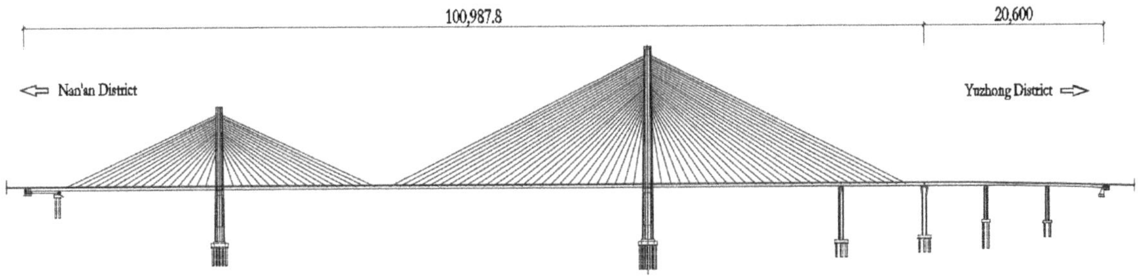

Figure 1. The bridge type layout of Nanjimen Yangtze River Track Special Bridge (unit: mm).

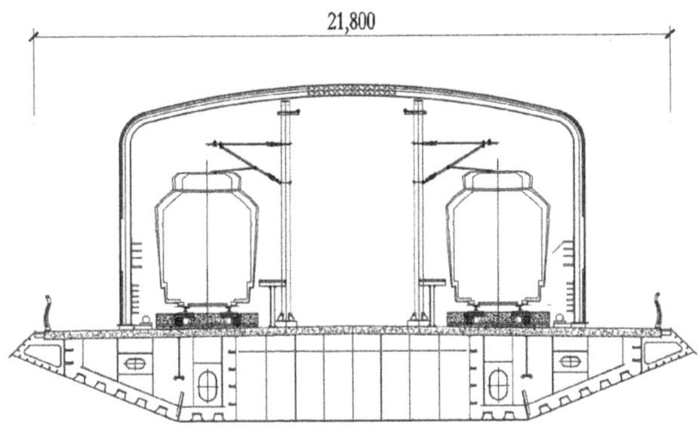

Figure 2. Cross-sectional view of the main beam (unit: mm).

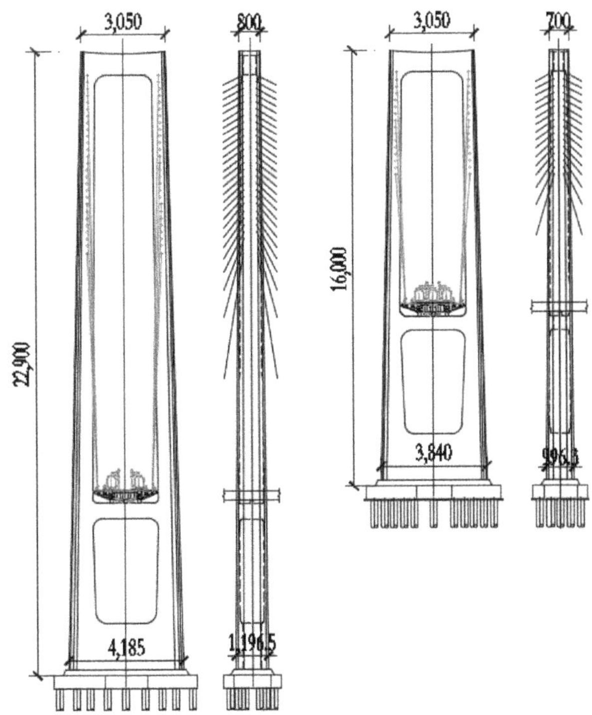

Figure 3. Cross-sectional view of the main tower (unit: mm).

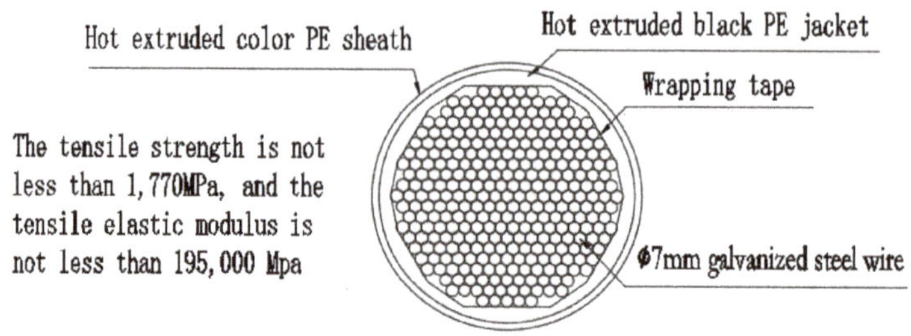

Figure 4. Cross-sectional view of the cable.

3.2. *Finite Element Analysis*

MIDAS/Civil was used to carry out the FE numerical analysis model of the Nanjimen Yangtze River Track Special Bridge (Figure 5). The model used beam elements to simulate the main tower, pier, and main beam, and truss elements to simulate the cables. The boundary conditions of the model included complete consolidation between the main tower and the pier, a rigid connection between the main tower and the stay cable, and tri-directional restraint at the boundary of the auxiliary pier. In this model, the shrinkage and creep of concrete were considered, and the material properties given in the design drawings were used. At the same time, according to the "Code for Design of Railway Bridges and culverts" (TB10002-2017) issued by the National Railway Administration, the time-dependent characteristics were considered, and the software was used to automatically calculate the most adverse effects of shrinkage and creep loads.

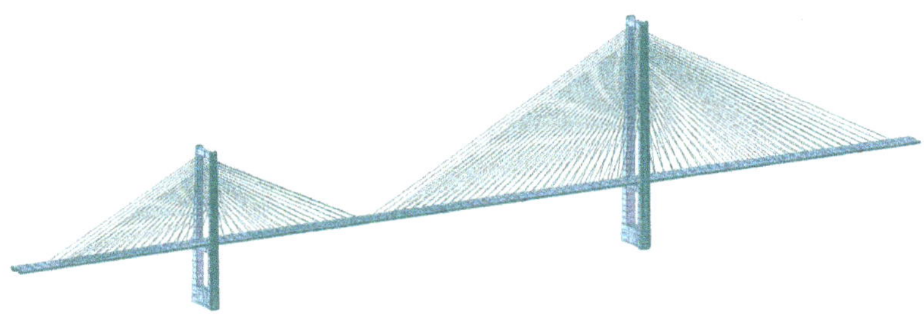

Figure 5. The FEM numerical analysis model of the Nanjimen Yangtze River Track Special Bridge.

3.3. *Model Establishment*

3.3.1. Selection of Parameters to Establish the Model

For the alignment analysis of large-span SCCB track cable-stayed bridges, it is necessary to select suitable random parameters and obtain the average value and coefficient of variation of these parameters. In this study, the random parameters and their statistical characteristics were selected considering the environmental factors, actual situation on the site, applicable standards and specifications, and other factors of the Nanjimen Yangtze River Track Special Bridge. The statistical characteristics of the random parameters are shown in Table 1, where X_1 is the strength of the main beam, X_2 is the mean annual humidity of the main beam environment, X_3 is the concrete age at the beginning of main beam shrinkage, X_4 is the strength of the main tower, X_5 is the mean annual humidity of the main tower environment, X_6 is the concrete age at the beginning of the main tower shrinkage, X_7 is the volume weight of the main beam, and X_8 is the volume weight of the main tower.

Table 1. Statistical characteristics of random parameters.

Statistical Characteristics	Strength of the Main Beam X_1/(N/mm^2)	Mean Annual Humidity of the Main Beam Environment X_2/(%)	Concrete Age at the Beginning of Main Beam Shrinkage X_3/(day)	Strength of the Main Tower X_4/(N/mm^2)	Mean Annual Humidity of the Main Tower Environment X_5/(%)	Concrete Age at the Beginning of the Main Tower Shrinkage X_6/(day)	Volume Weight of the Main Beam X_7/(N/mm^3)	Volume Weight of the Main Tower X_8/(N/mm^3)
Type of distribution	Normal distribution	Normal distribution	Normal distribution	Normal distribution	Normal distribution	Normal distribution	Normal distribution	Normal distribution
Average value	60	79	3	50	79	3	26	26
Coefficient of variation	0.15	0.20	0.22	0.14	0.12	0.26	0.11	0.11

3.3.2. Selection of Samples to Establish the Model

To study the long-term deformation of the entire bridge, six different time nodes were selected for analysis, namely, 1/2 year, 1 year, 2 years, 3 years, 5 years, and 10 years since the bridge completion. The vertical deformation of the mid-span position of the main span for these time nodes was studied. The structural influence values were denoted as $S_{1/2}$, S_1, S_2, S_3, S_5, and S_{10}, respectively, and the uniformity experiment was carried out according to the uniform design table of $U_8(8^8)$. Eight sample points were obtained by testing each set of data, with the deformation of the main beam defined as negative when vertically downward and positive when vertically upward. In this way, the sample points of the uniform design (Table 2) and the structural response data were obtained (Table 3).

Table 2. Sample points based on uniform design.

Serial Number	X_1/(N/mm^2)	X_2/(%)	X_3/(day)	X_4/(N/mm^2)	X_5/(%)	X_6/(day)	X_7/(N/mm^3)	X_8/(N/mm^3)
1	59.40	78.2	2.12	49.44	78.530	1.96	25.56	25.56
2	59.55	78.4	2.34	49.58	78.648	2.22	25.67	25.67
3	59.70	78.6	2.56	49.72	78.766	2.48	25.78	25.78
4	59.85	78.8	2.78	49.86	78.884	2.74	25.89	25.89
5	60.00	79.0	3.00	50.00	79.002	3.00	26.00	26.00
6	60.15	79.2	3.22	50.14	79.120	3.26	26.11	26.11
7	60.30	79.4	3.44	50.28	79.238	3.52	26.22	26.22
8	60.45	79.6	3.66	50.42	79.356	3.78	26.33	26.33

Table 3. Structural response data (mm).

Serial Number	$S_{1/2}$	S_1	S_2	S_3	S_5	S_{10}
1	−21.810	−37.231	−50.194	−61.070	−76.402	−98.043
2	−22.271	−34.782	−50.784	−61.656	−77.336	−99.256
3	−22.732	−35.310	−51.350	−62.803	−78.614	−100.498
4	−23.231	−35.639	−52.549	−64.103	−80.249	−101.795
5	−25.981	−39.213	−55.915	−67.192	−83.015	−104.833
6	−25.245	−38.547	−55.291	−66.585	−82.416	−104.202
7	−17.659	−27.846	−36.506	−41.716	−48.309	−55.856
8	−14.690	−24.974	−33.767	−39.047	−45.754	−52.834

3.4. Response Surface Analysis

3.4.1. Response Surface Model Fitting

By fitting the data in Tables 2 and 3 with the least square method, six response models were obtained, which are expressed as:

$$S_{1/2} = -237.4451 + 0.6211X_1 + 0.4659X_2 + 0.4235X_3 + 0.6655X_4 + 0.7896X_5 + 0.3584X_6 + 0.8470X_7 + 0.8470X_8 \quad (5)$$

$$S_1 = -403.5580 + 1.0634X_1 + 0.7976X_2 + 0.7251X_3 + 1.1394X_4 + 1.3518X_5 + 0.6135X_6 + 1.4501X_7 + 1.4501X_8 \quad (6)$$

$$S_2 = -638.1748 + 1.6983X_1 + 1.2737X_2 + 1.1579X_3 + 1.8196X_4 + 2.1589X_5 + 0.9798X_6 + 2.3159X_7 + 2.3159X_8 \quad (7)$$

$$S_3 = -883.0270 + 2.3753X_1 + 1.7814X_2 + 1.6195X_3 + 2.5449X_4 + 3.0194X_5 + 1.3703X_6 + 3.2390X_7 + 3.2390X_8 \quad (8)$$

$$S_5 = -1262.020 + 3.4276X_1 + 2.5707X_2 + 2.3370X_3 + 3.6724X_4 + 4.3571X_5 + 1.9774X_6 + 4.6740X_7 + 4.6740X_8 \quad (9)$$

$$S_{10} = -1879.095 + 5.1519X_1 + 3.8639X_2 + 3.5127X_3 + 5.5199X_4 + 6.5490X_5 + 2.9723X_6 + 7.0253X_7 + 7.0253X_8 \quad (10)$$

3.4.2. Response Surface Model Validation

To verify the statistical relationship between the structure and the selected random variables, the accuracy of the six fitted response surface models was tested. The results are shown in Table 4.

Table 4. Accuracy tests of response surface models.

Models	R^2	Models	R^2
$S_{1/2}$	0.8993	S_5	0.9455
S_1	0.8714	S_{10}	0.9425
S_2	0.8788	Average value	0.9091
S_3	0.9173		

As shown in Table 4, the average value of the six response surfaces is 0.9091. More than 90% is due to changes in the model parameters, indicating that the model has high fitting accuracy and is more reliable.

3.5. Deformation Data Analysis

3.5.1. Deformation Data Analysis since Closure of the Main Span

To verify the applicability of the method, the measured values, FEM calculated values and statistically calculated values (calculated by the MC-RSM method) of the vertical deformation data since closure of the main span were compared and analyzed. The measured deformation data was obtained from the health monitoring system (HMS) of Nanjimen Yangtze River Track Special Bridge. Using phase shift regression analysis, an empirical regression equation [28,29] was established to exclude the influence of temperature effects, and then compared with the calculation results. The results are shown in Figure 6.

As shown in Figure 6, the measured values of vertical deformation in the mid-span since closure of the main-span are relatively close to the FEM calculated values, with small fluctuations around the FEM calculated values. However, the engineering prediction values are low due to the fluctuation. At the same time, the statistically calculated values obtained by the MC-RSM method envelop the measured values. Although there is a gap between the statistically calculated and measured values, the statistically calculated values are more certain because they focus on the calculation and analysis of the sample data, and there is also some defect tolerance while maintaining accuracy and reliability. In addition, the statistically calculated values only consider some of the influencing factors and cannot include all the actual influencing factors, leading to partial deviation of the calculated results. If there were more factors studied, the statistically calculated results would be closer to the actual situation.

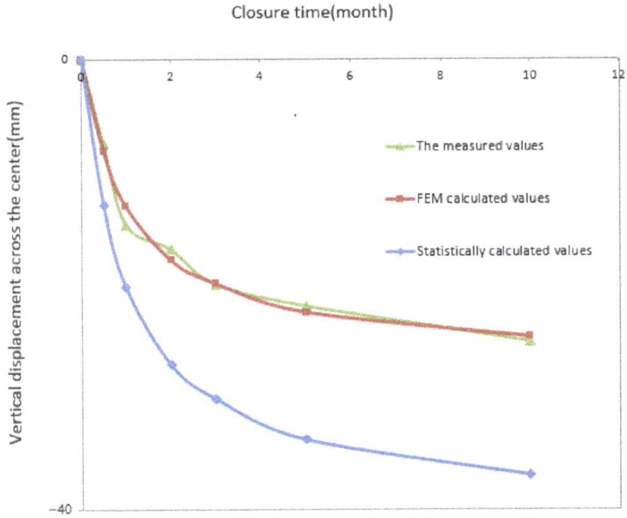

Figure 6. Comparison of vertical deformation at the mid-span of the main span.

3.5.2. Predictive Analysis of Deformation after Bridge Completion

When predicting deformation at the mid-span of the main span, the MC method was applied, selecting a sample number of 10,000. The shrinkage and creep of the Nanjimen Yangtze River Track Special Bridge were analyzed randomly based on the variability of the material parameters. The average value and standard deviation of vertical deformation at the mid-span of the main span were analyzed by selecting six time nodes from 1/2 year to 10 years after bridge completion. The results are shown in Table 5.

Table 5. Random analysis results for deformation at the mid-span of the main span.

Working Conditions		Design Expected Values (mm)	Measured Values (mm)	Statistical Characteristic Values	
				Average Values (mm)	Standard Deviation
Time after bridge completion	1/2 year	−21.33	−20.02	−21.70	3.54
	1 year	−33.56	−25.13	−34.19	4.77
	2 years	−47.28	/	−48.29	7.86
	3 years	−56.60	/	−58.02	10.40
	5 years	−69.46	/	−71.51	14.31
	10 years	−86.60	/	−89.66	20.52

Table 5 shows that the difference between the design expected values and the average values (obtained by sampling statistical analysis) is small, according to the parameters selected in this paper. It shows that the maximum vertical deformation at the mid-span of the main span can be well predicted. Furthermore, it can be seen that the standard value increases over time, indicating the discreteness of the deformation increment gradually increases.

The MC method obtained the probability distribution of vertical deformation at different time nodes, with the distribution results shown in Figure 7. The confidence levels under five representative percentages were selected, which were 35%, 55%, 75%, and 95%, respectively.

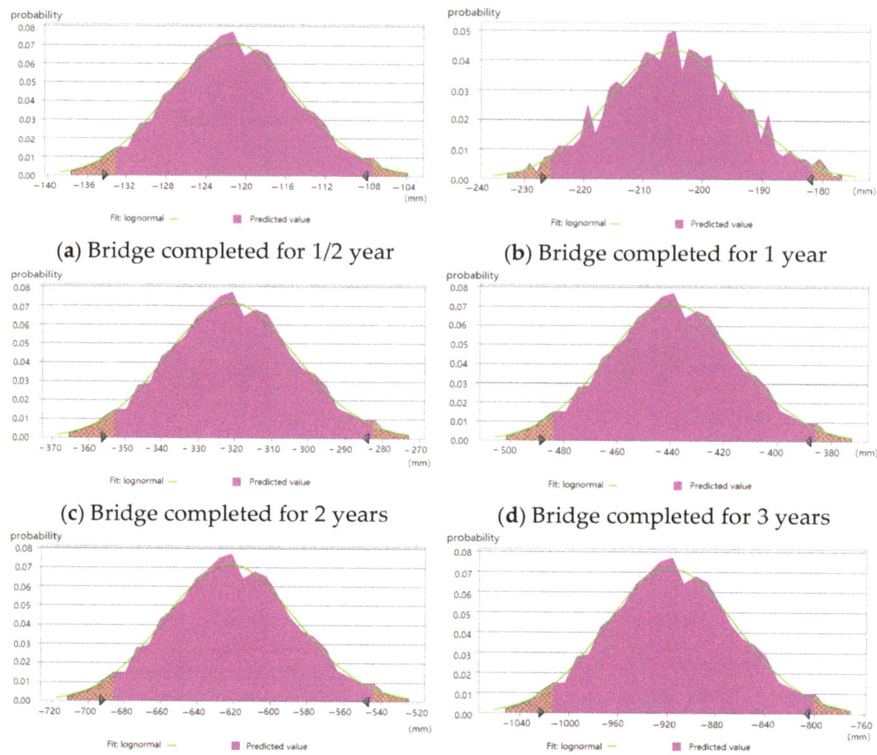

(**a**) Bridge completed for 1/2 year (**b**) Bridge completed for 1 year (**c**) Bridge completed for 2 years (**d**) Bridge completed for 3 years (**e**) Bridge completed for 5 years (**f**) Bridge completed for 10 years

Figure 7. Probability distribution of vertical deformation at different time nodes.

The vertical deformation predicted interval at each chosen time node for different random factors was plotted for these five different confidence levels. Figure 8 shows the results for the vertical deformation of different deformation intervals in 10 years.

Furthermore, the predicted values obtained by the FE calculation were compared with the confidence interval at the 95% confidence level. The predicted values for vertical deformation at the mid-span of the main span were obtained using MIDAS/Civil considering different time nodes after bridge completion. Under the same random parameters, the confidence intervals at the 95% confidence level were obtained by the MC method sampling, and the comparison results are shown in Figure 9.

Figure 9 illustrates that the predicted values obtained by the proposed model fall within the 95% confidence interval, demonstrating the high accuracy of the parameter distribution and analysis method. The 10-year deformation amplitude of the bridge calculated by the finite element software and the variation value of the lower limit of 95% confidence level are shown in Figure 10. It can be seen that the 10-year deformation interval calculated by the finite element software is (−86.5963, −21.3299) mm, and the variation value is 65.2664 mm. Using the uncertainty analysis method of MC-RSM, the variation interval of the lower limit of the 95% confidence level is (−133.64, −33.68) mm, and the variation amplitude of the lower limit of the 95% confidence level is 99.96 mm. The deformation difference between the two methods increased from 12.35 mm at half a year to 47.04 mm at 10 years. The deformation amplitude of the mid-span of the bridge at 10 years after completion (calculated by the FE method) and the lower limit of the 95% confidence level (predicted by the MC-RSM method) and the deformation difference obtained by the two methods in 10 years are shown in Figure 10. The vertical deformation at the mid-span of the main span is relatively discrete, resulting in small calculation values by the FE method, which is due to the influence of parameter randomness. It can be concluded that environmental factors, material factors, and other factors should be considered comprehensively when analyzing the alignment of large-span track SCCB cable-stayed bridges.

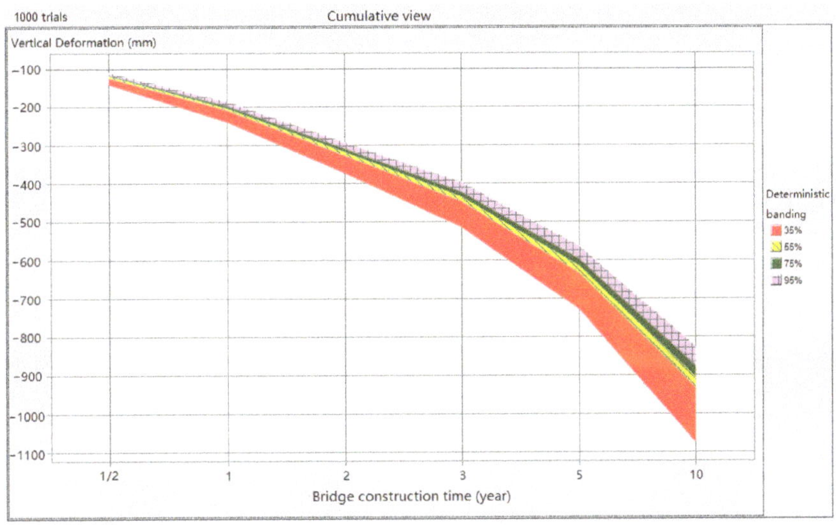

Figure 8. Vertical deformation predicted interval at different confidence levels.

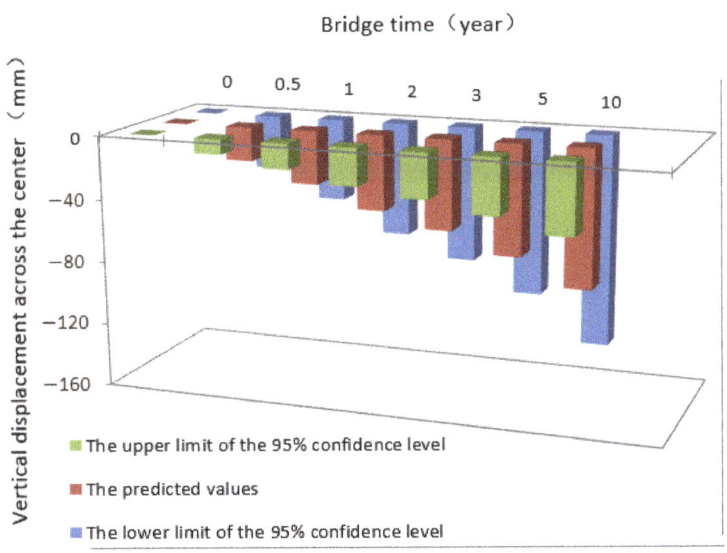

Figure 9. Prediction of vertical deformation at the mid-span of the main span (unit: mm).

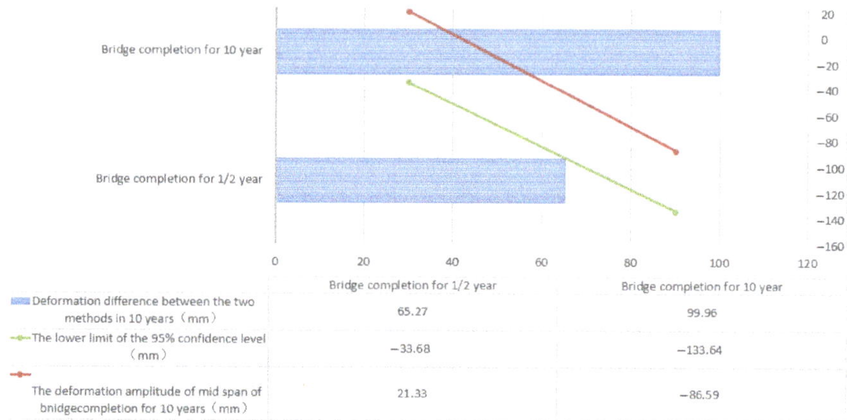

Figure 10. Comparison of 10-year deformation amplitude (unit: mm).

4. Conclusions

In this paper, an alignment prediction of a large-span track SCCB cable-stayed bridge was carried out by RSM considering the environmental factors, actual situation on site, applicable standards and specifications of the Nanjimen Yangtze River track Special Bridge. The deformation change of eight factors under the influence of randomness was discussed. The following main conclusions were obtained.

1. Combining different influencing factors with the alignment of the entire large-span track SCCB cable-stayed bridge, the RSM method was used to solve the problem of low accuracy of alignment prediction for large-span SCCB track cable-stayed bridges. Accurate alignment prediction was realized under the randomness of influencing factors.
2. The MC method was combined with the RSM method for analysis. The partial least squares method was used to verify the analysis according to the influencing factors combined with the uniform design method. The long-term deformation for different years after bridge completion was predicted. In addition, the finite element calculated value interval of (−86.5963, −21.3299) mm is included in the 95% confidence interval of (−133.64, −33.68) mm.
3. RSM can be used to analyze different influencing factors. In this paper, eight influencing factors, such as the strength of the main beam, were analyzed, to realize the alignment prediction of the bridge. In addition, other influencing factors could be analyzed and studied quickly and effectively through this method, reducing the test cost and time, and improving work efficiency.
4. RSM was used to predict and analyze the alignment of SCCB cable-stayed bridges, and the predicted values at different years over 10 years were obtained. Alignment prediction is an important consideration in the field of bridge health monitoring. The prediction method based on RSM provides a theoretical basis and engineering application reference for reasonable alignment control and state evaluation of bridges.

Due to the limited space, only the randomness of the eight influencing factors of the bridge was analyzed and studied in this paper, which resulted in the deviation between the predicted results and the actual results. To obtain more accurate results, further research on more influencing factors is required.

In addition, the time since the bridge completion is relatively short, so the deformation data for only one year after closure of the main span have been compared in this paper. As more extended deformation data becomes available, further verification is needed.

Author Contributions: Conceptualization, X.L. and X.C.; methodology, P.D.; software, P.D.; validation, X.L., X.C. and P.D.; formal analysis, X.L.; investigation, S.T. and H.L.; resources, P.D and S.T.; data curation, P.D. and S.T.; writing—original draft preparation, H.L.; writing—review and editing, X.L. and X.C.; visualization, H.L.; supervision, X.L.; project administration, X.L. All authors have read and agreed to the published version of the manuscript.

Funding: This work was supported by the Open Fund of the Chongqing Key Laboratory of Energy Engineering Mechanics and Disaster Prevention and Mitigation (EEMDPM2021201) and the Chongqing Institute of Science and Technology Research Grant Program (CKRC2021074).

Institutional Review Board Statement: Not applicable.

Informed Consent Statement: Not applicable.

Data Availability Statement: Apart from the original traffic data that were collected through WIM systems, all of the other data, models, and the code that was generated or used during the study are available from the corresponding author by request.

Conflicts of Interest: The authors declare no conflict of interest.

References

1. Ahad, A.P.; Xiong, F.; Yang, P.; Huang, D.; Reza, S.S. Comprehensive equation of local bond stress between UHPC and reinforcing steel bars. *Constr. Build. Mater.* **2020**, *262*, 119942.
2. Ahad, A.P.; Zhang, Z.R.; Mojdeh, A.P.; Xiong, F.; Zhang, L.L.; Yang, Q.H.; Siti, J.M. UHPC-PINN-parallel micro element system for the local bond stress–slip model subjected to monotonic loading. *Structures* **2022**, *46*, 570–597.
3. Xiong, S.Z. Structural Parameter Sensitivity Analysis of Cable-stayed Bridge with Composite Steel Beams. *China Foreign Highw.* **2016**, *36*, 159–162.
4. Xin, J.; Jiang, Y.; Zhou, J.; Peng, L.; Liu, S.; Tang, Q. Bridge deformation prediction based on SHM data using improved VMD and conditional KDE. *Eng. Struct.* **2022**, *261*, 114285. [CrossRef]
5. Lu, Z.Y.; Wang, X.M.; Zhao, B.J.; Ren, W.P. Construction alignment prediction of large span waveform steel web continuous rigid bridge based on MEC-BP agent model. *J. Chang. Univ. (Nat. Sci. Ed.)* **2021**, *41*, 53–62.
6. Zhou, S.X.; Deng, F.M.; Han, Z.; Yu, L.H.; Wu, L.Q. Construction alignment prediction for large-span continuous bridges based on optimal limit learning machine. *J. Railw.* **2019**, *41*, 134–140.
7. Chen, D.Y.; Yan, Q.S.; Zhu, S.Y. Gray prediction of spatial alignment of arch ribs of shaped arch bridges based on MGM(1,n) model. *J. South China Univ. Technol. (Nat. Sci. Ed.)* **2019**, *47*, 120–127.
8. Li, S.L.; Zhu, S.Y.; Xu, Y.L.; Chen, Z.W.; Li, H. Long-term condition assessment of suspenders under traffic loads based on structural monitoring system: Application to the Tsing Ma Bridge. *Struct. Control. Health Monit.* **2012**, *19*, 82–101. [CrossRef]
9. Li, S.L.; Wei, S.Y.; Bao, Y.Q.; Li, H. Condition assessment of cables by pattern recognition of vehicle-induced cable tension ratio. *Eng. Struct.* **2018**, *155*, 1–15. [CrossRef]
10. Li, S.; Xin, J.; Jiang, Y.; Wang, C.; Zhou, J.; Yang, X. Temperature-induced deflection separation based on bridge deflection data using the TVFEMD-PE-KLD method. *J. Civ. Struct. Health Monit.* **2023**. [CrossRef]
11. Zhang, H.; Li, H.; Zhou, J.; Tong, K.; Xia, R. A multi-dimensional evaluation of wire breakage in bridge cable based on self-magnetic flux leakage signals. *J. Magn. Magn. Mater* **2023**, *566*, 170321. [CrossRef]
12. Yang, H.; Ran, M.Y. Analysis and prediction of building energy consumption based on BPT-MLR model. *J. Huaqiao Univ. (Nat. Sci. Ed.)* **2023**, *3*, 1–9.
13. Li, Y.Q.; Liu, X.H.; Chen, Z.L.; Zhang, H. Model modification of cable-stayed bridge based on cable force response surface method. *J. China Railw. Soc.* **2021**, *43*, 168–174.
14. Ma, Y.P.; Liu, Y.J.; Liu, J. Multi-scale finite element model correction of steel tube concrete composite truss bridge based on response surface method. *China J. Highw. Transp.* **2019**, *32*, 51–61.
15. Xin, J.Z.; Zhou, J.T.; Yang, Y.; Zhong, J.H.; Niu, Y.F. Application of response surface method in long-term deformation prediction of cable-stayed bridge tower with large-span. *J. Jiangsu Univ. (Nat. Sci. Ed.)* **2016**, *37*, 367–372.
16. Liu, J.; Yan, C.; Fang, Q.Z.; Liu, C.Y. Reliability analysis of long-span bridge based on the combination of response surface method and JC method. *Bridge Constr.* **2022**, *52*, 32–38.
17. Tarhouni, I.; Frómeta, D.; Casellas, D.; Costa, J.; Maimí, P. Assessing the effect of the experimental parameters in the evaluation of the essential work of fracture in high-strength thin sheets. *Eng. Fract. Mech.* **2022**, *270*, 106949. [CrossRef]
18. Ye, C.L.; He, S.X.; Hou, Y.L. The Design of Simple Axial Tensile Testing Instrument and Experimental Study of Reasonable Sample Size. *Railw. Stand. Des.* **2016**, *60*, 9–15.
19. Ye, D.; Xu, Z.; Liu, Y.Q. Solution to the problem of bridge structure damage identification by a response surface method and an imperialist competitive algorithm. *Sci. Rep.* **2022**, *12*, 16495. [CrossRef]
20. Liu, J.; Wang, H.L.; Zhang, Z.G.; Lv, P. Research on damage recognition response surface model and optimization algorithm of cable-stayed bridge. *Railw. Constr.* **2016**, *15*, 2347.

21. Wang, Y.B.; Wang, L. Finite Element Model Modification of an Arch Bridge Based on Response Surface Method. In *IOP Conference Series: Materials Science and Engineering*; IOP Publishing: Bristol, UK, 2020; Volume 782.
22. Miao, C.Q.; Zhuang, M.L.; Dong, B. Environment—Environmental Factors; Data on Environmental Factors Reported by Researchers at Southeast University (Stress Corrosion of Bridge Cable Wire by the Response Surface Method). *Ecol. Environ. Conserv.* **2019**, *51*, 646–652.
23. Hong, X.J.; Zhang, X.S.; Guo, N.; Li, H.M. Application of optimized fractional order GM(1,1) model in bridge linear control. *J. Chongqing Jiaotong Univ. (Nat. Sci. Ed.)* **2022**, *41*, 65–70.
24. Jiang, Y.; Hui, Y.; Wang, Y.; Peng, L.; Huang, G.; Liu, S. A novel eigenvalue-based iterative simulation method for multi-dimensional homogeneous non-Gaussian stochastic vector fields. *Struct. Saf.* **2023**, *100*, 102290. [CrossRef]
25. Li, X.G.; Ding, P.; Zhou, J.T.; Chen, X.H.; Qi, Y. Linear prediction analysis of cable-stayed bridge of long-span rail transit considering parameter randomness. *Railw. Archit.* **2019**, *59*, 64–69.
26. Chen, Z.S.; KT, T.; Liu, S.M.; Zhou, J.T.; Zeng, Y. Application of a self-adaptive Kalman filter approach in alignment control for an extra long span rail transit cable-stayed bridge. *Struct. Infrastruct. Eng.* **2017**, *13*, 1186–1197. [CrossRef]
27. Chen, Z.S.; Zhou, J.T.; Kim, T.T.; Hu, G.; Li, Y.; Wang, X. Alignment control for a long span urban rail-transit cable-stayed bridge considering dynamic train loads. *Sci. China Technol. Sci.* **2016**, *59*, 1759–1770. [CrossRef]
28. Veselov, I.V.; Gamlitskiy, Y.A. Engineering Method of Modeling and Optimization of Chemical-Technological Processes. *Russ. Sci. J.* **2017**, *5*, 56–65.
29. Liu, S.M.; Liu, F.P.; Liu, L. Bridge Structural Safety Assessment Technology Based on Statistic Analysis. *Appl. Mech. Mater.* **2012**, *1802*, 2486–2491. [CrossRef]

buildings

MDPI

Article

Statistical Analyses of the Non-Uniform Longitudinal Temperature Distribution in Steel Box Girder Bridge

Wenda Ma [1,2], Bo Wu [1,2,*], Dingsong Qin [3], Bin Zhao [3] and Xianyi Yang [1,2]

1 State Key Laboratory of Mountain Bridge and Tunnel Engineering, Chongqing Jiaotong University, Chongqing 400074, China; 611200080001@mails.cqjtu.edu.cn (W.M.); xianyi@yahoo.com (X.Y.)
2 School of Civil Engineering, Chongqing Jiaotong University, Chongqing 400074, China
3 China Railway First Group Bridge Engineering Co., Ltd., Chongqing 404100, China; 13983219281@163.com (D.Q.); 13217028191@163.com (B.Z.)
* Correspondence: bo.wu@cqjtu.edu.cn

Abstract: The frequently conventional assumption that bridge temperature is uniformly distributed on long-span bridges could lead to uncertainty when analyzing temperature effects. This study investigated the surface temperature of steel box girders on a long-span suspension bridge, emphasizing the distribution characteristics in the longitudinal (spanwise) direction. The girder surface temperature distribution was monitored using the long-term structural health monitoring system (SHMS). First, the probability density functions (PDF) of the girder surface temperature were analyzed. The results showed that the PDFs had bimodal characteristics and could be well-fitted using the weighted superposition of two normal distributions. Meanwhile, there was an obvious difference between the PDFs of the measuring points at different longitudinal sections of the bridge, which is inconsistent with the assumption that the temperature was uniformly distributed in the longitudinal direction. Subsequently, the longitudinal distributions of the girder surface temperature were statistically analyzed, and polynomial functions were introduced to fit the distribution curves along the left and right sides of the mid-span. A correlation analysis was then performed, highlighting the variability in temperature in the longitudinal direction. Additionally, the longitudinal temperature distribution pattern could be summarized as (i) the highest in the mid-span, the lowest in the tower, and increasing along the side span; (ii) there were also significant differences between the left and right sides of the mid-span. Finally, the time- and space- distributions of the temperature were studied, and a contour map was displayed. The results showed that the girder surface temperature had significant three-dimensional spatial characteristics and was not only non-uniformly distributed in space but also in time. This work is useful for a more accurate analysis of temperature effects on long-span bridges.

Keywords: bridge temperature; non-uniform temperature field; longitudinal temperature distribution; steel box girder; long-span bridge

Citation: Ma, W.; Wu, B.; Qin, D.; Zhao, B.; Yang, X. Statistical Analyses of the Non-Uniform Longitudinal Temperature Distribution in Steel Box Girder Bridge. *Buildings* **2023**, *13*, 1316. https://doi.org/10.3390/buildings13051316

Academic Editor: Andrea Benedetti

Received: 12 April 2023
Revised: 8 May 2023
Accepted: 16 May 2023
Published: 18 May 2023

1. Introduction

Long-span bridge structures can be affected by various environmental factors during their long-term operations, among which the temperature effect is particularly significant [1–3]. In recent years, numerous studies on bridge temperature effects have been carried out [4–7], mostly considering uniform temperature fields. Some scholars [8,9] have studied the influence of boundary factors such as solar radiation on the bridge temperature field. Some scholars [10–13] studied the time-varying temperature distribution of bridges and established a temperature gradient distribution model. There have also been numerous scholars [14–20] who have studied the long-term measured cross-section temperatures of a large number of bridges with different cross-section forms, observed transverse temperature gradients and proposed two-dimensional temperature field calculation models.

It is worth noting that the above studies are based on the assumption that the temperature is uniformly distributed along the longitudinal span direction of the bridge. However, in actual engineering, the temperature field of large-scale structures, such as long-span bridges, is usually non-uniform. It has been shown that the stresses and deformations caused by non-uniform temperature distribution can be comparable to those caused by static and live loads, which may also cause cracks and excessive deformations [21–25].

Actually, the existing standards [26–30] have neglected the influence of the longitudinal non-uniform temperature distribution on bridges, assuming that the temperature field can be represented by one-point measuring data, i.e., a uniform distribution field. Although this assumption simplifies the analysis of bridges affected by temperature in the two-dimensional plane, it may overestimate the temperature effects at some spanwise locations and underestimate them at others. Moreover, the nonlinear effects caused by temperature effects are neglected by the uniform-distribution assumption [31,32]. Therefore, the temperature effects cannot be accurately evaluated when adopting the above assumption. Additionally, the inability to accurately assess the temperature effects of large-span bridges may bring uncertainty to the bridge during the design, construction and operation stages, thus affecting the structural safety of the bridge [33,34].

Therefore, the longitudinal temperature distribution is important for the accurate analysis of bridge temperature effects. Several studies have started to pay attention to this topic. Abid et al. [35] investigated the temperature distribution of bridge segmental models under the influence of solar radiation and air temperature variations through numerical simulations. Based on the health monitoring data of the Aizhai Bridge, Hu et al. [36] found that there were temperature differences at different measuring points along the longitudinal direction of the bridge span. Gu et al. [37] proposed a vertical and lateral temperature gradient model for different longitudinal positions along the bridge span. Liu et al. [38] studied the non-uniform longitudinal distribution temperature of concrete-filled steel tubular bridges due to the change in component inclination. Although the above studies considered the variability of temperature distribution, it is limited to a single or several measuring points, and the conclusions may not be well applied to actual bridges. In this context, the longitudinal temperature distribution on the whole bridge and its statistical characteristics was insufficient. Moreover, the role of the non-uniform longitudinal distributions of temperature was not analyzed in depth.

Therefore, this work studies the longitudinal temperature distribution of steel box girders based on the field-measured temperature data. The main objective was to provide a more accurate description of the longitudinal temperature field of long-span bridges.

The paper is organized as follows. In Section 2, the bridge and its health monitoring system are introduced. In Section 3, the probability density characteristic analysis is carried out by selecting typical measuring points in the longitudinal direction of the bridge, and the probability density curves are fitted; next, the temperature of all measurement points is statistically analyzed, and longitudinal distribution curves are fitted; A correlation analysis was then performed, highlighting variability in the longitudinal direction. Finally, the frequency domain features were analyzed, and the space-time contour maps were given to study the mechanisms and processes of the non-uniform distribution of temperature.

2. Description of the Bridge and Its Monitoring System

The Egongyan rail bridge is a long-span suspension bridge located in southwestern China and across the Yangtze River. Figure 1 is a picture of this bridge, and Figure 2 shows its elevated view. It is a five-span self-anchored suspension bridge with double towers and double plane cables. The middle three spans are the main bridge, and the extended span with a span of 50 m is the approach bridge. The overall length of the bridge is 1650.5 m, and the main span is 600 m. In addition, the main girder is a closed single-box five-chamber steel box girder structure [39–41].

Figure 1. Egongyan rail bridge.

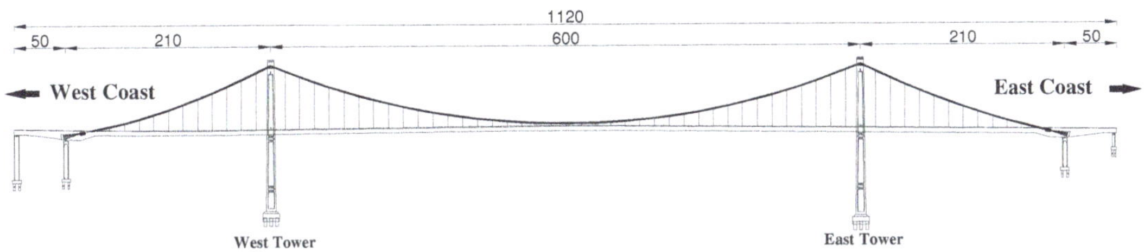

Figure 2. Elevation layout of the main bridge (Unit: m).

To monitor the operation status, a structural health monitoring system (SHMS) was installed on the Egongyan rail bridge [42] and has been in operation since 2020. A total of 267 sensors were installed at different parts of the bridge to collect various types of structural and environmental information. Among them, the temperature measuring points on the steel box girder of the main bridge were arranged symmetrically about the mid-span, as shown in Figure 3. A total of 88 temperature sensors were installed at 11 sections (S1–S11) along the main bridge, and there were also eight measuring points (T1~T8) inside the surface of each section. Taking the mid-span section (S6); for instance, a plan view of the measuring point at each section is shown in Figure 3.

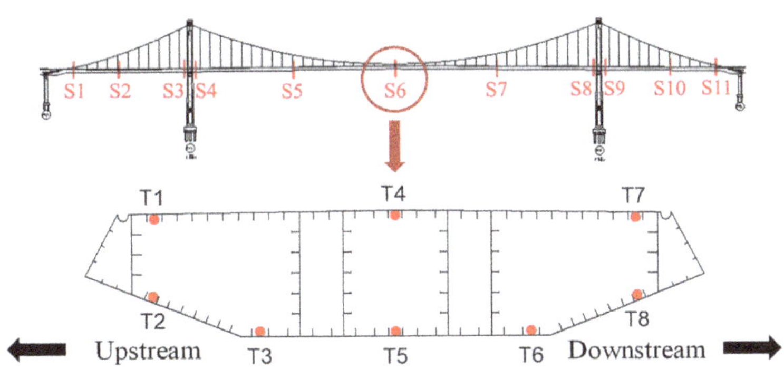

Figure 3. The layout of the temperature monitoring points. (S: Section number; T: Number of temperature sensors arranged on each section).

The surface temperature of the box girder was monitored using a JMZX-212HAT surface intelligent digital tandem strain gauge with built-in temperature sensors as the testing instrument [43], as shown in Figure 4. The main technical parameters of the sensor were as follows: Operating ambient temperature: −40 °C~+80 °C; Temperature measurement range: −20 °C~+80 °C; Temperature resolution: 0.1 °C; Temperature measurement accuracy: ±0.5 °C.

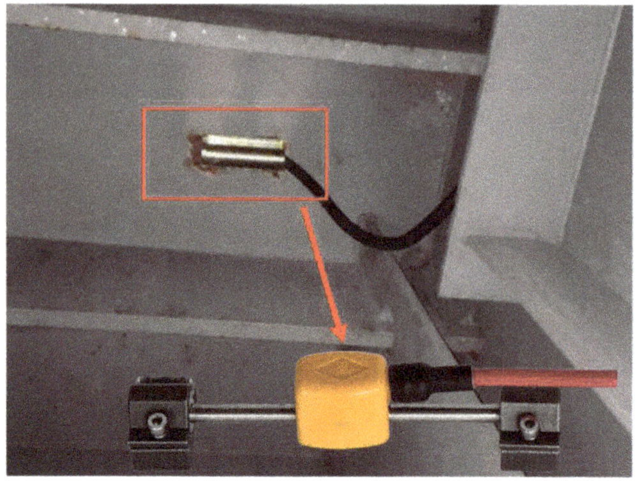

Figure 4. JMZX-212HAT surface intelligent digital string strain gauge.

3. Probability Density Characteristics of Temperature

To display the annual variations in the girder surface temperature, the measuring point T4 at different longitudinal positions S2, S6 and S10 was chosen, for instance. Figure 5 shows the monitoring data from 1 September 2020 to 31 August 2021. It can be observed that the annual temperature changes all appeared to have obvious periodic patterns, and this could be approximately described by sinusoidal functions, with lower temperatures from December to February and higher temperatures from June to August. In addition to the low-frequency periods, the temperature data also contained high-frequency fluctuations. High-frequency fluctuations may be due to the effect of daily temperature changes. In addition, it was found that the overall temperature value of S6 and its variance amplitude were significantly larger than those of S2 and S10, and the difference was more obvious in summer. This was a preliminary indication of the non-uniformity of the temperature distribution along the longitudinal direction of the bridge, which is discussed later.

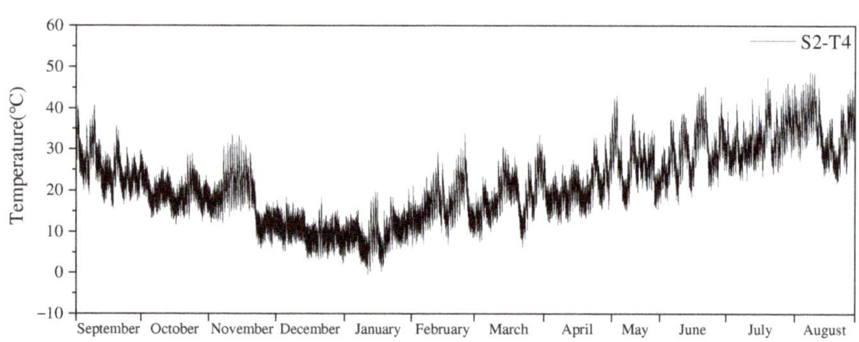

Figure 5. *Cont.*

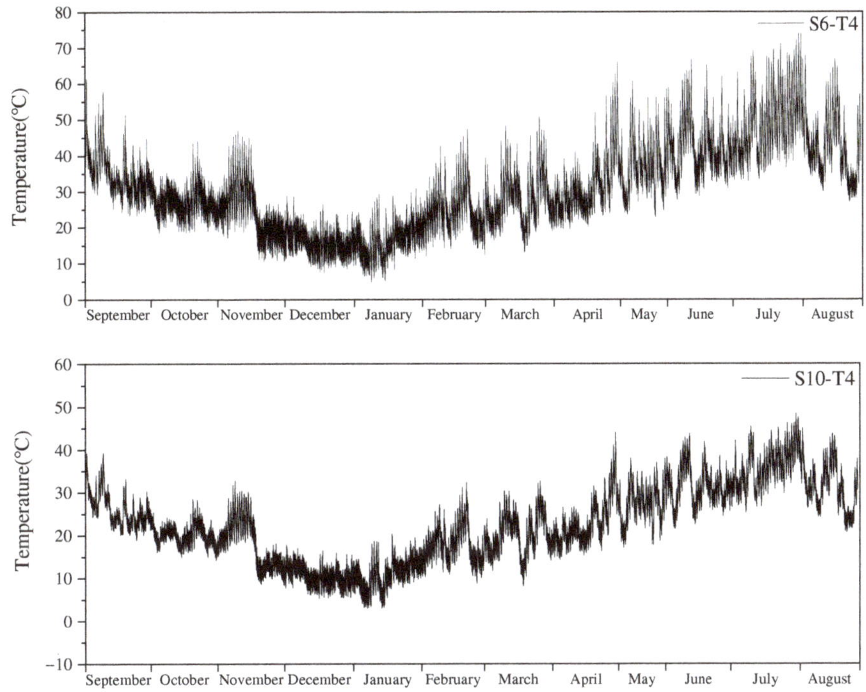

Figure 5. Annual time-history of the girder surface temperature. (S-T: Sensor number T The section number S).

In previous studies, the Gaussian distribution model [44–46] has been commonly used for fitting the probability density function (PDF) of the temperature. However, it is obvious in Figure 6a that the PDF cannot be properly fitted by the Gaussian distribution model due to bimodal characteristics. By comparing the fitting results of several models, the weighted superposition of two normal distributions was selected for fitting the current data, with its mathematical expression as follows:

$$f(x) = \gamma f_1(x) + (1-\gamma) f_2(x) \tag{1}$$

$$f_1(x) = \frac{1}{\sqrt{2\pi}\sigma_1} e^{-\frac{(x-\mu_1)^2}{2\sigma_1^2}} \tag{2}$$

$$f_2(x) = \frac{1}{\sqrt{2\pi}\sigma_2} e^{-\frac{(x-\mu_2)^2}{2\sigma_2^2}} \tag{3}$$

where x denotes the surface temperature of the steel box girder, $f(x)$ denotes the probability density model of x, $f_i(x)$ $(i = 1,2)$ denotes the normal distribution function with the mean value μ_i and the standard deviation σ_i, γ is the weight of the two normal distribution functions with $0 \leq \gamma \leq 1$.

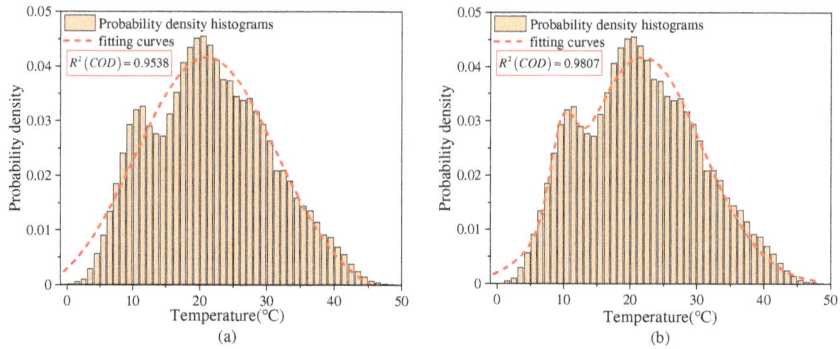

Figure 6. Fitting of the PDFs of temperature at S2-T4 using different models: (**a**) Gaussian distribution model; (**b**) Superposition of two normal distribution models.

The goodness-of-fit for the different models is shown in Figure 6. The R^2 (coefficient of determination, COD) of the superposition of two normal distribution models was significantly higher than that of the Gaussian distribution model. Therefore, it was reasonable to choose the superposition of two normal distribution models to describe the probability distribution of temperature.

The probability density histograms of measuring points S2-T4, S6-T4 and S10-T4 and their fitted curves are shown in Figure 7. It can be seen that the fitted curves were in good agreement with the measured ones. Meanwhile, the fitting results all passed the K-S test with a significance level $\alpha = 0.05$, indicating that the fitting curves could accurately describe the probability density characteristics of the surface temperature of the steel box girder. The results show that the probability density curves of each measuring point have bimodal characteristics. This was due to the bridge being located in the subtropical humid monsoon climate zone and its experience of a transition period between seasons. The temperatures at the two peaks represent the temperature during the spring-summer and summer-fall transitions and during the fall-winter and winter-spring transitions, respectively.

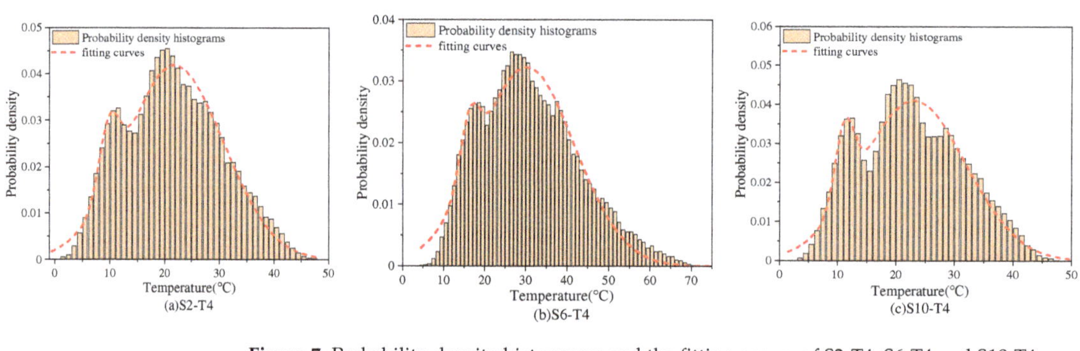

Figure 7. Probability density histograms and the fitting curves of S2-T4, S6-T4 and S10-T4.

To investigate the difference in temperature distribution along the longitudinal direction of the bridge, the probability densities of T4 and T5 (on the top and bottom plates, respectively) and at different longitudinal sections S2, S6 and S10 were fitted based on the abovementioned model. The estimated floating parameters are shown in Table 1, and the corresponding PDFs are given in Figure 8.

Table 1. Estimated fitting parameters for the probability density models of typical measuring points.

Measuring Points	Fitting Parameters				
	γ	σ_1	σ_2	μ_1	μ_2
S2-T4	0.9360	8.9207	1.8087	21.8373	9.8373
S6-T4	0.9437	11.6900	2.3217	30.1064	16.2538
S10-T4	0.9209	8.9665	1.6595	23.0285	11.1745
S2-T5	0.9486	9.1765	1.6817	21.7870	10.6727
S6-T5	0.9301	9.9613	2.2033	29.4177	16.9026
S10-T5	0.9278	9.0896	1.6930	22.6721	11.1846

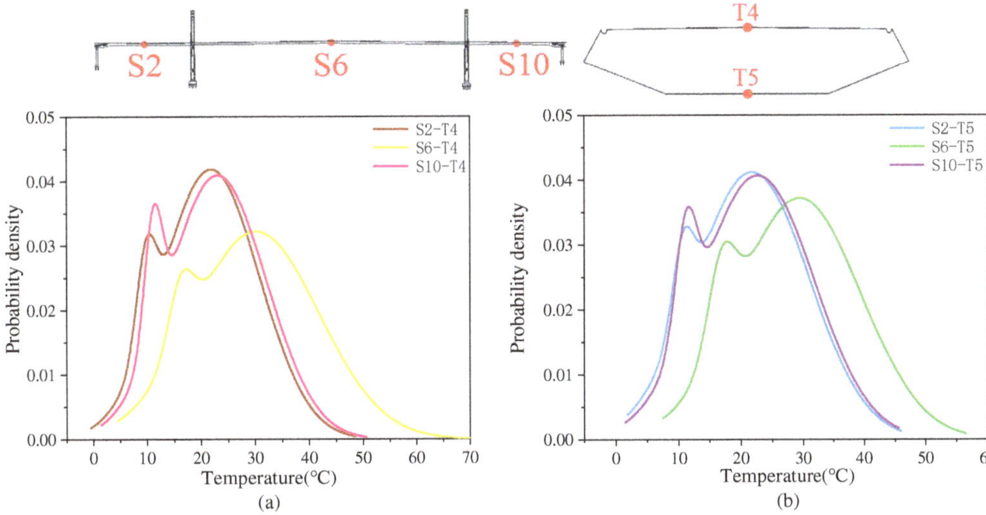

Figure 8. Fitted PDFs of the girder surface temperatures at different longitudinal sections. (**a**) Top plate measuring points; (**b**) Bottom plate measuring points.

From Table 1 and Figure 8, it can be seen that there was a significant difference between the PDFs of the mid-span and side-span for both the top and the bottom plates. Concerning the two peaks in the PDFs, the measuring point S6-T4 was mainly concentrated around 30 °C and 16 °C, while S10-T4 was mainly concentrated around 23 °C and 11 °C, indicating that there was a significant difference in the temperature distribution along the longitudinal direction of the bridge, and the temperature in the mid-span was significantly higher than that in the side span. Moreover, PDFs were not the same for the side-span measuring points, which were symmetrical along the bridge centerline, indicating a non-uniform longitudinal temperature distribution in the steel box girder. This phenomenon shows that the assumption of uniform temperature used in previous studies and specifications [14,20,30] is not reasonable.

4. Statistical Analyses of Temperature Distribution along the Bridge

4.1. Temperature Longitudinal Distributions

To further analyze the longitudinal gradient of the bridge girder surface temperature, the annual temperature and annual temperature difference were statistically analyzed by selecting one year of the temperature time history data from 11 measurement points (S1–S11) in the longitudinal direction. The statistical analyses included the following four aspects: annual maximum temperature, annual minimum temperature, annual average temperature and annual maximum temperature difference for each measurement point. The annual maximum temperature difference was the difference between the annual maximum temperature and the annual minimum temperature.

By taking the top and bottom measuring points, including T4 and T5; for instance, the annual temperature and statistical analyses along the longitudinal direction of the main bridge are shown in Figure 9. An analysis of Figure 9 shows that the distribution of the temperature field along the longitudinal direction was non-uniform. The average temperature was approximately symmetrical about the mid-span. On the other hand, the annual average and maximum temperature curves showed the same patterns: the temperature was highest in the middle of the span and lower near the bridge tower while also rising again from there to the side span. In addition, the variation trends of the annual maximum temperature difference were similar to that of the annual maximum temperature. Although there was some variation in the minimum temperature between the different measurement points, this variation was not very significant compared to the maximum temperature. In addition, compared to the bottom plate, the temperature variation range from the top plate was larger. Measured data show that many cities in China have experienced extremely high temperatures over recent years. For example, the maximum temperature in Chongqing reached 43 °C in 2022, and the temperature of the steel box girder, in this case, may show a peak. We will continue to follow up on the study.

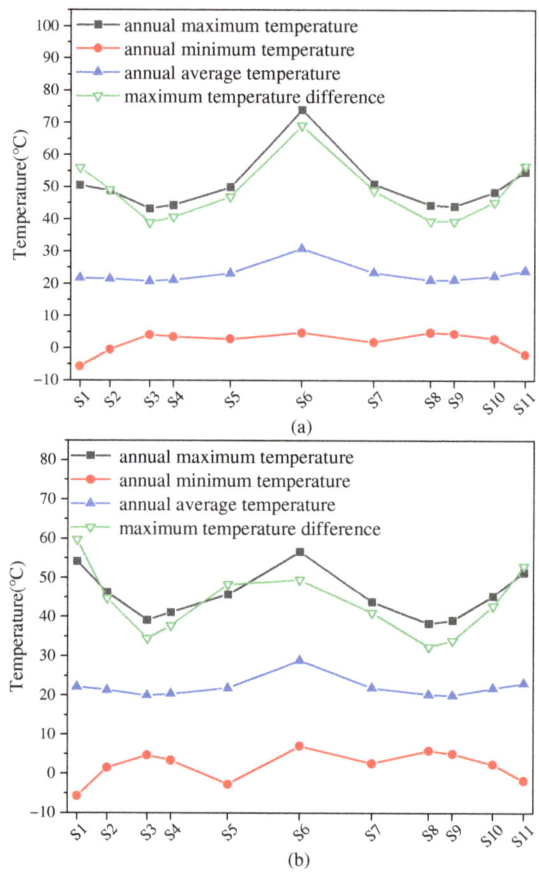

Figure 9. Distribution of annual temperature statistic values along the longitudinal direction. (**a**) Top plate measuring points T4; (**b**) Bottom plate measuring points T5.

To further investigate the longitudinal temperature distribution pattern, the annual measured data for all eight temperature measuring points (T1~T8) in each section were statistically analyzed, and the annual average temperature of each measuring point is shown in Figure 10. It can be seen that the surface temperature of the steel box girder had different distributions between the top and bottom plates, as well as the left and right sides of the mid-span. To clarify the distribution pattern, the polynomial fitting of the longitudinal temperature distribution curve was performed below.

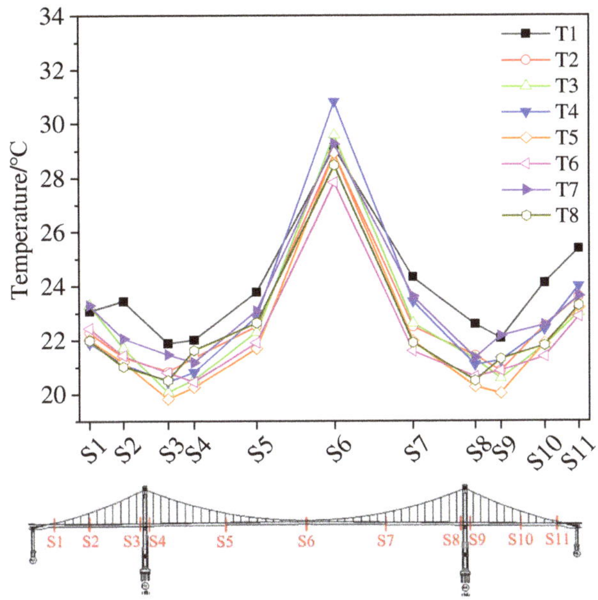

Figure 10. Longitudinal distributions of annual average temperature.

First, the measured temperature of the top and bottom plates was, respectively, time-averaged. Then, the average temperature values of different measuring points on the left and right sides of the mid-span (S6) were separately fitted using polynomial curves $y = Ax^2 + Bx + C$, where y denoted the temperature value, x denoted the Longitudinal distance from the mid-span, and the three parameters of the equation were A, B, and C. In order to conveniently compare the fitted formulas, the horizontal axis on the left side was set as x_1 and the horizontal axis on the right side was set as x_2. The fitting results are shown in Figure 11. Meanwhile, the equations of the fitted curves are listed in Table 2. The temperature values at any longitudinal location on the steel box girder could be subsequently estimated from these fitted curve equations. By referring to the practice of the specifications and other studies [8,26,30,33] on the horizontal and vertical temperature gradient patterns, in our future research work, we should incorporate data from other similar bridges to obtain the longitudinal temperature distribution patterns of large-span suspension bridges.

Table 2. Longitudinal distribution of temperature fitting equation.

Position	Fitted Curves Equation
Left side of the mid-span (top plate)	$y = 8.21 \times 10^{-5}{x_1}^2 - 5.25 \times 10^{-2}x_1 + 29.60$
Right side of the mid-span (bottom plate)	$y = 8.91 \times 10^{-5}{x_2}^2 - 5.31 \times 10^{-2}x_2 + 29.74$
Left side of the mid-span (top plate)	$y = 9.24 \times 10^{-5}x^2 - 5.58 \times 10^{-2}x + 28.65$
Right side of the mid-span (bottom plate)	$y = 9.24 \times 10^{-5}x^2 - 5.49 \times 10^{-2}x + 28.65$

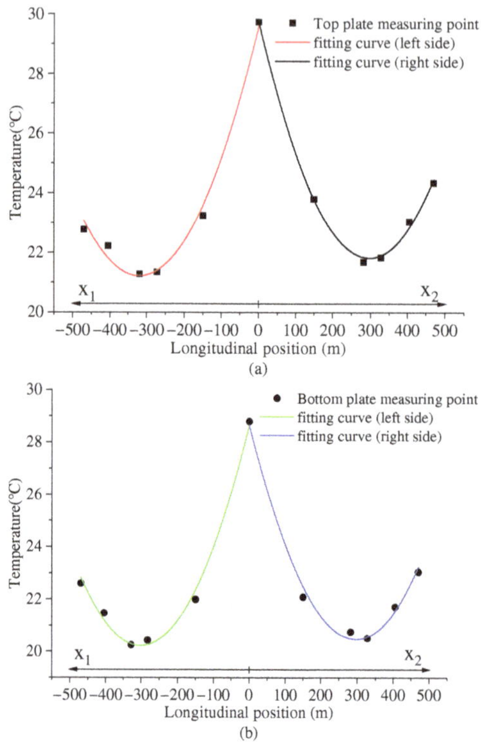

Figure 11. Fitting curves of the temperature's longitudinal distribution (**a**) Top plate; (**b**) Bottom plate.

From the fitted curves and the fitting parameters, it could be seen that: (1) For the top plate temperature, the two fitting curves along the left and right spans had significantly different parameters, with higher temperatures on the right side than on the left side, possibly due to variations in geographic position and material properties. (2) For the bottom plate temperature, the two fitting curves along the left and right spans had very similar parameters, indicating that the temperature of the bottom plate was longitudinally distributed and symmetric along the mid-span.

4.2. Temperature Longitudinal Correlation Analysis

The analysis of statistical values, such as the annual mean and annual maximum temperatures, showed that there was a significant temperature gradient along the longitudinal direction of the bridge. To investigate the connection between surface temperatures at different longitudinal locations, correlation analysis was necessary. In order to evaluate the non-uniformity of the longitudinal temperature distribution, for each section measurement point (T1~T8), correlation analysis was performed with the same measurement points at other longitudinal positions, respectively. The position of S6 (mid-span) was taken as the reference point, where the correlation coefficient [47] could be defined as:

$$R_x(y) = \frac{Cov[x(t, y_1), x(t, y_2)]}{\sqrt{Cov[x(t, y_1)]} \cdot \sqrt{Cov[x(t, y_2)]}} \quad (4)$$

where, x denoted the surface temperature of the steel box girder, $Cov[]$ was the operator of covariance, t corresponded to time, and y_1 and y_2 represented different longitudinal measuring points.

Figure 12 shows the correlation analysis results and reveals that: (1) On the main span, the correlation coefficient decreased as the distance from S6 increased. This indicated that S6 had a greater effect on the temperature of S5 and S7 compared to S4 and S8. (2) The lowest correlation was found at the bridge tower. This is probably a result of the variability in the spatial distribution of cross-sections due to differences in the duration and intensity of the solar radiation received. The solar radiation at the bridge tower would be absorbed more by the tower, so the correlation coefficient of the S4 and S8 cross-sections was the lowest. (3) The sudden increase in the correlation coefficients of cross-sections S3 and S9 at the side span position indicated that solar radiation had a more dominant effect than the shading effect on the bridge tower. (4) The temperature correlation coefficient was generally higher on the right side of the mid-span than on the left side, meaning that the temperature was more uniform on the right side than on the left side. In addition, the upstream and downstream sides showed different temperature correlation characteristics. These differences may be related to the differences in bridge orientation, section form and material properties.

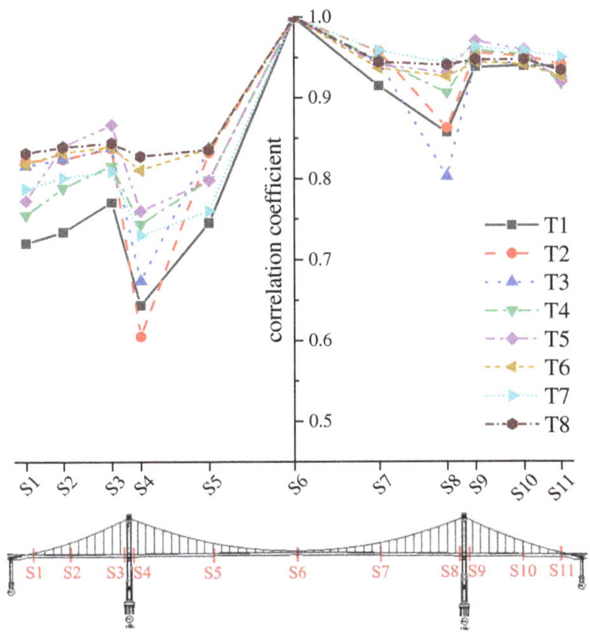

Figure 12. Correlation analysis of the measurement points in the mid-span section with those in other locations.

The longitudinal correlation analysis showed again that the longitudinal temperature distribution was non-uniform. These results were more comprehensive than previous studies with limited point comparisons. Therefore, using single-point measurement data to represent the temperature field may lead to overestimating or underestimating the temperature effect at different bridge spanwise locations. As a result, the temperature effects of large-span bridges could not be assessed accurately.

5. Mapping Longitudinal Temperature Variation Contours

5.1. Power Spectral Density

Power Spectral Densities (PSDs) refer to the spectral energy distribution per unit of time. PSDs focus on various features of the signal in the frequency domain, intending to extract useful signals in frequency domains that are drowned in noise. Based on Fourier transform, the PSDs of the annual temperatures of the girder surface temperature could be calculated. Taking a typical section S6 for an example, the results of the top and bottom plates are shown in Figure 13.

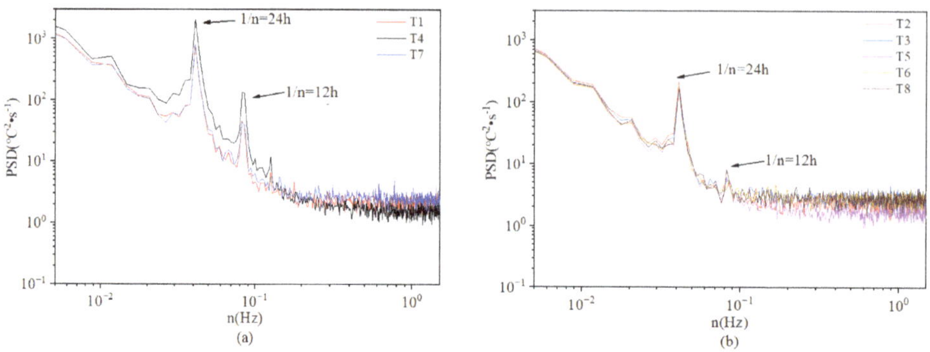

Figure 13. The PSD of the measured annual temperature. (**a**) Top plate measurement points; (**b**) Bottom plate measurement points.

There are two prominent peaks that can be observed in the PSD of Figure 13. The period of the first peak was 24 h, which was identical to the daily period. The period of the second peak was 12 h, which was induced by the transmission from day to night. The results of PSD analysis show that the daily temperature could be considered a representative sample of the annual temperature. Moreover, the daily temperature could provide a more detailed view of the temperature change process compared to the annual temperature. Therefore, the temperature in one day can be selected to investigate the mechanisms and processes of longitudinal temperature gradients.

In addition, the PSD generally decreased with the increase in frequency and tended to become a horizontal line in the high-frequency region. The general decreasing trend resulted from a reduction in the high-frequency components of temperature. When this frequency was high enough, the temperature did not change in the corresponding short period. Thus, the PSD was almost horizontal in the high-frequency region. These results indicate that the surface temperature of the steel box girder was mainly controlled by low-frequency components, and the high-frequency effect was less affected.

5.2. Contours of Daily Temperature

To further analyze the generation mechanisms and processes of the longitudinal temperature gradient, taking into account the apparent daily periodicity of the temperature data, the data for one day were selected for analysis. The following is an analysis of the longitudinal temperature distribution pattern based on daily temperature curves. A total of 11 longitudinal sections were selected, and three measurement points on each of the top and bottom surfaces were taken for analysis. To better observe their regularity, the hourly temperature was averaged to obtain the time history of the measured daily temperature curves, as shown in Figure 14.

Figure 14 points out that, as summarized in the previous section, the temperature distribution along the longitudinal direction was always non-uniform. The temperature statistic values varied non-uniformly in spatial locations, and the temperature transients varied non-uniformly in space as well. Thus, the non-uniform longitudinal temperature gradient contained both spatial non-uniformity and temporal non-uniformity. Although previous studies have investigated the effect of temperature at different longitudinal locations on the temperature gradient of the cross-section through finite points, they have not investigated the mechanism of this non-uniform temperature field in depth.

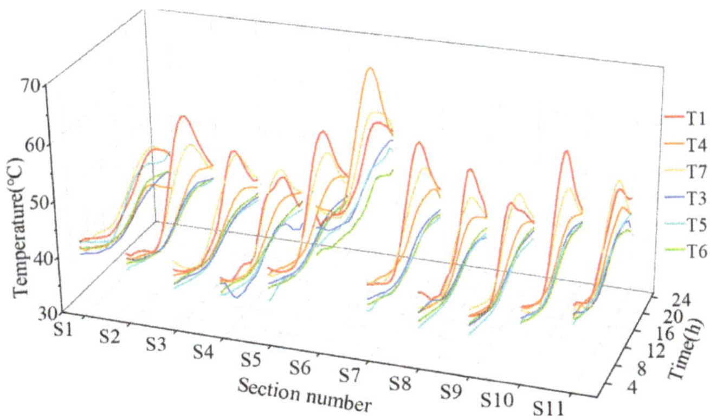

Figure 14. Daily temperature-time histories for different measuring points at 11 sections.

Meanwhile, Sections S3, S4 and S8, S9 are, respectively, located on both sides of the east and west main towers. The amplitude of daily temperature variation in these cross-sections was smaller than that of other adjacent measurement points. This phenomenon is similar to the findings of Figures 10 and 12. Additionally, this was mainly due to the shading effect of the main tower, which reduced the temperature of this section. In addition, the temperature of the mid-span S6 section was the highest because it was directly affected by solar radiation. By comparing the temperature-time histories of the top and bottom plates between different sections, it is found that there was a difference in the longitudinal temperature distribution between the top and bottom surfaces.

If the surface temperature of any position on the box girder could be obtained, the accuracy of the temperature effect analysis could be improved. Therefore, Figure 15 shows the daily temperature contour maps of the top and bottom plate temperatures, respectively. The example time instant was chosen at 17:00 because the girder surface temperature at that time was the highest value in one day, and the longitudinal temperature distribution of the steel box girder had obvious non-uniformity.

It can be seen from Figure 15 that: (1) The longitudinal temperature contour on the top plate indicates that the high-temperature area slightly deviated to the right of the mid-span. This is because the bridge is east–west oriented, and the bridge deck on the right side of the mid-span was exposed to solar radiation for a longer time, so the temperature would be slightly higher than the bridge deck on the left side of the mid-span. (2) The longitudinal temperature distribution of the bottom plate was almost symmetrical about the mid-span because the bottom plate was not directly affected by solar radiation but was mainly affected by the environmental temperature. (3) The transverse temperature difference of the top plate was more obvious than that of the bottom plate because of solar radiation, and the temperature distribution had obvious three-dimensional spatial distribution characteristics.

After analyzing the temperature distribution characteristics on other days, we could also obtain a similar conclusion. The longitudinal distribution characteristics of daily and annual temperature were similar, while daily temperature also had a three-dimensional spatial distribution.

To summarize, the temperature of the long-span bridge structure was non-uniformly distributed not only in space but also in time. The non-uniformity of the longitudinal temperature distribution could be caused by the non-uniformity of the material and is also related to the environmental location where the measurement points are located. When performing temperature effect analysis, the temperature field of large-scale bridge structures is usually assumed to be a one-dimensional temperature field, which erases the longitudinal and transverse temperature gradients, resulting in large deviations between the calculated results and the real temperature field of the bridge. Further research into the effects of these factors on the longitudinal temperature distributions could be useful in the design stage of long-span bridges to effectively reduce the risk caused by the temperature effects. Therefore, during the refinement of the temperature effect analysis for long-span bridges, the longitudinal distribution of temperature non-uniformity needed to be considered.

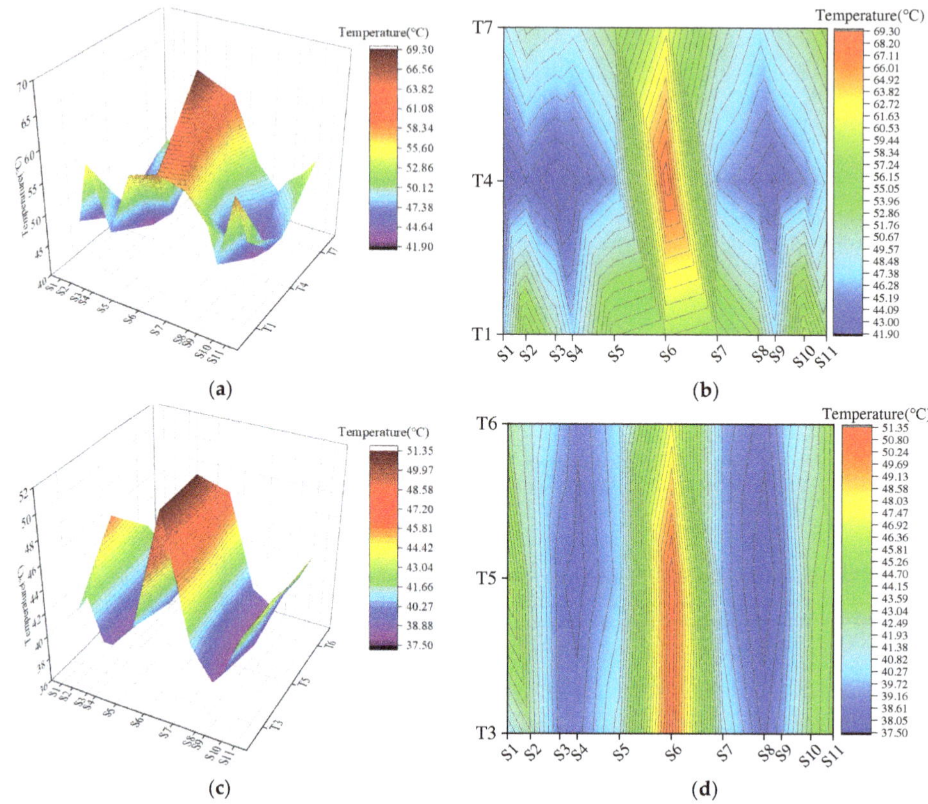

Figure 15. Contour maps of measured temperature at 17:00. (**a**) Top plate temperature contours; (**b**) Projection of (**a**); (**c**) Bottom plate temperature contours; (**d**) Projection of (**c**).

6. Conclusions

This study investigated the longitudinal distribution characteristics of the box girder surface temperature on a long-span suspension bridge. The probability density statistics, statistical values, cross-correlations coefficients, power spectrum densities, and time-space contours are detailed and analyzed. The main findings are as follows:

(1) The annual temperature's probability density curves at different longitudinal measuring points of the bridge all have bimodal characteristics, which could be fitted by the weighted sum of two normal distributions. This bimodal distribution was mainly caused by the transitions between different seasons.

(2) The statistical analyses show that the distribution of the girder surface temperature along the longitudinal direction was non-uniform. Moreover, the equations of the longitudinal distribution curves were obtained by polynomial fitting, and the distribution pattern was the highest in the mid-span, the lowest in the bridge tower, and increased along the side span. Therefore, the design phase should consider the non-uniform distribution of temperature.

(3) Using the correlation analysis between the temperature measured at the mid-span and other longitudinal sections, it was found that the correlation coefficient in the main span

gradually decreased farther away from the mid-span section, with the lowest at the tower section. While in the side span, the correlation coefficient increased as the measurement points moved away from the tower. This revealed that the girder surface temperature was influenced by solar radiation and also the shedding effects of the tower.

(4) There were two prominent peaks in the frequency domain of the annual temperature, corresponding to the 24 h and 12 h time periods, respectively. This meant that the daily temperature could be considered a representative sample of the annual temperature.

(5) By comparing the daily time histories of temperature at different longitudinal sections, it was found that the temperature field was significantly three-dimensional and was not only non-uniformly distributed in space but also non-uniformly distributed in time.

(6) According to the time-space contour maps of temperature, we can gain insight into the underlying mechanisms of the generation of this non-uniformity distribution. Combined with the statistical analyses, the statistical values of the bridge temperature at any section could be obtained using the contour map.

In conclusion, the surface temperature of the steel box girder was non-uniformly distributed along the longitudinal direction of the bridge. Therefore, in analyzing the temperature effects, the traditional uniform-distribution assumption could lead to inestimable deviations from the real conditions. This work is helpful for a more accurate analysis of temperature effects on long-span bridges and can also provide a reference for the longitudinal distribution of temperature fields on other similar bridges.

Author Contributions: Conceptualization, W.M. and B.W.; Formal analysis, W.M. and B.W.; Methodology, W.M. and B.W.; Data curation, D.Q. and B.Z.; Validation, D.Q. and B.Z.; Investigation, D.Q. and X.Y.; Writing—original draft, W.M.; Writing—review and editing, B.W.; Project administration, B.W. and X.Y.; Funding acquisition, B.W. and X.Y. All authors have read and agreed to the published version of the manuscript.

Funding: This work was supported by the National Natural Science Foundation of China (Grant No. 51978111), Chongqing Technology Innovation and Application Development Special Key Project (Grant No. CSTB 2022TIAD-KPX0205), Chongqing Transportation Science and Technology Project (Grant No. 2022-01), Natural Science Foundation of Chongqing, China (Grant No. cstc2021jcyj-bshX0061), China Postdoctoral Science Foundation (Grant No. 2022MD713699), Special Funding of Chongqing Postdoctoral Research Project (Grant No. 2021XM1016) and Chongqing Zhongxian Science and Technology Plan Project (Grant No. zxkyxm202202) are greatly acknowledged.

Data Availability Statement: Data are available on request due to restrictions.

Conflicts of Interest: The authors declare no conflict of interest.

References

1. Wang, J.F.; Zhang, J.T.; Xu, R.Q.; Yang, Z.X. Evaluation of Thermal Effects on Cable Forces of a Long-Span Prestressed Concrete Cable–Stayed Bridge. *J. Perform. Constr. Facil.* **2019**, *33*, 04019072. [CrossRef]
2. Honarvar, E.; Sritharan, S.; Rouse, J.M.; Meeker, W.Q. Probabilistic Approach to Integrating Thermal Effects in Camber and Stress Analyses of Concrete Beams. *J. Bridge Eng.* **2020**, *25*, 04020010. [CrossRef]
3. Zhou, Y.; Xia, Y.; Chen, B.; Fujino, Y. Analytical Solution to Temperature-Induced Deformation of Suspension Bridges. *Mech. Syst. Signal Process.* **2020**, *139*, 106568. [CrossRef]
4. McClure, R.M.; West, H.H.; Hoffman, P.C. Observations from Tests on a Segmental Bridge. *Transp. Res. Rec.* **1984**, *2*, 60–67.
5. Riding, K.A.; Poole, J.L.; Schindler, A.K.; Juenger, M.C.; Folliard, K.J. Temperature Boundary Condition Models for Concrete Bridge Members. *ACI Mater. J.* **2007**, *104*, 379–387.
6. Tayşi, N.; Abid, S. Temperature Distributions and Variations in Concrete Box-Girder Bridges: Experimental and Finite Element Parametric Studies. *Adv. Struct. Eng.* **2015**, *18*, 469–486. [CrossRef]
7. Li, S.; Xin, J.; Jiang, Y.; Wang, C.; Zhou, J.; Yang, X. Temperature-Induced Deflection Separation Based on Bridge Deflection Data Using the TVFEMD-PE-KLD Method. *J. Civ. Struct. Health Monit.* **2023**, *13*, 781–797. [CrossRef]
8. Xue, J.; Lin, J.; Briseghella, B.; Tabatabai, H.; Chen, B. Solar Radiation Parameters for Assessing Temperature Distributions on Bridge Cross-Sections. *Appl. Sci.* **2018**, *8*, 627. [CrossRef]
9. Honorio, T.; Bary, B.; Benboudjema, F. Evaluation of the Contribution of Boundary and Initial Conditions in the Chemo-Thermal Analysis of a Massive Concrete Structure. *Eng. Struct.* **2014**, *80*, 173–188. [CrossRef]

10. Zhang, F.; Shen, J.; Liu, J. Effect of Encased Concrete on Section Temperature Gradient of Corrugated Steel Web Box Girder. *Adv. Struct. Eng.* **2021**, *24*, 2321–2335. [CrossRef]
11. Zhou, L.; Xia, Y.; Brownjohn, J.M.W.; Koo, K.Y. Temperature Analysis of a Long-Span Suspension Bridge Based on Field Monitoring and Numerical Simulation. *J. Bridge Eng.* **2016**, *21*, 04015027. [CrossRef]
12. Yang, L.I.U.; Hai-ping, Z.; Yang, D.; Nan, J.; Jian-ren, Z. Temperature Field Characteristic Research of Steel Box Girder for Suspension Bridge Based on Measured Data. *China J. Highw. Transp.* **2017**, *30*, 56.
13. Shi, T.; Zheng, J.; Deng, N.; Chen, Z.; Guo, X.; Wang, S. Temperature Load Parameters and Thermal Effects of a Long-Span Concrete-Filled Steel Tube Arch Bridge in Tibet. *Adv. Mater. Sci. Eng.* **2020**, *2020*, 9710613. [CrossRef]
14. Abid, S.R.; Tayşi, N.; Özakça, M. Experimental Analysis of Temperature Gradients in Concrete Box-Girders. *Constr. Build. Mater.* **2016**, *106*, 523–532. [CrossRef]
15. Peiretti, H.C.; Parrotta, J.E.; Oregui, A.B.; Caldentey, A.P.; Fernandez, F.A. Experimental Study of Thermal Actions on a Solid Slab Concrete Deck Bridge and Comparison with Eurocode 1. *J. Bridge Eng.* **2014**, *19*, 04014041. [CrossRef]
16. Hedegaard, B.D.; French, C.E.; Shield, C.K. Investigation of Thermal Gradient Effects in the I-35W St. Anthony Falls Bridge. *J. Bridge Eng.* **2013**, *18*, 890–900. [CrossRef]
17. Lee, J.-H.; Kalkan, I. Analysis of Thermal Environmental Effects on Precast, Prestressed Concrete Bridge Girders: Temperature Differentials and Thermal Deformations. *Adv. Struct. Eng.* **2012**, *15*, 447–459. [CrossRef]
18. Ding, Y.; Zhou, G.; Li, A.; Wang, G. Thermal Field Characteristic Analysis of Steel Box Girder Based on Long-Term Measurement Data. *Int. J. Steel Struct.* **2012**, *12*, 219–232. [CrossRef]
19. Wang, G.; Ding, Y.; Wang, X.; Yan, X.; Zhang, Y. Long-Term Temperature Monitoring and Statistical Analysis on the Flat Steel-Box Girder of Sutong Bridge. *J. Highway Transp. Res. Dev.* **2014**, *8*, 63–68. [CrossRef]
20. Zhang, C.; Liu, Y.; Liu, J.; Yuan, Z.; Zhang, G.; Ma, Z. Validation of Long-Term Temperature Simulations in a Steel-Concrete Composite Girder. *Structures* **2020**, *27*, 1962–1976. [CrossRef]
21. Xia, Y.; Xu, Y.-L.; Wei, Z.-L.; Zhu, H.-P.; Zhou, X.-Q. Variation of Structural Vibration Characteristics versus Non-Uniform Temperature Distribution. *Eng. Struct.* **2011**, *33*, 146–153. [CrossRef]
22. Xin, J.; Jiang, Y.; Zhou, J.; Peng, L.; Liu, S.; Tang, Q. Bridge Deformation Prediction Based on SHM Data Using Improved VMD and Conditional KDE. *Eng. Struct.* **2022**, *261*, 114285. [CrossRef]
23. Zhang, H.; Li, H.; Zhou, J.; Tong, K.; Xia, R. A Multi-Dimensional Evaluation of Wire Breakage in Bridge Cable Based on Self-Magnetic Flux Leakage Signals. *J. Magn. Magn. Mater.* **2023**, *566*, 170321. [CrossRef]
24. Zeng, Y.; He, H.; Qu, Y.; Sun, X.; Tan, H.; Zhou, J. Numerical Simulation of Fatigue Cracking of Diaphragm Notch in Orthotropic Steel Deck Model. *Materials* **2023**, *16*, 467. [CrossRef] [PubMed]
25. Zeng, Y.; Qiu, Z.; Yang, C.; Haozheng, S.; Xiang, Z.; Zhou, J. Fatigue Experimental Study on Full-Scale Large Sectional Model of Orthotropic Steel Deck of Urban Rail Bridge. *Adv. Mech. Eng.* **2023**, *15*, 16878132231155272. [CrossRef]
26. AASHTO. *AASHTO LRFD Bridge Design Specifications*; American Association of State Highway and Transportation Officials: Washington, DC, USA, 2012.
27. *BS 5400: Part 2*; Concrete and Composite Bridges. Specification for Loads. British Standard Institution: London, UK, 1978.
28. *EN 1991-1*; Eurocode 1—Actions on Structures—Part 1-5: General Actions—Thermal Actions. CEN (European Committee for Standardization): Brussels, Belgium, 2004.
29. *AS5100. 2-2004*; Bridge Design—Part 2. Design Load. Standards Australia: Sydney, NSW, Australia, 2004.
30. *JTG D60-2015*; General Specifications for Design of Highway Bridges and Culverts. China Communications Press: Beijing, China, 2015.
31. Niu, Y.; Wang, Y.; Tang, Y. Analysis of Temperature-Induced Deformation and Stress Distribution of Long-Span Concrete Truss Combination Arch Bridge Based on Bridge Health Monitoring Data and Finite Element Simulation. *Int. J. Distrib. Sens. Netw.* **2020**, *16*, 1550147720945205. [CrossRef]
32. Taeb, A.; Ooi, P.S. Comparison of Field Behavior with Results from Numerical Analysis of a Geosynthetic Reinforced Soil Integrated Bridge System Subjected to Thermal Effects. *Transp. Res. Rec.* **2020**, *2674*, 294–306. [CrossRef]
33. Wang, D.; Liu, Y.; Liu, Y. 3D Temperature Gradient Effect on a Steel-Concrete Composite Deck in a Suspension Bridge with Field Monitoring Data. *Struct. Control. Health Monit.* **2018**, *25*, e2179. [CrossRef]
34. Jiang, Y.; Hui, Y.; Wang, Y.; Peng, L.; Huang, G.; Liu, S. A Novel Eigenvalue-Based Iterative Simulation Method for Multi-Dimensional Homogeneous Non-Gaussian Stochastic Vector Fields. *Struct. Saf.* **2023**, *100*, 102290. [CrossRef]
35. Abid, S.R. Three-Dimensional Finite Element Temperature Gradient Analysis in Concrete Bridge Girders Subjected to Environmental Thermal Loads. *Cogent Eng.* **2018**, *5*, 1447223. [CrossRef]
36. Hu, J.; Wang, L.; Song, X.; Sun, Z.; Cui, J.; Huang, G. Field Monitoring and Response Characteristics of Longitudinal Movements of Expansion Joints in Long-Span Suspension Bridges. *Measurement* **2020**, *162*, 107933. [CrossRef]
37. Gu, B.; Zhou, F.Y.; Gao, W.; Xie, F.Z.; Lei, L.H. Temperature Gradient and Its Effect on Long-Span Prestressed Concrete Box Girder Bridge. *Adv. Civ. Eng.* **2020**, *2020*, 5956264. [CrossRef]
38. Liu, J.; Liu, Y.; Zhang, C.; Zhao, Q.; Lyu, Y.; Jiang, L. Temperature Action and Effect of Concrete-Filled Steel Tubular Bridges: A Review. *J. Traffic Transp. Eng.* **2020**, *7*, 174–191. [CrossRef]
39. Xin, J.; Zhou, C.; Jiang, Y.; Tang, Q.; Yang, X.; Zhou, J. A Signal Recovery Method for Bridge Monitoring System Using TVFEMD and Encoder-Decoder Aided LSTM. *Measurement* **2023**, 112797. [CrossRef]

40. Morgese, M.; Wang, C.; Ying, Y.; Taylor, T.; Ansari, F. Stress–Strain Response of Optical Fibers in Direct Tension. *J. Eng. Mech.* **2023**, *149*, 04023037. [CrossRef]
41. Tang, Q.; Xin, J.; Jiang, Y.; Zhou, J.; Li, S.; Chen, Z. Novel Identification Technique of Moving Loads Using the Random Response Power Spectral Density and Deep Transfer Learning. *Measurement* **2022**, *195*, 111120. [CrossRef]
42. Wang, C.; Ansari, F.; Wu, B.; Li, S.; Morgese, M.; Zhou, J. LSTM Approach for Condition Assessment of Suspension Bridges Based on Time-Series Deflection and Temperature Data. *Adv. Struct. Eng.* **2022**, *25*, 3450–3463. [CrossRef]
43. Zhang, H.; Ma, X.; Jiang, H.; Tong, K.; Zheng, Y.; Zhou, J. Grading Evaluation of Overall Corrosion Degree of Corroded RC Beams via SMFL Technique. *Struct. Control. Health Monit.* **2023**, *2023*, 6672832. [CrossRef]
44. Wu, B.; Zhou, J.; Li, S.; Xin, J.; Zhang, H.; Yang, X. Combining Active and Passive Wind Tunnel Tests to Determine the Aerodynamic Admittances of a Bridge Girder. *J. Wind. Eng. Ind. Aerodyn.* **2022**, *231*, 105180. [CrossRef]
45. Wu, F.; Zhou, J.; Xin, J.; Zhang, H.; Zhao, N.; Yang, X. Wind Damage Estimation of Roof Sheathing Panels Considering Directionality: Influences of Both Correlations of Directional Wind Speeds and Multiple Response Coefficients in Each Direction. *J. Wind. Eng. Ind. Aerodyn.* **2023**, *236*, 105396. [CrossRef]
46. Tao, T.; He, J.; Wang, H.; Zhao, K. Efficient Simulation of Non-Stationary Non-Homogeneous Wind Field: Fusion of Multi-Dimensional Interpolation and NUFFT. *J. Wind. Eng. Ind. Aerodyn.* **2023**, *236*, 105394. [CrossRef]
47. Li, H.; Zhang, L.; Wu, B.; Yang, Y.; Xiao, Z. Investigation of the 2D Aerodynamic Admittances of a Closed-Box Girder in Sinusoidal Flow Field. *KSCE J. Civ. Eng.* **2022**, *26*, 1267–1281. [CrossRef]

buildings

MDPI

Article

Research on Optimal Arch Rib Inclination of Large Span Highway CFST through Arch Bridge

Zengwu Liu [1,2], Yuexing Wu [3], Chengwei Wang [2,4], Yonghui Fan [2], Chao Luo [2,*] and Shaorui Wang [2]

1 School of Transportation and Civil Engineering, Shandong Jiao Tong University, Changqing District, Jinan 250357, China; 611180086015@mails.cqjtu.edu.cn
2 State Key Laboratory of Mountain Bridge and Tunnel Engineering, Chongqing Jiaotong University, Chongqing 400074, China; cwang268@uic.edu (C.W.); fyh1995@mails.cqjtu.edu.cn (Y.F.); ruiruiplace@163.com (S.W.)
3 School of Civil Engineering, Hunan City University, Yiyang 413000, China; wuyuexing@hncu.edu.cn
4 Smart Sensors and NDT Laboratory, Department of Civil, Materials, and Environmental Engineering, University of Illinois at Chicago, W. Taylor Street, Chicago, IL 60607, USA
* Correspondence: luochao@mails.cqjtu.edu.cn

Abstract: To investigate the reasonable range of the inclination angle of arch ribs, a spatial finite element method was employed based on a concrete-filled steel tube (CFST) basket-handle through an arch bridge with a span of 360 m. A spatial finite element model was established using Midas/Civil software, which was verified with actual bridge data. The effects of different arch rib inclination angles were investigated under static loads. The structural natural frequencies, linear elastic stability coefficients, internal forces, and displacements were comprehensively considered to determine the reasonable range of the inclination angle. The results show that when the inclination angle ranges between 8° and 10°, the first, third, and sixth natural frequencies of the structure are increased. It effectively improves the lateral and torsional stiffness of the arch ribs while ensuring optimal out-of-plane stability of the arch ribs. Compared with the parallel arch, the stability is improved by 20.2%. The effects of angle variation on displacement and internal force of the arch ribs were not significant. Considering all indicators, the optimal range of the inclination angle for the arch ribs of 300-m-level highway CFST arch bridges is suggested to be 8~10°.

Keywords: steel tube concrete truss arch; highway basket-handle arch bridge; arch rib inclination angle; finite element method; lateral stability

Citation: Liu, Z.; Wu, Y.; Wang, C.; Fan, Y.; Luo, C.; Wang, S. Research on Optimal Arch Rib Inclination of Large Span Highway CFST through Arch Bridge. *Buildings* **2023**, *13*, 1415. https://doi.org/10.3390/buildings13061415

Academic Editors: Simon X. Yang, Elena Ferretti, Jingzhou Xin, Yan Jiang and Bo Wu

Received: 25 April 2023
Revised: 23 May 2023
Accepted: 29 May 2023
Published: 30 May 2023

1. Introduction

In mountainous regions of China, arch bridges are widely used due to their advantages of high stiffness, high bearing capacity, good seismic performance, and low cost [1,2]. The number of existing concrete-filled steel tube (CFST) arch bridges in China has exceeded 400, and their spans continue to break records. However, there are few CFST basket-handle arch bridges with spans above 300 m [3–5]. Therefore, the research on the large-span basket-handle arch bridge is relatively limited, especially on the mechanical performance affected by inclination angle [1,6].

As the span of steel-concrete composite arch bridges increases, the problem of lateral stability becomes more prominent. The basket-handle arch not only has an elegant appearance but also has good lateral stability. Table 1 shows the statistics of some highway CFST basket-handle arch bridges. The inclination angle of the arch ribs is between 4.6° and 13°. Existing literature [7–9] has shown that if an improper inclination angle of the basket-handle arch ribs is used, the lateral stiffness decreases. In addition, for truss structures, changes in structural form may lead to a redistribution of member axial forces, thereby affecting the safety of the structure. Scholars have proposed methods to estimate the axial forces on the members of the structure. Therefore, when determining the optimal inclination angle of the arch ribs, the effects on the mechanical performance, such as axial forces and bending moments of the structure, need to be considered [10,11]. Many scholars have studied the effects of inclination angle on the structural mechanical performance of CFST arch bridges. Yang et al. [12] studied a 220-m railway CFST arch bridge with varying inclination angles. It was found that when the inclination angle ranges from 0° to 10°, the modal frequencies of lateral and torsional vibrations of the arch ribs increase with inclination. However, the vertical and longitudinal vibration frequencies remain almost constant. Yun et al. [13] used the finite element analysis method to study the influence of the arch ribs rise-to-span ratio and inclination angle on the structural vibration mode and stability. A mid-supporting steel-concrete composite arch bridge was employed in this study. It was found that the inclination angle of the arch ribs should not exceed 10°. Wang et al. [14] conducted a self-vibration characteristic analysis of a 240-m railway CFST arch bridge. In this study, the lateral vibration frequency was found to be more sensitive than the other two directions when the inclination angle varied from 7.5° to 9.5°. Wei et al. [15] conducted a parameter analysis of several existing CFST arch bridges and built a CFST arch bridge with a span of 105 m. The study showed that the linear elastic stability safety factor gradually increased as the inclination angle of the arch ribs increased within 0–15° and reached an optimal state at 9°. Zeng et al. [16] studied the structural stability of a steel box arch rib bridge with varying arch rib inclination angles from 0° to 12° using the finite element analysis method. The results showed that the out-of-plane stability safety factor increased first and then decreased with the increase in inclination angle. The out-of-plane stability safety factor reached its peak at 10°. However, the in-plane stability safety factor decreased with the increase in the inclination angle. Therefore, it is necessary to comprehensively analyze and determine the optimal inclination angle. Zhao et al. [17] studied the variation of internal forces at key positions of the arch ribs when the inclination angles were 0°, 7°, and 15°. A flying-swallow-shaped CFST cable-stayed arch bridge with a main span of 360 m was employed in this study. The results showed that the internal forces of the arch ribs increased with the increase in the inclination angle, especially the bending moment at the crown. Huang et al. [18] examined the variation of internal forces and vertical displacement of a 260-m CFST arch bridge when the inclination angle of the arch rib ranged from 0° to 12°. The results showed that with an increase in the inclination angle of the arch rib, the bending moment and horizontal thrust force at the abutments decreased significantly, reaching a minimum of 8°. The increase in inclination angle beyond 8° did not result in a significant change in the internal forces or vertical displacement. Xu et al. [19] used a railway CFST arch bridge as an example to investigate the influence of the arch rib inclination angle on the seismic performance of the structure. The study found that an inclination angle of 3.5~4° not only reduced the displacement and axial force of the arch rib but also avoided the problem of excessive growth of tensile stress in the concrete of the arch rib, which could result in inadequate strength. Wang et al. [20] determined the optimal inclination angle of the arch rib by considering the effects of the rise-to-span ratio, width-to-span ratio, and number of transverse braces. A centrally supported dumbbell-shaped steel-concrete composite basket-handle arch bridge was employed in this study. The study found that the optimal inclination angle of the arch rib was negatively correlated with the rise-to-span ratio and not significantly correlated with the number of transverse braces. Ji et al. [21] investigated the variation patterns of the linear elastic and ultimate bearing capacities of a large-span railway steel-concrete composite lever arch bridge with a tube-shaped structure. The study found that the structural stability of the arch bridge would increase first and then decrease as the arch rib inclination angle increased. The optimal value of the inclination angle was determined in this paper as well. Pan et al. [22] conducted a study on the influence of the inclination angle on the lateral stability of a tied arch bridge using the finite element method. Through parameter analysis, the study obtained an approximate expression for the reasonable inclination angle of the double-rib arch. The studies of the aforementioned scholars have achieved rich results. However, the majority of studies were focused on railway bridges. Related studies on highway bridges are relatively few, especially for bridges with spans larger than 300 m. Furthermore, the evaluation is not comprehensive enough. The relationship between multiple factors of the structure and the arch rib inclination angle needs to be systematically studied. Further research is required to comprehensively determine the reasonable range of inclination angles for the arch rib.

Table 1. Partial basket-handle arch bridges were built in China.

Serial Number	Bridge Name	Span/m	Arch Rib Section Form	Inclination Angle
1	The Yarlung Zangbo River	430	Four limb truss	4.6°
2	Yellow River Extra Large Bridge	380	Four limb truss	8°
3	Guangxi Shawei Zuojiang Bridge	360	Four limb truss	10°
4	Lancang River Grand Bridge	342	Four limb truss	6.8
5	Anhui Taiping Lake Bridge	336	Four limb truss	10°
6	Zhejiang Sanmenkou Cross Sea Bridge	270	Dumbbell shaped	8°
7	Zhejiang Tongwamen Bridge	238	Dumbbell shaped	8.5°
8	Jinghang Canal Grand Bridge	235	Four limb truss	10°
9	Menghua Railway Longmen Yellow River Bridge	202	Four limb truss	6°
10	Huayudong Bridge	180	Four limb truss	10°
11	Jiangning Grand Bridge	128	Dumbbell shaped	9°
12	Hujiawan Grand Bridge	112	Dumbbell shaped	9°
13	Dongtiaoxi Grand Bridge	112	Dumbbell shaped	13°
14	Longmen Yellow River Bridge	202	Four limb truss	6°

The objective of this paper is to study the influence of the arch rib inclination angle on the large span highway CFST basket-handle arch bridge (i.e., 300 m level). The mechanical performance of the structure under different rib inclination angles was investigated. A reasonable range of arch rib inclination angles for large-span CFST arch bridges was proposed to enhance their safety and provide a reference for the design of similar bridges in the future. Based on the world's largest CFST basket-handle arch bridge with a span of 360 m—the Shaowei Zuojiang Extra-large Bridge—this paper employs Midas Civil to establish a full-bridge finite element model. Firstly, the accuracy of the finite element was verified by using the measured vertical displacement values of the main arch during the installation of the main beam and bridge pavement. Then, different arch rib inclination angles were simulated in the FE model. The structural vibration characteristics, linear elastic stability, internal forces, and displacement under static loads with different inclination angles were analyzed. The relationship between the arch rib inclination angles and the structural mechanical performance was systematically studied.

2. Bridge Background and Finite Element Model

The Shawei Zuojiang Bridge is a through-arch bridge with varying cross-section CFST truss ribs. It is designed as a basket-handle arch bridge. It has a main span of 360 m (an effective span of 340 m) and a rise-to-span ratio of 1/4.533. The arch axis follows a catenary curve, and the arch axis coefficient is m = 1.55. The two arch ribs are inclined 10 degrees towards the centerline of the bridge, as shown in Figure 1a. The transverse spacing between the arch feet is 38 m. Each arch rib is constructed from a four-tube truss with a varying cross-section. The cross section has a height of 7 m at the arch crown section and 12 m at the arch foot section, and a rib width of 3.2 m. Both upper and lower chord members of the arch rib are made of steel pipes with a diameter of 1200 mm and a wall thickness of 24–32 mm, filled with C60 self-compacting concrete to compensate for shrinkage. The rectangular cross-section is formed by connecting the chord members of the main arch rib with 720 mm-diameter connecting pipes and two vertical web members with a diameter of 610 mm. The chord members of the main arch rib are all made of Q345 steel.

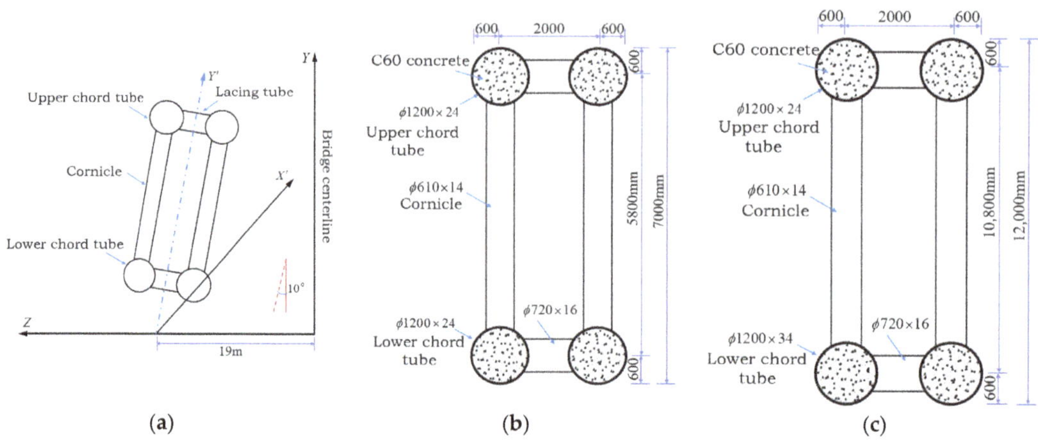

Figure 1. Arch rib structure diagram. (**a**) Schematic diagram of arch rib inclination angle; (**b**) arch rib arch crown section; and (**c**) arch rib and arch foot section.

The two arch ribs are connected by "X"-braces. There are 12 "X"-braces arranged along the upper chord and 10 along the lower chord, with the steel pipes of the "X"-braces measuring Φ900 × 18 mm. The cross-sectional diagrams of the main arch rib at the crown and foot are shown in Figures 1b and 1c, respectively, and the overall layout is shown in Figure 2. The cross-section of the bridge deck is shown in Figure 3. The bridge deck has a concrete slab and steel grid beams supporting it. The steel grid beam consists of 7 longitudinal beams and 24 transverse beams. The transverse beam is made of 20-mm-thick steel plates and is located under each suspender. The longitudinal beams are I-beams made of Q345 steel. The thickness and the width of the top and bottom flanges of the longitudinal beams are 20 mm and 600 mm, respectively. The thickness of the web plate is 16 mm. The No. 2 and No. 6 beams in Figure 3 have a height of 1924 mm. The No. 1 and No. 7 beams have a height of 1113 mm. The No. 3 and No. 5 beams have a height of 1274 mm. The No. 4 beam has a height of 1400 mm. A C40 concrete slab with a thickness of 140 mm is located at the top of the grid beams. The steel grid beams and concrete slab are connected by a steel plate with a thickness of 100 mm and shear nails. The entire bridge has 24 pairs of suspenders, which are composed of steel strands with a tensile strength of 1860 MPa. This type of through-arch bridge has a clear load path. The dead load and live load are transmitted to the arch ribs through the suspenders and then to the abutment at the arch foot.

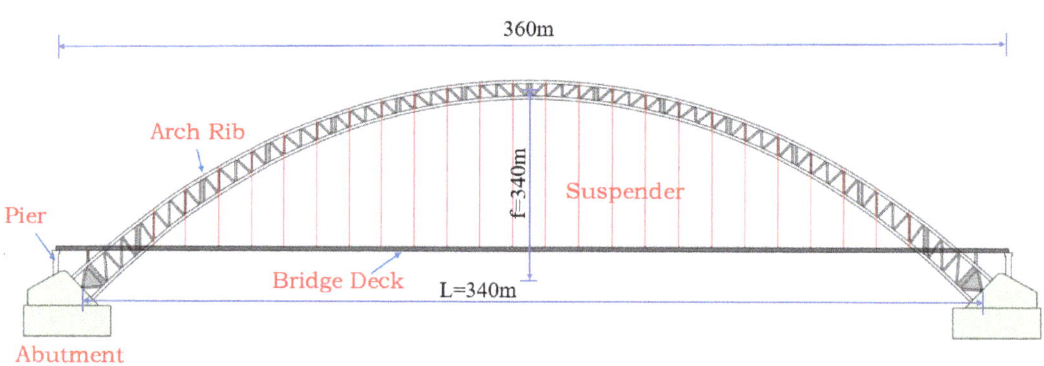

Figure 2. Elevation of Shawei Zuojiang Bridge.

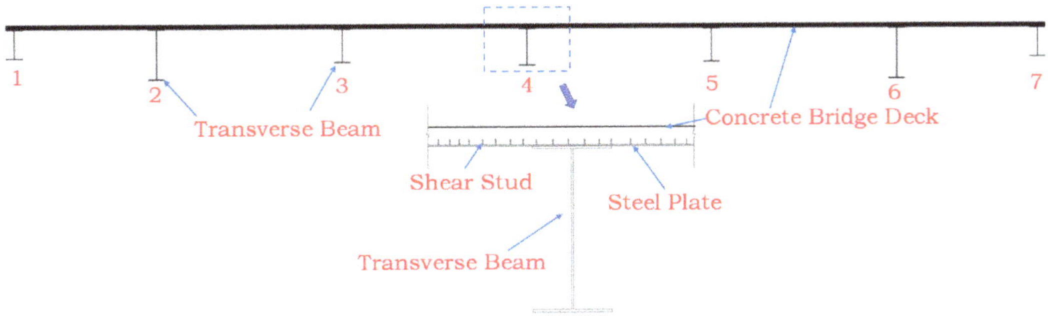

Figure 3. Cross section of the main beam.

MIDAS/CIVIL is software for structural and spatial finite element model building. It was wildly used in bridge construction, temperature, dynamics, and mechanical performance analysis [23,24].

In this study, the finite element software MIDAS/CIVIL was used to implement the spatial finite element modeling of the bridge. The appropriate spatial elements were selected based on the actual stress conditions of each component. The corresponding constraints were applied according to the boundary conditions. The following assumptions were made in the modeling process: (1) all section deformations comply with the cross-sectional assumption; (2) no slippage occurs between steel tubes and concrete; and (3) linear elastic theory is adopted and nonlinear effects are not considered.

The arch ribs and the steel grid beams were modeled using beam elements, the bridge slab was modeled using plate elements, and the suspenders were modeled using truss elements. The entire bridge has a total of 3930 nodes and 5714 elements, including 4942 beam elements, 724 plate elements, and 48 truss elements. The abutments were fixed. The suspenders and arch ribs were rigidly connected using elastic connections. The supports at two ends of the main girder were modeled as simply supported. The second-stage load, such as the pavement and parapet, was simulated using uniformly distributed loads applied to the main girder. The types and quantities of elements used in the model are shown in Table 2. The finite element model is shown in Figure 4.

Table 2. Element type and quantity.

	Element Type	Quantity
Arch Rib	Beam	3080
Steel lattice beam	Beam	1862
Bridge deck	Plate Element	724
Suspender	Truss Element	48

To verify the accuracy of the finite element model, the measured vertical displacement of the arch ribs during construction was compared with the finite element values. Reflectors are arranged at the center of the top chord of the arch rib at the key sections (i.e., L/8, 2L/8, 3L/8, 4L/8, 5L/8, 6L/8, and 7L/8). A Leica total station was used to measure the vertical displacement of the arch rib before and after the installation of the bridge deck. The resolution of the total station is 0.1 mm. To avoid the impact of solar radiation on structural deformation, measurements were made in the early morning with similar atmospheric temperatures. Meanwhile, the displacements of ribs induced by bridge deck installation were calculated by the FE model. Two strategies were used during the calculation (i.e., the geometric nonlinearity strategy and the linear elastic strategy). Figure 5a is the on-site construction drawing, and Figure 5b is a comparison between the measured values and calculated values of the vertical displacement at key sections of the arch rib. The deviation is shown in Table 3 as well. In both strategies, the calculated vertical displacement of the key section of the arch rib matches the measured value. The geometric nonlinearity strategy has slightly better performance than the linear elastic strategy, but it is not significant.

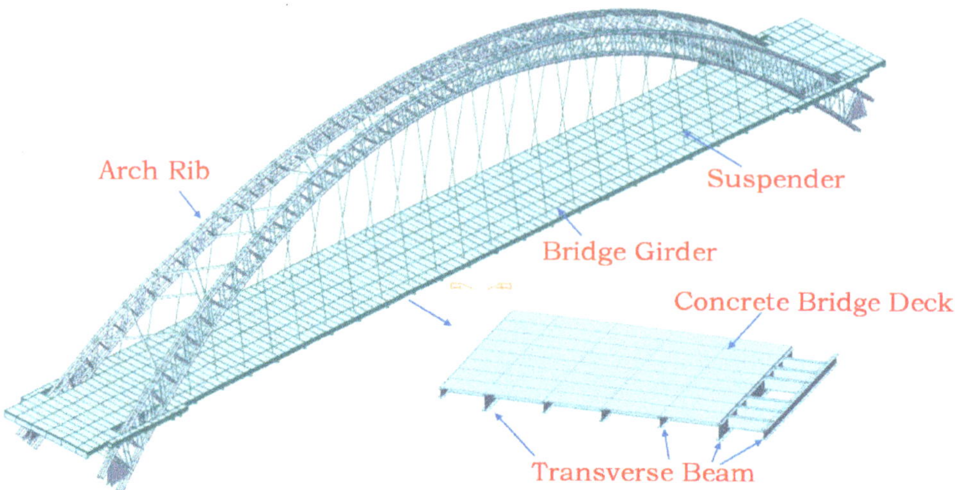

Figure 4. Finite element model diagram.

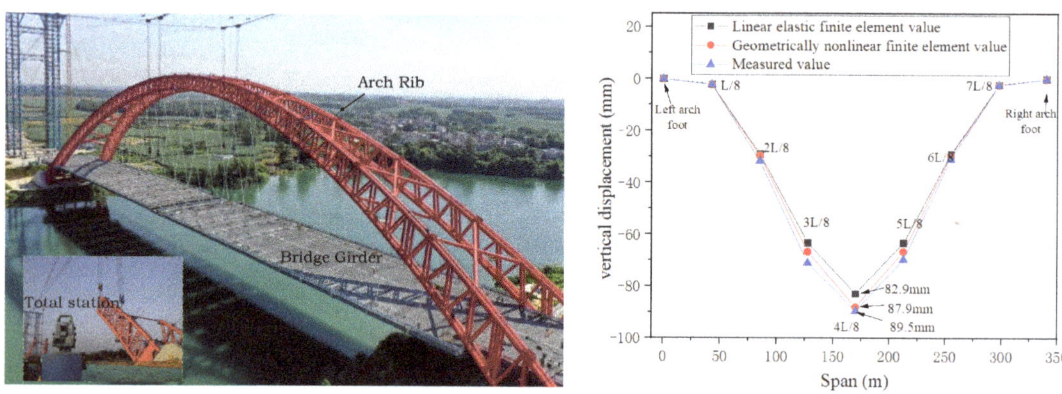

(**a**) (**b**)

Figure 5. (**a**) Construction drawing of bridge deck installation Site. (**b**) Comparison of vertical displacement of key sections of arch rib.

Table 3. The deviation between calculated values and measured values of vertical displacement at key sections of arch ribs (mm).

	L/8	2L/8	3L/8	4L/8	5L/8	6L/8	7L/8
(3)—(1)	−0.1	−2.6	−7.6	−6.6	−6.3	−1.7	−0.2
(3)—(2)	0.0	−2.1	−4.3	−1.6	−3.1	−1.3	−0.1

In the table, (1) is the linear elastic finite element value, (2) is the geometrically nonlinear finite element value, and (3) is the measured value.

3. Analysis of Reasonable Value of Arch Rib Inclination Angle

For the basket-handle through the arch bridge, the inclination angle of the arch rib cannot infinitely increase due to the geometrical limit. The width and height of the bridge deck, the arch rise, and other factors are shown in Figure 6. The relationship between the inclination angle of the arch rib and these parameters can be expressed as follows:

$$\tan\theta = \frac{B_1 - B_2}{2(f - h)} \tag{1}$$

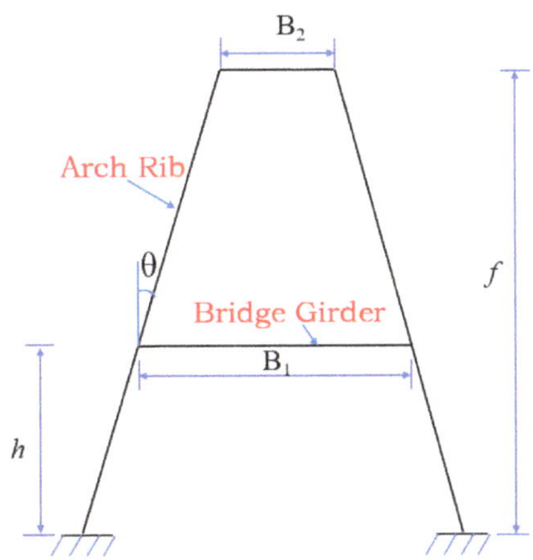

Figure 6. Schematic diagram of arch rib inclination calculation.

The arch rib inclination angle is:

$$\theta = \arctan(\frac{B_1 - B_2}{2(f - h)}) \tag{2}$$

where θ—Arch rib inclination angle;
B_1—The bridge deck width between the arch ribs;
B_2—The clear distance between arch ribs at the crown;
f—Rise of the arch;
h—The height from the bridge deck to the center of the arch foot.

When the inclination angle is too large, the arch ribs at the top of the arch will intersect, thereby limiting the increase in the inclination angle. On the other hand, a too-small inclination angle will not be able to take advantage of the benefits of the basket-handle arch. A typical basket-handle arch adjusts the inclination angle of the arch ribs by keeping the bridge deck width constant. For the Shawei Zuojiang Bridge, B_1 = 28 m, f = 75 m, h = 16.2 m. When the inclination angle is zero, $\theta = 0°$, it is a parallel arch. The inclination angle reaches its maximum value when $B_2 = 0$, which can be calculated by the formula.

$$\theta = \arctan(\frac{28 - 0}{2(75 - 16.2)}) = 13.393°$$

Therefore, the range of inclination angles for arch ribs is 0~13°. To analyze the mechanical performance with different rib inclination angles, 0°, 1°, 2°, 3°, 4°, 5°, 6°, 7°, 8°, 9°, 10°, 11°, 12°, and 13° were employed in this study. The natural vibration characteristics, linear elastic stability coefficient, internal force, and displacement of arch ribs under static load were calculated for structures at each selected angle. The optimal range of inclination angle for arch ribs was then determined by considering all the mechanical performances above comprehensively.

3.1. Analysis of the Influence of Natural Vibration Characteristics

To evaluate the natural frequencies and modes of vibration of the bridge under different arch rib inclinations, the structural self-weight and second stage loads (i.e., the weight of the bridge pavement and the parapet) were considered as the loads. As a result, low-frequency modes contain the majority of the energy, while high-order modes have a negligible impact on the structure's vibration. For the parallel arch (with an inclination angle of 0°), the first six mode shapes and their natural frequencies are shown in Table 4. The first three mode shapes are shown in Figures 7–9.

Table 4. Natural frequency and mode direction of the first six modes.

Modal	Natural Frequency (HZ)	Vibration Mode Direction
1	0.32	Symmetrical transverse bending of the main arch
2	0.53	Antisymmetric vertical bending of the main beam and arch rib
3	0.55	Antisymmetric transverse bending of the main arch
4	0.64	Symmetrical transverse bending of the main beam and arch rib
5	0.73	Symmetrical vertical bending of the main beam and arch rib
6	0.86	Symmetrical torsion of the main beam and arch rib

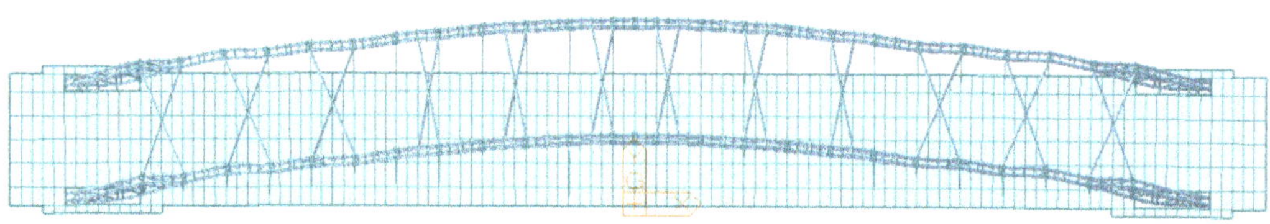

Figure 7. Symmetrical transverse bending of the main arch.

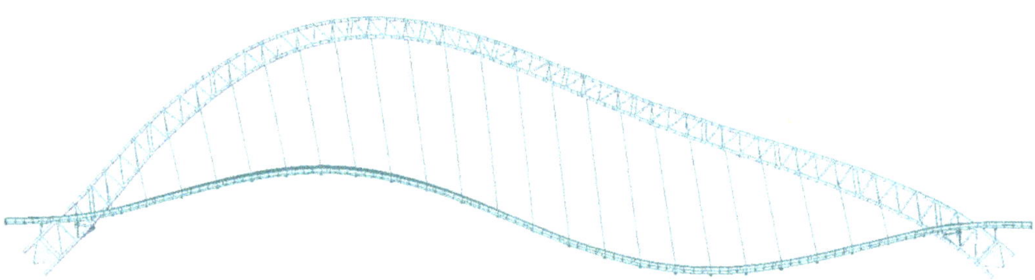

Figure 8. Antisymmetric vertical bending of the main beam and arch rib.

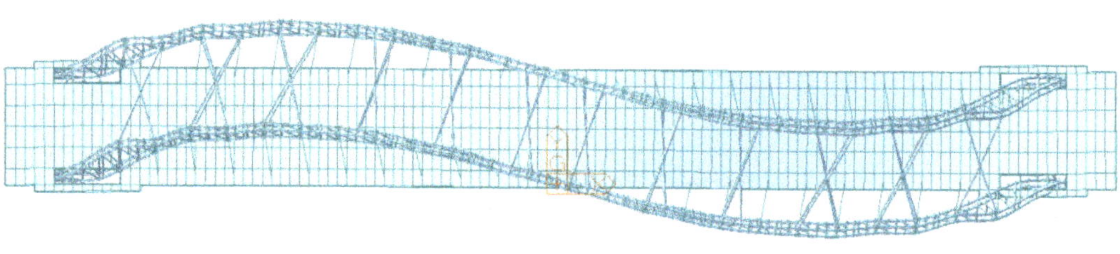

Figure 9. Antisymmetric transverse bending of the main arch.

According to the analysis in Table 4 and Figures 7–9, the first six modal frequencies of the bridge have a maximum value of 0.8586, indicating that it is a flexible structure. The first two modes were dominated by lateral bending and vertical bending. Torsion mode was ranked as the sixth mode. It indicated that the structure has a relatively weak lateral stiffness, moderate vertical stiffness, and a relatively strong torsional stiffness. To investigate the modal frequency variation with respect to different inclination angles, the modal frequencies under different inclination angles were compared with those of the parallel arch. The percentage difference of each modal frequency with respect to the parallel arch model is shown in Figure 10.

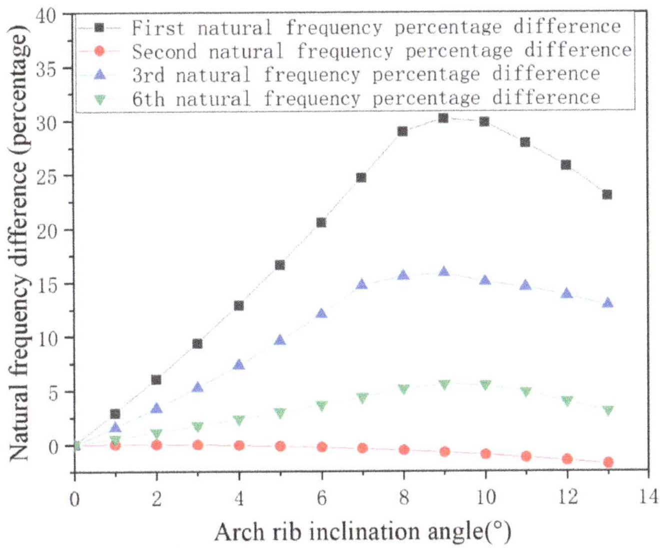

Figure 10. Natural frequency difference of different arch rib angles and parallel arches.

According to Figure 10, the changing rate varied with different modes while increasing the inclination angle. For the first, third, and sixth modes, the curve shows a pattern of increasing first and then decreasing. The natural frequency of the first, third, and sixth modes reached a maximum value at 8° to 10°, indicating the optimal inclination angle for transverse and torsional stiffness improvement. For the second mode, the natural frequency gradually decreases with an increase in the inclination angle. However, the decrease was small—less than 2% at the inclination angle of 8° to 10°. It indicated that the increase in the inclination angle of the arch rib reduced the vertical stiffness of the arch rib, but the effect was small.

3.2. Linear Elastic Stability Analysis

The linear elastic stability coefficient (λ) is an important indicator for evaluating the stability and safety of structures. The definition of λ can be found in Cao et al. [9], Xu et al. [19], and Wang et al. [25]. Due to the complexity of the structure, it is difficult to obtain its elastic instability limit load using analytical methods. However, using finite-element numerical analysis methods can yield reliable results [25]. Midas Civil employed the subspace iteration method to calculate λ and the buckling shape of each buckling mode by inputting the load conditions, number of buckling modes, and convergence conditions of structural buckling. The change in inclination angle leads to a change in buckling capacity. Under certain loads, the linear elastic stability coefficient λ increases with the increase in buckling capacity. Therefore, λ can be used as an indicator of structural stability. The dead load was employed as the load condition herein because it is the dominant load on the linear stability of the structure. In practice, only the first buckling mode is important because it has the smallest linear elastic stability coefficient. The shape and λ of the first buckling mode were calculated for both in-plane and out-of-plane buckling. The λ buckling shape of the parallel arch ribs is shown in Table 5. The corresponding buckling shape diagrams are shown in Figures 11 and 12.

Table 5. First-order stability coefficient and instability mode of a parallel arch.

Working Condition	Modal	Linear Elastic Stability Coefficient λ	Buckling Shape
Dead Load	Out-of-plane 1st bucking mode	7.447	Antisymmetric transverse bending of the main arch
	In plane 1st bucking mode	15.69	Antisymmetric vertical bending of the main arch

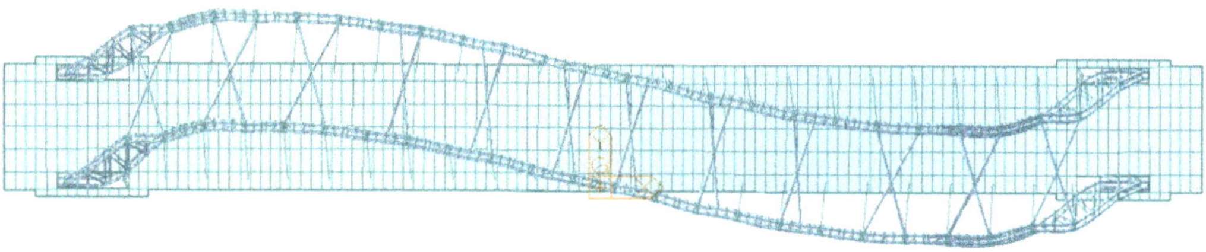

Figure 11. Out-of-plane buckling shape of a parallel arch plane.

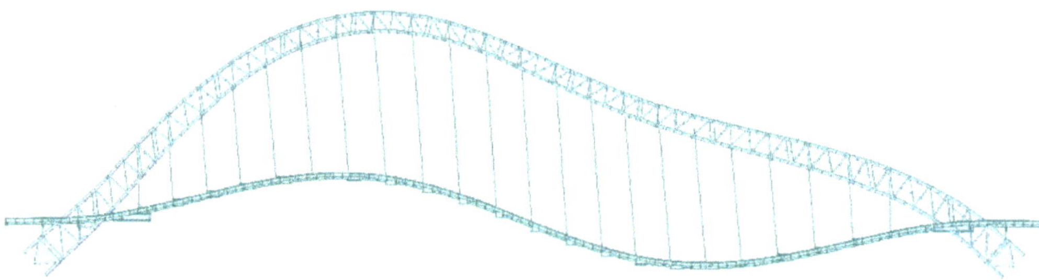

Figure 12. In-plane buckling shape of a parallel arch plane.

According to Table 5 and Figures 11 and 12, the first buckling modes of the parallel arch are the anti-symmetric transverse bending and the anti-symmetric vertical bending for out-of-plane buckling and in-plane buckling, respectively. The linear elastic stability coefficient of the out-of-plane buckling is smaller than that of the in-plane buckling for the parallel arch. It indicated that the lateral stiffness of the arch rib is weaker than the vertical stiffness of the parallel arch. The percentage difference of λ with respect to the parallel arch under different arch rib inclination angles is shown in Figure 13.

According to Figure 13, when the inclination angle of the arch rib ranges from 0° to 13°, the λ for the out-of-plane and in-plane directions showed different changing patterns with the increase of the inclination angle. The λ of out-of-plane increased at the beginning and then decreased with the increase in inclination angle. It reached its peak at an inclination angle of 9° with a 20.2% improvement and then sharply decreased after the inclination angle exceeded 10°. The λ of in-plane mode showed a negative correlation with the inclination angle. The decrease was small (3%) when the inclination angle was less than 10°. However, the λ of in-plane mode decreases significantly when the inclination angle exceeds 10°. The results indicated that the optimal inclination angle of the arch rib is between 8° and 10°. It effectively improves the out-of-plane stability of the structure with an insignificant decrease in in-plane stability. Therefore, it is recommended to use an appropriate inclination angle of 8° to 10° for the arch rib while ensuring the optimal first-order stability safety factors for both the out-of-plane and in-plane directions.

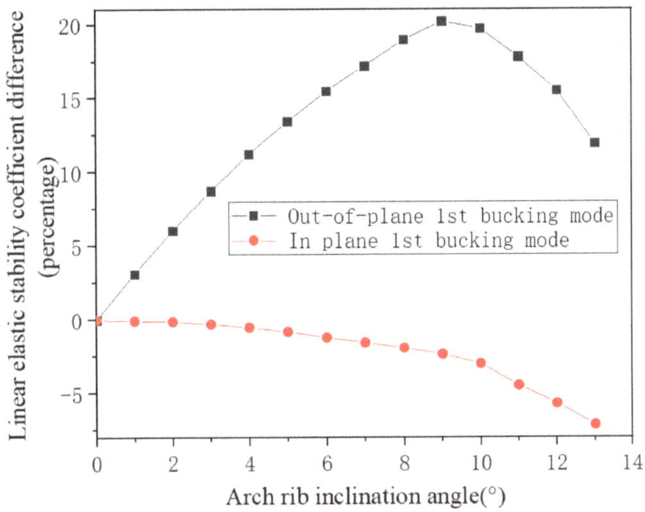

Figure 13. Percentage difference of the linear elastic stability coefficient with respect to parallel arch.

In order to consider the impact of geometric nonlinearity on the stability of the structure, the λ of the parallel arch was calculated using a geometric nonlinear model. The results are shown in Table 6.

Table 6. Stability comparison.

	λ of Out-of-Plane Mode	λ of In-Plane Mode
Linear elastic model	7.447	15.69
Geometric nonlinear model	7.335	14.77
Difference percentage (%)	−1.50	−5.86

From Table 6, it can be observed that, considering the geometric nonlinearity of the structure, the λ is smaller with respect to that of the linear elastic model. However, the difference is not significant, with 1.5% for out-of-plane mode and 5.86% for in-plane mode. Therefore, the geometric nonlinearity of the structure is no longer considered.

3.3. Static Performance Analysis

Deformation and internal forces are macroscopic manifestations of the stress state of bridges [26,27], and the internal forces of an arch bridge are mainly affected by the dead load. As the span increases, the proportion of internal forces generated by the dead load becomes larger. Therefore, the analysis of the effect of dead loads on the internal forces of arch bridges is essential [28,29]. The dead load-induced internal forces and displacements of the arch rib were calculated under different inclination angles. As a result, the arch rib is a symmetrical structure, and the arch foot, L/8, 2L/8, 3L/8, and 4L/8 (arch top) were selected as the key sections for displacement analysis. The key sections for internal force selection were the arch foot, L/4, and 4L/8 (arch top). For a parallel arch, the vertical displacements of the arch rib at the arch foot (L/8, 2L/8, 3L/8, and 4L/8) are 0, −16.5 mm, −83 mm, −162.8 mm, and −208.8 mm, respectively. Figure 14 shows the difference in displacement for different arch rib inclination angles with respect to a parallel arch.

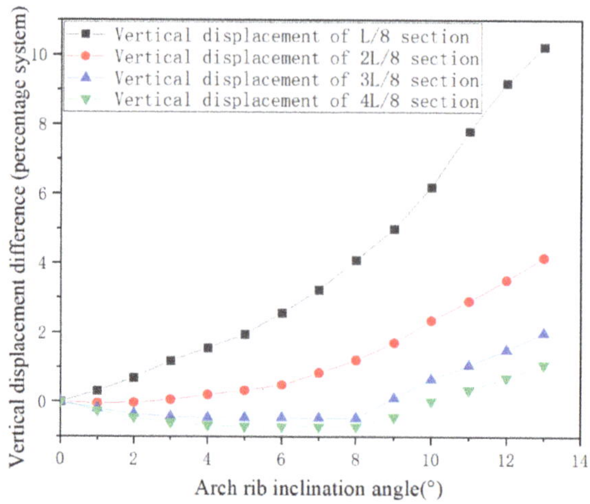

Figure 14. Vertical displacement difference with respect to parallel arch at key sections.

According to Figure 14, with the increase in the inclination angle, the vertical displacements of the L/8 and 2L/8 sections show a continuous increase. When the inclination angle is less than 10°, the vertical displacement changes slowly, but it increases rapidly in later stages. Vertical displacement at L/8 shows the largest percentage increase of 10% at the inclination angle of 13°. However, the vertical displacement value of the L/8 section is relatively small. The 3L/8 and 4L/8 sections show a trend of first decreasing and then increasing. The vertical displacement percentage difference reached its smallest for both 3L/8 and 4L/8 sections when the inclination angle was 8°. It indicated that when the inclination angle of the arch rib is 8°, the vertical displacement of the 3L/8 section and the 4L/8 section of the arch rib is the smallest, and the percentage difference is less than 5%. As far as the vertical displacement of the arch rib is concerned, the influence of the inclination angle on the vertical displacement of the structure is not significant.

For the parallel arch, the internal forces of the key sections of the arch ribs are shown in Table 7. The axial force unit in the table is kN and the moment unit is kN·m. The internal force differences of different inclinations with respect to a parallel arch are shown in Figure 15.

Table 7. Internal force values of key sections of parallel arches.

Internal Force	Position	Arch Foot Section	2L/8 Section	4L/8 Section
Axial Force (kN)	Upper chord	−13,995.48	−18,325.64	−22,800.5
	Lower chord	−22,804.11	−21,297.89	−13,871.3
Bending Moment (kN·m)	Upper chord	124.67	−164.95	487.35
	Lower chord	−984.09	−87.74	670.42

From Table 7 and Figure 15, the axial force of each key section of the arch rib shows a different trend with the increase of the inclination angle. The axial forces at the upper chord arch top section and the lower chord 2L/8 section decrease with the inclination angle when the inclination angle is less than 10° and then increase. The axial force of the other key sections of the arch rib shows an increasing trend, but the percentage difference is less than 5%, and the change in axial force is relatively small. The bending moment of different sections also shows different trends. The bending moment of the upper chord arch foot section increases with the arch rib inclination angle at first and then decreases. It reached a peak when the inclination angle was 10° with a 30% larger parallel arch. The effects of inclination angle on bending moments at other key sections are not significant.

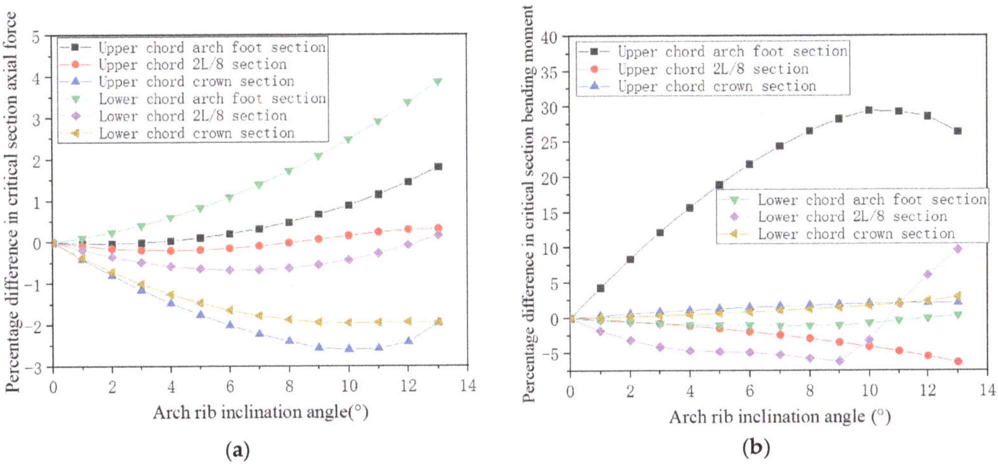

Figure 15. Internal force difference of arch ribs at different arch rib angles with respect to parallel arch. (**a**) Axial force and (**b**) bending moment.

4. Conclusions

Based on the world's largest CFST basket-handle arch bridge, this paper conducted a study on the effect of changes in arch rib inclination angle on the mechanical performance of large CFST basket-handle arch bridges. Different arch rib inclination angles were analyzed in terms of structural natural frequency, linear elastic stability, internal forces, and displacement under static load. The relationship between mechanical performance and arch rib inclination angle was systematically studied, and the following main conclusions were obtained:

(1) According to the analysis of the vibration characteristics, it is recommended that the reasonable range of the rib inclination angle be 8° to 10°, with the optimal angle being 9°. At this angle, not only does the second-order natural frequency of the structure change slightly, but the first, third, and sixth-order natural frequencies are also increased, which can effectively improve the rib's lateral and torsional stiffness.

(2) Based on the analysis of the structural elastic stability coefficient, it is suggested that the reasonable range of the arch rib inclination angle is 8~10°, and the optimal angle is 9°. This not only ensures the best out-of-plane stability of the arch rib, with a stability improvement of 20.2% compared to a parallel arch but also ensures that the decrease in in-plane stability is within 3%.

(3) Based on the analysis of static performance, the variation of rib inclination angle has little effect on rib displacement and internal force. When the rib inclination angle varies within the range of 0° to 13°, the difference ratio of rib displacement is within 5%. The bending moment difference ratio of the upper chord foot section reaches 30% at an inclination angle of 10°, but the bending moment value is relatively small. Considering the relationship between rib displacement and internal force with rib inclination angle, it is suggested that the rib inclination angle be between 9° and 10°.

(4) Based on the above analysis of the structural mechanics performance under different rib inclination angles, it is recommended that the optimal range of inclination angle for the arch rib should be 8–10° for the 300-m-class CFST arch bridge.

In the future, further research will be conducted by considering material nonlinearity, geometric nonlinearity, and material nonlinear coupling effects. Additionally, CFST arch bridges with different span lengths will be investigated as well.

Author Contributions: Z.L. and Y.W., methodology and finite element analysis; C.W. and Y.F., drawing and translation; C.L. and S.W., review and editing. All authors have read and agreed to the published version of the manuscript.

Funding: The support from the Guangxi Key R&D Plan (Guike AB22036007), the National Natural Science Foundation Youth Fund Project (51608080), the Chongqing Natural Science Foundation General Project (CSTC2021jcyj-msxm2491), The Science and Technology Innovation Project of the Chongqing Education Commission for the Construction of the Chengdu Chongqing Double City Economic Circle (KJCXZD2020032), and the Special Key Project for Technological Innovation and Application Development in Chongqing (CSTB2022TIAD-KPX0205) are greatly acknowledged.

Data Availability Statement: Data is available on request due to restrictions.

Conflicts of Interest: The authors declare no conflict of interest.

Abbreviations

Symbol	Illustrate
L	Calculated span
f	Calculate rise-to-span ratio
m	arch-axis coefficient
θ	Arch rib inclination angle
λ	Structural stability safety factor
N	axial force
M	bending moment

References

1. Chen, B.C.; Wei, J.G.; Zhou, J. Application status and Prospect of concrete filled steel tubular arch bridges in China. *J. Civ. Eng.* **2017**, *50*, 12. (In Chinese) [CrossRef]
2. Hu, C.F.; Huang, Y.M. In-plane nonlinear elastic stability of pin-ended parabolic multi-span continuous arches. *Eng. Struct.* **2019**, *190*, 435–446. [CrossRef]
3. Chen, B.C.; Liu, J.P. Summary of World Arch Bridge Construction and Technical Development. *J. Transp. Eng.* **2020**, *20*, 27–41. (In Chinese) [CrossRef]
4. Zheng, J.; Wang, J. Concrete-Filled Steel Tube Arch Bridges in China. *Engineering* **2017**, *4*, 143–155. [CrossRef]
5. Chen, B.C.; Wang, T.L. Overview of Concrete Filled Steel Tube Arch Bridges in China. *Pract. Period. Struct. Des. Constr.* **2009**, *14*, 70–80. [CrossRef]
6. Rajeev, S.; John, P.D.; Varkey, M.V. Study of concrete filled steel tubular arch bridge: A review. *Appl. Mech. Mater.* **2016**, *857*, 261–266. [CrossRef]
7. Hu, J.H. Analysis of Basic Dynamic Characteristics of a Steel Tube Concrete Basket Arch Bridge. *J. Nanchang Univ. (Eng. Ed.)* **2019**, *41*, 144–148+156. (In Chinese)
8. Zhang, Q.M.; Zhou, G. Research on the influence of the lateral inclination angle of the arch ribs on the stability of large-span basket arches. *Bridge Constr.* **2007**, *37*, 32–34. (In Chinese)
9. Cao, Z.Z.; Feng, Y.T.; Shen, F.J. Spatial Stability Analysis of Long Span Steel Box Basket Arch Bridge. *Bridge Constr.* **2011**, *205*, 43–47. (In Chinese)
10. Bonopera, M.; Chang, K.C.; Chen, C.C.; Lin, T.K.; Tullini, N. Bending tests for the structural safety assessment of space truss members. *Int. J. Space Struct.* **2018**, *33*, 026635111880412. [CrossRef]
11. Maes, K.; Peeters, J.; Reynders, E.; Lombaert, G.; De Roeck, G. Identification of axial forces in beam members by local vibration measurements. *J. Sound Vib.* **2013**, *332*, 5417–5432. [CrossRef]
12. Yang, Y. Design and Research on Long Span Concrete Filled Steel Tubular Arch Bridges for Heavy Haul Railways. *Railw. Stand. Des.* **2018**, *62*, 107–111+186. (In Chinese) [CrossRef]
13. Yun, D.; Liu, H.; Zhang, S.M. Natural vibration characteristics and stability of large-span through steel tube concrete arch bridges. *J. Jilin Univ. (Eng. Ed.)* **2013**, *43*, 86–91. (In Chinese) [CrossRef]
14. Wang, L.; Peng, R.H.; Guo, X.R. The effect of arch rib inclination angle on the dynamic response of steel tube concrete basket arch bridges. *J. Railw. Sci. Eng.* **2012**, *9*, 25–29. (In Chinese) [CrossRef]
15. Wei, J.G.; Chen, J.W.; Xie, Z.T. Construction and analysis of a steel tube concrete dumbbell shaped cross-section basket shaped standard arch bridge. *J. Fuzhou Univ. (Nat. Sci. Ed.)* **2019**, *47*, 663–668. (In Chinese) [CrossRef]
16. Zeng, D.R.; Zhang, Q.M. The influence of the inclination angle of the arch rib of a basket arch bridge on lateral stability. *J. Chongqing Jiaotong Univ.* **2006**, *25*, 4–8. (In Chinese) [CrossRef]
17. Zhao, Y.Y.; Lao, W.Q.; Feng, R. Analysis of the influence of internal inclination angle on the mechanical properties of steel tube concrete basket arches. *Highw. Transp. Technol.* **2007**, *24*, 56–58+89. (In Chinese)

18. Huang, P.M.; Ren, X.; Li, W.J. The influence of arch rib inclination angle on the static and dynamic behavior of steel tube concrete arch bridges. *J. Chang. Univ. (Nat. Sci. Ed.)* **2009**, *29*, 51–55. (In Chinese)
19. Xu, H.; Wu, Y.T.; Xie, W.W. Research on Reasonable Arch Rib Inclination of CFST Arch Bridges for 400 m Railway. *J. Railw. Sci. Eng.* **2021**, *18*, 1203–1212. (In Chinese)
20. Wang, D.; Yao, X.L.; Yan, W.F. Analysis of the Influence of the Optimal Inclination Angle of the Arch Ribs of a Through Type Basket Arch Bridge. *Transp. Sci. Eng.* **2022**, *38*, 60–64+71. (In Chinese) [CrossRef]
21. Ji, R.C.; Shi, M.X. Stability Analysis of Long Span Railway Steel Pipe Concrete Tied Arch Bridges. *Vib. Impact* **2011**, *30*, 87–91. (In Chinese) [CrossRef]
22. Pan, S.S.; Zhang, Z.; Huang, C.L. Research on the influence of the inclination angle of the X-shaped double rib arch on lateral stability. *J. Dalian Univ. Technol.* **2008**, *06*, 845–850. (In Chinese) [CrossRef]
23. Li, S.; Xin, J.; Jiang, Y.; Wang, C.; Zhou, J.; Yang, X. Temperature-induced deflection separation based on bridge deflection data using the TVFEMD-PE-KLD method. *J. Civ. Struct. Health Monit.* **2023**, *13*, 781–797. [CrossRef]
24. Liu, Z.; Zhou, S.; Zou, K.; Qu, Y. A numerical analysis of buckle cable force of concrete arch bridge based on stress balance method. *Sci. Rep.* **2022**, *12*, 12451. [CrossRef]
25. Wang, Q.; Wang, C.S.; Yu, X. Local Stability Test and Numerical Analysis of Steel Bridge Towers. *J. Chang. Univ. (Nat. Sci. Ed.)* **2008**, *127*, 67–72. (In Chinese)
26. Xin, J.; Jiang, Y.; Zhou, J.; Peng, L.; Liu, S.; Tang, Q. Bridge deformation prediction based on SHM data using improved VMD and conditional KDE. *Eng. Struct.* **2022**, *261*, 114285. [CrossRef]
27. Xin, J.; Zhou, C.; Jiang, Y.; Tang, Q.; Yang, X.; Zhou, J. A signal recovery method for bridge monitoring system using TVFEMD and encoder-decoder aided LSTM. *Measurement* **2023**, *214*, 112797. [CrossRef]
28. Geng, Y.; Ranzi, G.; Wang, Y.T. Out-of-plane creep buckling analysis on slender concrete-filled steel tubular arches. *J. Constr. Steel Res.* **2018**, *140*, 174–190. [CrossRef]
29. Liu, C.Y.; Hu, Q.; Wang, Y.Y. In-plane stability of concrete-filled steel tubular parabolic truss arches. *Int. J. Steel Struct.* **2018**, *18*, 1306–1317. [CrossRef]

buildings

MDPI

Article

Working Stress Measurement of Prestressed Rebars Using the Magnetic Resonance Method

Junfeng Xia [1,2], Senhua Zhang [1,2,*], Leng Liao [1,3], Huiling Liu [1,2] and Yisheng Sun [4]

1 State Key Laboratory of Mountain Bridge and Tunnel Engineering, Chongqing Jiaotong University, Chongqing 400074, China; jfxia@mails.cqjtu.edu.cn (J.X.); lengliao@cqjtu.edu.cn (L.L.); 631601030132@mails.cqjtu.edu.cn (H.L.)
2 School of Civil Engineering, Chongqing Jiaotong University, Chongqing 400074, China
3 School of Materials Science and Engineering, Chongqing Jiaotong University, Chongqing 400074, China
4 Chongqing Rail Transit Design and Research Institute Co., Ltd., Chongqing 401122, China; 17783093703@163.com
* Correspondence: shzhang@mails.cqjtu.edu.cn

Abstract: Prestressed rebars are usually used to apply vertical prestress to concrete to prevent web cracking. The reduction of working stress will affect the durability of the structure. However, the existing working stress detection methods for prestressed rebars still need to be improved. To monitor the working stress of rebars, a magnetic resonance sensor was introduced to carry out experimental research. The correlation between rebar stress and the sensor's induced voltage was theoretically analyzed using the magnetoelastic effect and magnetic resonance theory. A working stress monitoring method for prestressed rebars based on magnetic resonance was proposed. Working stress monitoring experiments were carried out for 16 mm, 18 mm, and 20 mm diameter rebars. The results showed that the induced voltage peak-to-peak value and the rebar prestress were nonlinearly correlated under different working conditions. Correlations between the characteristic indicators and the rebar working stress were obtained using nonlinear and linear fit. The cubic polynomial segmented fit outperformed the gradient overall linear fit, with the goodness of fit R^2 greater than 0.96. The average relative error values of working stress monitoring were less than 5% under different working conditions. This provides a new method for working stress measurement of vertical prestressed rebars.

Keywords: working stress; rebar; monitoring; magnetoelastic effect; magnetic resonance

Citation: Xia, J.; Zhang, S.; Liao, L.; Liu, H.; Sun, Y. Working Stress Measurement of Prestressed Rebars Using the Magnetic Resonance Method. *Buildings* **2023**, *13*, 1416. https://doi.org/10.3390/buildings13061416

Academic Editor: Roberto Capozucca

Received: 21 April 2023
Revised: 24 May 2023
Accepted: 29 May 2023
Published: 30 May 2023

1. Introduction

The prestressed concrete bridge is widely used in bridge construction because of its advantages of sizeable structural stiffness, smooth driving, and low maintenance cost [1]. Vertical prestressed rebar is used to provide vertical compressive stress to the reinforcement by post-tensioned method. The effect of vertical prestressed rebar can make the shear load capacity of the structure significantly increase by 95% [2]. However, the elongation of vertical prestressed rebar is slight during vertical prestressing tensioning in construction. Therefore, the prestress loss caused by rebar retraction is significant [3]. Furthermore, the loss of vertical prestress has an important influence on the principal tensile stress of the box girder web [4]. Once the vertical prestress is lost and the web cracks, the bridge structure's safety and durability will be affected [5–7]. Therefore, the vertical prestressed rebar working stress must be accurately monitored to ensure the structure's safety.

To avoid prestress detection affecting the structure's durability, nondestructive testing methods are usually used [8]. Commonly used methods are the strain method, electromagnetic resonance method, the stiffness method, the ultrasonic guided wave method, the eddy current method, and the magnetoelastic method. The strain method is based on the stress–strain relationship. The test is carried out by pasting electronic strain gauges or embedded sensors, and then converting the stress. Sawicki [9] successfully identified the stress of rebars by using strain gauges installed on rebars and distributed optical fiber sensors for strain detection. However, this method is susceptible to temperature and has low durability. The electromagnetic resonance method connects the steel strand analog inductor to the oscillation circuit and calculates the stress by measuring the oscillation frequency of the circuit with a frequency meter. Cui [10] measured the stress of prestressed concrete beams by the electromagnetic resonance method and obtained the functional relationship between resonance frequency increment and stress increment. However, rebars cannot simulate the inductance due to their different configurations, so this method is unsuitable for rebar stress monitoring. The stiffness method measures the frequency of the anchorage zone of the exposed section of the rebar and then infers the magnitude of the prestress [11]. Zhong has conducted much research on the measurement of rebar tension by the stiffness method and achieved specific results [12,13]. However, this method is susceptible to boundary conditions and is only suitable for stress measurement during construction [14]. The ultrasonic guided wave method measures stress by the acoustoelastic effect. Chen [15] used single-source high-frequency cylindrical guided waves to improve the accuracy of the ultrasonic guided wave method. The ultrasonic guided wave method has a rapid energy attenuation rate due to the bonding between rebar and concrete, so the method's reliability needs to be improved. It can be seen that the existing methods are not suitable for rebar stress monitoring, or the detection accuracy needs to be further improved. Therefore, the monitoring of working stress of rebar still needs further study.

The magnetoelastic effect indicates that the magnetism of ferromagnetic material changes with its stress. Based on the magnetoelastic effect, scholars have conducted much research and proposed the eddy current and magnetoelastic methods. The eddy current method realizes stress monitoring through the relationship between sensor impedance and stress [16]. To improve the eddy current sensor's performance, Xiu [17] designed a sleeve structure to reduce the loss of magnetic field and provide a higher permeability path. Alonso [18] used an eddy current sleeve structure and phase shift measurements to detect the stress in iron-based materials. Liang [19] found that the magnetoelastic method is more suitable for stress monitoring than the eddy current method.

The magnetoelastic method uses the magnetoelastic effect to monitor the change of magnetization intensity to obtain the magnitude of stress. Due to its advantages of noncontact, high sensitivity, and robustness, the method is considered a promising nondestructive stress monitoring method [20]. According to their different structural forms, magnetoelastic sensors can be divided into U-type sensors, permanent magnet magnetization sensors, and sleeve sensors. Joh [21] designed a U-shaped sensor to measure the magnitude of prestress. Deng [22] used the static magnetization of the permanent magnet to replace the magnetization of the coil. However, the U-shaped sensor and the permanent magnet magnetization sensor are unsuitable for monitoring vertical prestress due to the irregular excitation structure and large size. The sleeve magnetoelastic sensor uses the coil as the excitation element and the sensing element. The monitoring object is used as the coil core. This method has the advantages of a clear magnetic circuit and less magnetic field leakage [23]. To optimize the sleeve sensor, Duan [24] proposed an intelligent elastomagnetic (EME) sensor by replacing the secondary coil with a laminated composite magnetic sensor. Due to its large size and high functional requirements, the sensor is mainly used for cable force detection. Zhang [25] simplified the primary coil and induction unit of the EM sensor into self-inductive coils, then proposed a magnetoelastic inductance method using weak magnetic excitation. The magnetoelastic inductance method has the advantages of reducing the sensor size and reducing the power supply demand. However, the low sensitivity of this method affects the accuracy of stress monitoring. Therefore, the sensor needs to be optimized.

Kurs [26] first proposed the magnetic resonance theory, which improves the transmission efficiency of the coil-based energy transmission system. To improve the working performance of the sensor, magnetic resonance theory is introduced into the sensor field. Hughes [27] studied the enhancement effect of resonant coupling on eddy current sensors and improved the sensitivity of corrosion damage detection. The magnetoelastic sensor is also composed of the coil as the main component. To improve the sensor's sensitivity, Zhang [28] introduced the magnetic resonance theory into the magnetoelastic effect method. He proposed the resonance enhanced magnetoelastic method (REME) and verified the feasibility of this method for monitoring the stress of steel strands. However, as a hot-rolled low-carbon steel structure, the rebar's section form, initial magnetization state, and stress–strain relationship differ from those of the steel strand, resulting in different stress identification. In addition, the advantage of small size of the magnetic resonance sensor meets the pre-embedded requirement of vertical prestressed rebar and can be applied in post-tensioned pipeline [29]. Therefore, monitoring the working stress of the rebar by REME needs further study.

Based on the existing research, this paper combined the magnetoelastic effect, electromagnetic induction law, and magnetic resonance effect. A working stress monitoring method for vertical prestressed rebar was proposed using the magnetic resonance sensor. Firstly, the relationship between sensor induced voltage and rebar stress was analyzed. Then, working stress monitoring experiments under different working conditions were carried out on rebars with different diameters. According to the experiment results, the nonlinear relationship between the induced voltage peak-to-peak values and the prestress was analyzed. Based on the experiment data, the correlation between the characteristic indicator and the rebar working stress was obtained by nonlinear fit and linear fit. According to the relationship, the working stress was accurately evaluated, and the feasibility of the proposed method was verified.

2. Theory

According to the Joule effect and the magnetization theory of ferromagnetic material, there is a functional relationship between the stress of rebar and the change in magnetic permeability [30,31]. In Equation (1), μ is the permeability of rebar, μ_0 is the vacuum permeability, λ_s is the axial deformation constant, M_s is the saturation magnetization, K_u is the uniaxial magnetic anisotropy constant, H_R is the excitation magnetic field, and θ_0 is the angle between the magnetic field and the easy magnetization axis [32].

$$\sigma = E\frac{3\lambda_s M_s}{2K_u}(\mu - \mu_0)H_R \sin^2\theta_0 \cos\theta_0 \tag{1}$$

A magnetic resonance sensor [28] was used to monitor the working stress of the rebar. The sensor's two coils are the excitation and induction coils. The coil is wound on the PVC skeleton, as shown in Figure 1. The equivalent circuit diagram [28] of the magnetic resonance sensor is shown in Figure 1. L_T and L_R are the inductance of the excitation coil and induction coil, respectively. C_T and C_R are the excitation and induction coil's compensation capacitors, respectively. u_{CT} and u_{CR} are the voltage of the compensation capacitor of the excitation coil and the induction coil. R_T and R_R are the internal resistance of the excitation coil and the induction coil, respectively. The voltage source is AC power, and the input voltage is u_{in}. The millivoltmeter is regarded as a load connected in series with an induction coil, and its equivalent resistance is R_L.

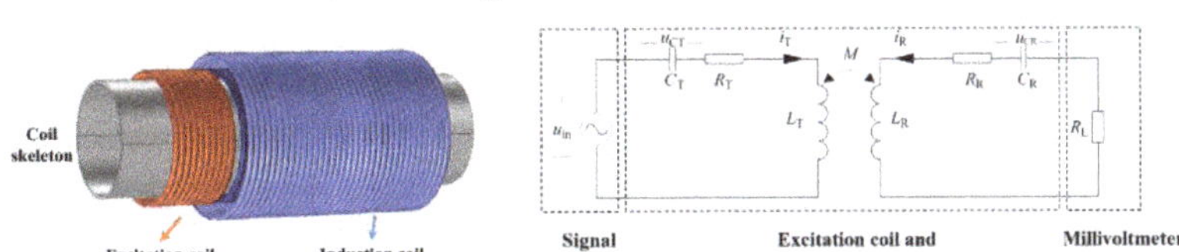

Figure 1. The magnetic resonance sensor and equivalent circuit diagram schematic.

According to Kirchhoff's voltage law [33], the self-impedance of the excitation coil and the induction coil is Z_T and Z_R, respectively, as shown in Equations (2) and (3). The loop current I_R of the induction coil is shown in Equation (4), where j is the imaginary part of the complex number, U_{in} is the effective value of u_{in}, ω is the angular frequency of u_{in}, and M is the mutual inductance between the excitation coil and the induction coil. According to the coupled mode equation of LC coupled circuit [34], the relationship between coupling coefficient κ and mutual inductance M can be expressed as Equation (5); ω_0 is the resonant frequency.

$$Z_T = R_T + j\omega L_T + \frac{1}{j\omega C_T} \quad (2)$$

$$Z_R = R_R + R_L + j\omega L_R + \frac{1}{j\omega C_R} \quad (3)$$

$$\dot{I}_R = \frac{-j\omega M\dot{U}_{in}}{Z_T Z_R + (\omega M)^2} \quad (4)$$

$$M = \frac{2\kappa\sqrt{L_R L_T}}{\omega_0} \quad (5)$$

A rebar with a cross-sectional area of A_{iron} is placed in a magnetic resonance sensor. A_{air} is the cross-sectional area of the nonmagnetic material between the coil and the rebar. The voltage source provides alternating current for the excitation coil. Under the action of alternating current, the excitation coil generates an excitation magnetic field [35,36]. The excitation coil and the induction coil are resonantly coupled. An excitation magnetic field of the magnetized rebar is generated in the induction coil. The magnetic field is expressed as H_R, which has a functional relationship with the coupling coefficient κ, as shown in Equation (6). N_R is the number of turns of the induction coil. l_R is the effective magnetic circuit length of the induction coil. According to electromagnetic induction law, the induced voltage of the induction coil can be obtained by the magnetic flux in the area around the coil [37], as shown in Equation (7), where Φ is the magnetic flux around the area of the induction coil, and t is the time.

$$H_R = \frac{N_R \dot{I}_R}{l_R} = \frac{-j\omega N_R 2\kappa\sqrt{L_1 L_2}\dot{U}_{in}}{\left[Z_T Z_R + \left(\frac{2\omega\kappa\sqrt{L_1 L_2}}{\omega_0}\right)^2\right] l_R \omega_0} \quad (6)$$

$$u_{CR} = N_R \frac{d\Phi}{dt} = N_R \frac{d(\mu H_R A_{iron} + \mu_0 H_R A_{air})}{dt} \quad (7)$$

Combined with the electric power calculation formula, the excitation coil's input power P_{in} and the millivoltmeter's output power P_o as the load can be calculated, respectively. The results are shown in Equations (8) and (9). The transmission efficiency can be obtained as shown in Equation (10). $X_T = \omega L_T - 1/\omega C_T$, $X_T = \omega L_R - 1/\omega C_R$.

$$P_{in} = \frac{U_{in}^2}{R_R} = \frac{\left\{R_T\left[(R_R + R_L)^2 + X_R{}^2\right] + \omega^2 M^2 (R_R + R_L)\right\} U_{in}{}^2}{[R_T(R_R + R_L) - X_T X_R + \omega^2 M^2] + [R_T X_R + (R_R + R_L) X_T]^2} \quad (8)$$

$$P_o = I_R{}^2 R_L = \frac{\omega^2 M^2 R_L U_{in}{}^2}{[R_T(R_R + R_L) - X_T X_R + \omega^2 M^2] + [R_T X_R + (R_R + R_L) X_T]^2} \quad (9)$$

$$\begin{aligned} \eta &= \frac{P_o}{P_{in}} \times 100\% \\ &= \frac{\omega^2 M^2 R_L}{R_T\left[(R_R + R_L)^2 + X_R{}^2\right] + \omega^2 M^2 (R_R + R_L)} \times 100\% \end{aligned} \quad (10)$$

When the induction coil resonates, $X_R = 0$, the transmission efficiency reaches the maximum, and the measured induced voltage is the highest. In the working stress monitoring experiment, the rebar is used as the core of the induction coil. The change of permeability of rebar caused by working stress also causes the induction coil's inductance change. After the inductance changes, the resonant frequency of the induction coil changes, as shown in Equation (11). When the resonant frequency of the induction coil deviates from the initial resonant frequency, the coil coupling coefficient κ and the sensor induced voltage are significantly reduced, thereby improving the sensitivity of the rebar working stress monitoring.

$$\omega_0 = \frac{1}{2\pi\sqrt{L_R C_R}} = \frac{1}{2\pi\sqrt{C_R}} \frac{1}{\sqrt{(\mu A_{iron} + \mu_0 A_{air})\frac{N_R{}^2}{l_R}}} \tag{11}$$

The above relationship is solved simultaneously to explore the internal relationship among stress, magnetism, and electricity. The change in stress will lead to the change of permeability of the rebar. The relationship between induced voltage and permeability can be simplified from Equations (7)–(12), where $f(u_{CR})$ is the function of induced voltage u_{CR} representing permeability μ. The induced voltage u_{CR} is related to the coupling coefficient κ. For a specific rebar and sensor, the relationship between the sensor's induced voltage and the rebar's working stress can be expressed as Equation (13); $g(u_{CR})$ is the function of the induced voltage u_{CR} representing the working stress σ.

$$\mu = \frac{\int u_{CR} dt - N_R H_R \mu_0 A_{air}}{N_R H_R A_{iron}} = f(u_{CR}) \tag{12}$$

$$\sigma = h[f(u_{CR}) - \mu_0] = g(u_{CR}) \tag{13}$$

Through the above derivation, it can be found that the rebar working stress is related to the sensor's induced voltage. Therefore, the induced voltage of the magnetic resonance sensor can be used to evaluate the working stress of the rebar. To verify the feasibility of the magnetic resonance monitoring method (REME) for rebar working stress monitoring, rebar working stress monitoring experiments were carried out.

3. Experiment Design

Vertical prestressed tendons generally use rebar. To explore the relationship between rebar stress and sensor induced voltage, this paper uses the magnetic resonance sensor to carry out working stress monitoring experiments on rebar under different working conditions.

3.1. Experiment Equipment

A rebar working stress monitoring system was built to carry out the experiment, as shown in Figure 2. The experiment system comprised a universal testing machine, magnetic resonance sensor, signal generator, power amplifier, millivoltmeter, and computer. The maximum tensioning load of the universal testing machine is 100 tons. The universal testing machine was used to tension the rebar to different stress levels. In this experiment, a magnetic resonance sensor was used for working stress monitoring. The signal type for data analysis was induced voltage. The induced voltage peak-to-peak value was chosen as the electrical characteristic value characterizing the variation of the magnetic properties of the rebar with stress. The signal generator was connected to the excitation coil. The millivoltmeter was considered as a resistor connected to the induction coil. The signal generator generated an alternating excitation signal as a sine wave. The power amplifier was used to amplify the excitation signal power. The effective value of the induced voltage was measured by the millivoltmeter during the experiment. The value was transmitted and saved in the computer for further processing.

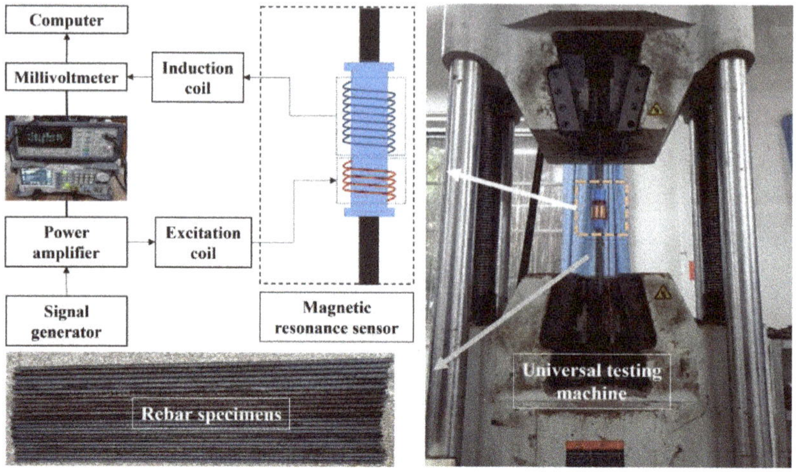

Figure 2. Rebar working stress monitoring system.

3.2. Sensor and Specimen Preparation

A PVC tube with an outer diameter of 40 mm was used as the magnetic resonance sensor skeleton. The excitation and induction coil were wound with 0.25 mm-diameter enameled wire. The total number of turns of the excitation coil was 40 turns, and 1 layer was wound. The total number of turns of the induction coil was 1400 turns, and 10 layers were wound. The yield strength of HRB400 rebar is 400 MPa. HRB400 rebar is widely used in engineering projects. Rebar is a common ferromagnetic material with magnetoelastic effect [38,39]. The length of the sensor was 80 mm. The specimens were made of rebars with the yield strength of 400 MPa. To verify the applicability of the working stress monitoring method to different diameters of rebars, the specimens' diameters were made of 16 mm, 18 mm, and 20 mm. In actual engineering, the working stress of the rebar is lower than the yield strength. Therefore, the maximum design stresses are 50%, 70%, and 90% of the yield strength, respectively. There were 6 specimens of each diameter and a total of 18 specimens. The specimens were divided into three groups according to their diameter. Each rebar diameter yielded at 89 kN, 109 kN, and 155 kN in tension, respectively. The specimens were numbered as shown in Table 1, with D being the diameter of the rebar and P being the maximum stress-to-yield strength ratio. To ensure the reproducibility of the experiment results, two specimens with the same stress conditions were set up, numbered T1 and T2.

Table 1. Specimen number and loading procedure.

Group	Rebar Diameter (mm)	Excitation Frequency (kHz)	Excitation Voltage (V)	Maximum Stress (Yield Strength Ratio) (%)	Specimen Number
1	16	32.97 ± 0.8	6.40 ± 0.4	50 70 90	D16-P50-T1/T2 D16-P70-T1/T2 D16-P90-T1/T2
2	18	32.60 ± 0.9	7.12 ± 0.4	50 70 90	D18-P50-T1/T2 D18-P70-T1/T2 D18-P90-T1/T2
3	20	32.56 ± 0.9	7.89 ± 0.3	50 70 90	D20-P50-T1/T2 D20-P70-T1/T2 D20-P90-T1/T2

3.3. Loading Procedure

In the experiment, the magnetic resonance sensor was fixed in the middle of the rebar to avoid the magnetic field's influence at the rebar's end. The universal testing machine stretched the specimens. Loading and unloading were carried out with 3 kN as the starting and ending points to avoid instrument errors. The working stress of rebar does not reach its yield strength. Therefore, to ensure that no plastic deformation of the rebar occurs, the maximum stress levels were designed to be 50%, 70%, and 90% of the yield strength, and the step size was 10% of the yield strength. In practical engineering, the prestressed rebar will be initially tensioned to reduce the prestress loss. Therefore, the experiment was conducted with pretreatment of the rebars to simulate the initial tensioning during the construction phase. Then, the specimens were loaded and unloaded using the universal testing machine. The loading stage simulated the prestress application during the construction phase. The unloading stage was used to simulate the working stress during the operation phase. The loading and unloading speeds were both 0.2 kN/s.

During the experiment, the excitation coil was excited with the initial resonant frequency of the induction coil (with rebar inside). Then, the induction coil resonated with the excitation coil. The excitation frequency and excitation voltage of each specimen are shown in Table 1. It can be seen that different diameters of rebars' excitation frequency and excitation voltage had some differences, but those of the same diameter were more stable. To ensure adequate deformation of the rebar and stability of the loading and test systems, the load was held for 30 s after each loading to the specified tension (each tension level). After the induced voltage was stabilized, the induced voltage peak-to-peak value of the induction coil was measured. The peak-to-peak induction voltage (*Vpp*) was repeated seven times, and the average value was taken to reduce the measurement error.

4. Experimental Results and Discussion

To study the relationship between the induced voltage peak-to-peak value and working stress, the loading and unloading experiment results with the maximum design stress of rebars with diameters of 16 mm, 18 mm, and 20 mm being 50%, 70%, and 90% of yield strength, respectively, were analyzed.

4.1. The Evolution Law of Induced Voltage with Working Stress

Due to the different diameters of the specimens, the tensile force during the data analysis was converted into stress to facilitate the control variable. To compare different groups of rebars with the same stress level, the stress to yield strength ratio was taken as the abscissa and expressed by T_p. Considering the different initial magnetization states of different specimens, their initial induced voltage peak-to-peak values after pretreatment were different. Therefore, the starting point of each group of data was excluded from the initial value, and the increment of induced voltage peak-to-peak value (ΔVpp) was used as an indicator. As shown in Figure 3, ΔVpp and rebar working stress are nonlinearly correlated. According to the magnetoelastic effect, the magnetization strength of the rebar changes when the stress changes. During the elastic stage, the force-induced magnetization is theoretically reversible. Therefore, during the unloading stage, the elastic strain recovery makes the reversible magnetization intensity recover. However, as shown in Figure 3, the induced voltage peaks did not fully recover when the rebar was unloaded to its starting value. This was because the plastic deformation generated by the rebar fabrication was not completely eliminated in the pretreatment stage. The magnetic domain structure was irreversibly rotated during loading, which resulted in irreversible magnetization.

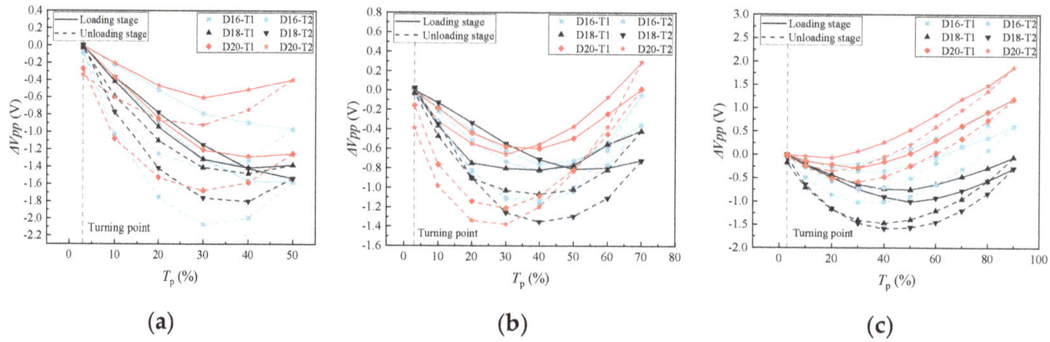

Figure 3. The ΔVpp for each stress level measured during loading and unloading: (**a**) design stress level of P50; (**b**) design stress level of P70; (**c**) design stress level of P90.

The corresponding ΔVpp-T_p curves were similar for each specimen. Therefore, the design stress range of 90% yield strength in each group of specimens was selected for further analysis. As shown in Figure 4, the relationship of the ΔVpp-σ was similar for the same stressing process for different rebars. During the loading stage, the ΔVpp decreased and then increased with the increase of working stress. During the unloading stage, the ΔVpp decreased and then increased with the decrease of working stress. The corresponding ΔVpp-σ curves in the loading and unloading stages were different. The same stress level in loading and unloading corresponded to two different ΔVpp. This was due to the hysteresis of the rebar as a ferromagnetic material after loading and unloading [40].

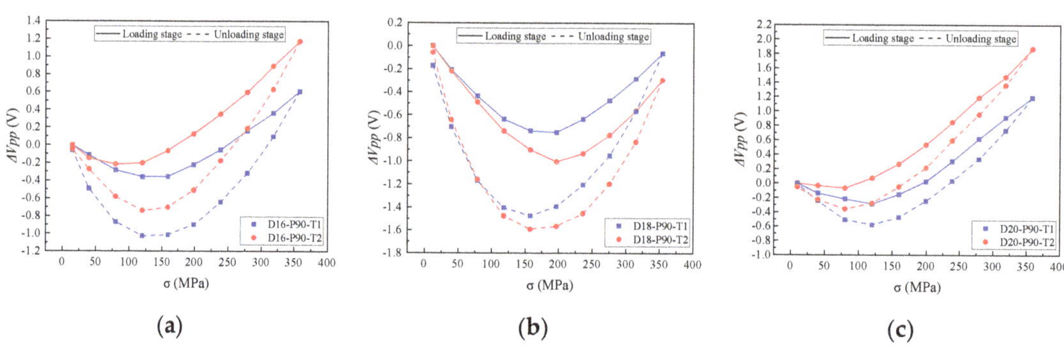

Figure 4. The ΔVpp corresponding to the design stress condition of P90 in the loading and unloading stages: (**a**) Group 1; (**b**) Group 2; (**c**) Group 3.

The working stress loss stage corresponded to the unloading stage. Further analysis of the unloading stage was performed. The maximum stress level was taken as the starting point for comparison purposes. The starting point of each specimen was removed from the initial value. The increment of induced voltage peak-to-peak (ΔVpp) was used to characterize the working stress of the rebar, as shown in Figure 5.

There was a similar relationship between the ΔVpp and working stress for rebars with different diameters. For specimens with the same diameter, due to the different composition and processing technology of different rebars, the force-induced magnetization law of rebars was different. Therefore, the reversible magnetization of each specimen was different, which made the ΔVpp of different rebars different under the same working stress level. However, under different working conditions, the ΔVpp-σ curve was similar. In the unloading stage, the ΔVpp decreased first and then increased with the decrease of working stress. From the perspective of magnetic domain theory, it can be seen that the working stress had a more substantial influence on magnetization than the excitation magnetic field at a greater working stress level. Therefore, magnetization would increase with the increase of stress at a greater stress level [41].

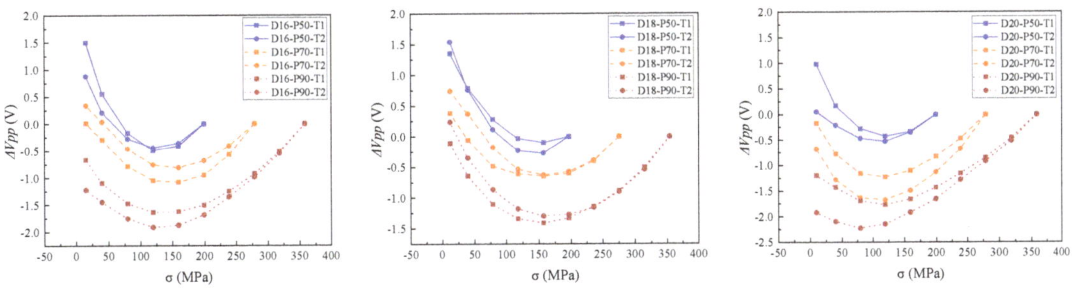

(a) (b) (c)

Figure 5. Comparison of the same stress level of the ΔVpp under three design stress conditions for each group of specimens: (**a**) Group 1; (**b**) Group 2; (**c**) Group 3.

Comparing Figure 5a–c, it can be seen that under the same working stress level, the greater the design stress of different rebars in the same group, the lower the ΔVpp corresponding to the specimen. This was because when the design stress increased, the elastic strain generated by the rebar during the prestressing process increased, reducing the rebar's effective area. In addition, the more extensive range increased the magnetization range of the rebar. Therefore, in the unloading stage, the rebar simulated the working stress loss; when the working stress was lost to the same stress level, the ΔVpp measured by the specimen with high design stress was less. For each group of specimens, the turning point of the ΔVpp-σ curve was different, but it was concentrated at 135 ± 25 MPa. For the same group, the distribution of turning points was more concentrated. For example, the turning point of Group 2 was 157.19 MPa. In the same design stress of the same group, except for D20-P90-T1 and D20-P90-T2, the turning point of the ΔVpp-σ curve of other repetitive tests was the same stress level.

From the above analysis, it can be seen that the induced voltage peak-to-peak value was nonlinearly related to working stress. Therefore, to evaluate the working stress of prestressed rebar using the induced voltage peak-to-peak value, the mapping relationships between characteristic indicators and working stress were established by nonlinear fit and linear fit.

4.2. Characteristic Indicators for the Evaluation of Working Stress

4.2.1. Relationship between Working Stress and the ΔVpp

Due to the measurement under different working conditions, the changing trend between the stress of prestressed rebar and the ΔVpp was basically the same. Therefore, a representative ΔVpp-σ curve was selected from three diameters for further analysis. Because the design stress of 90% yield strength included the stress process of 50% and 70% yield strength design conditions, this paper selected specimens D16-P90-T1, D18-P90-T1, and D20-P90-T1 for discussion. In working stress monitoring, the working stress was unknown and needed to be evaluated based on the measured ΔVpp. Therefore, the ΔVpp was used as the abscissa and the stress converted by tension was used as the ordinate, which was recorded as Method 1. The σ-ΔVpp curves of D16-P90-T1, D18-P90-T1, and D20-P90-T1 are shown in Figure 6.

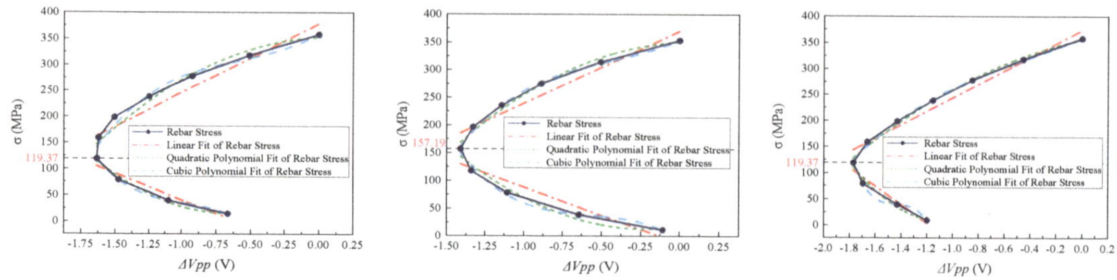

(a) (b) (c)

Figure 6. The ΔVpp-σ curves and fitted curves of three specimens: (**a**) D16-P90-T1; (**b**) D18-P90-T1; (**c**) D20-P90-T1.

As shown in Figure 6, during the unloading stage, the ΔVpp decreased first and then increased with the decrease of rebar working stress. Therefore, when the increase of the ΔVpp was observed, it could be considered that the working stress of the rebar had dropped to a low stress level relative to the design prestress. However, all three specimens had a ΔVpp corresponding to two different rebar prestress levels, and the mapping relationship between the ΔVpp and working stress could not be established. Therefore, the corresponding relationship between working stress and the ΔVpp variation under each stress level was discussed in sections.

The whole unloading stage bounded by the turning point can be divided into two sections: the high stress section and low stress section. Since the importance of the two sections was the same, it was necessary to evaluate the fit effect as a whole. The Taylor expansion of Equation (13) was carried out, the higher order term after the third order was ignored, and Equation (14) was obtained. The first, second, and third orders of the corresponding relationship between the ΔVpp and rebar working stress were discussed separately, as shown in Equation (15). The turning points of the corresponding curves of each specimen in Figure 6 were 119.37 MPa, 157.19 MPa, and 119.37 MPa, respectively. Taking the turning point as the dividing line, the three specimens were fitted to obtain the corresponding linear, quadratic, and cubic fit curves. Therefore, the goodness of fit (R^2) was used to evaluate the fit effect. The R^2 of each specimen was calculated based on two segmented data.

$$\sigma \approx g(0) + g'(0)(u) + \frac{g''(0)}{2!}(u)^2 + \frac{g'''(0)}{3!}(u)^3 \tag{14}$$

$$\begin{aligned} &\sigma \approx a(u) + b \\ &\sigma \approx a(u)^2 + b(u) + c \\ &\sigma \approx a(u)^3 + b(u)^2 + c(u) + d \end{aligned} \tag{15}$$

The R^2 is shown in Figure 7, demonstrating that as the order of fit increased, the R^2 approached one. The R^2 of cubic polynomial fit was higher than that of quadratic polynomial fit and linear fit, and its R^2 reached 0.98 on average. The cubic polynomial R^2 of the specimen D20-P90-T2 was as high as 0.99781, which was close to 1. In addition, when the order increased from three to four, there was little room for improvement in the R^2. Considered comprehensively, the cubic polynomial was selected for piecewise fit to explore the correlation between working stress and the ΔVpp. To verify the feasibility of using cubic polynomial fit to determine the correlation between ΔVpp and working stress, the ΔVpp data of each specimen were fitted by cubic polynomial, and the R^2 was shown as follows.

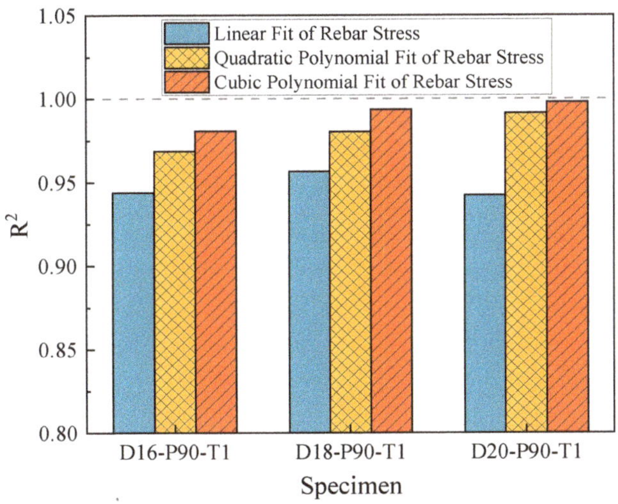

Figure 7. The goodness of fit R^2 of different fit methods of three specimens.

As shown in Figure 8, the R^2 of D18-P50-T2 was at least 0.96928. The R^2 of each specimen was more significant than 0.96, indicating a high degree of compliance with the cubic polynomial fit of the line between σ-ΔVpp. Therefore, the working stress of the rebar could be determined from the σ-ΔVpp curve.

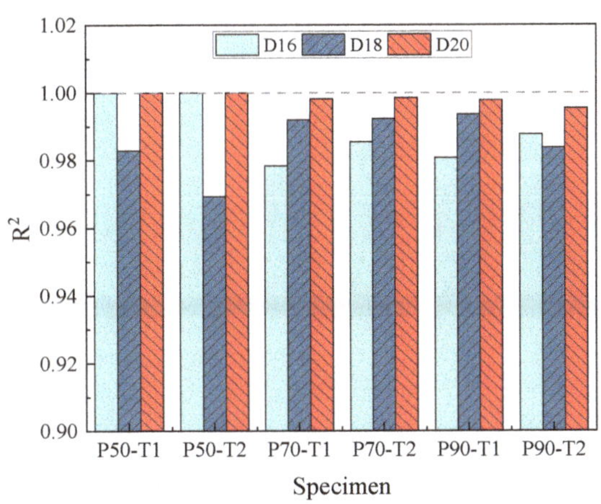

Figure 8. The cubic polynomial R^2 of the relationship between the ΔVpp and working stress change under different working conditions.

4.2.2. Relationship between Working Stress and dΔVpp

As shown in Figure 5, the gradient of the σ-ΔVpp curve ($d\Delta Vpp$) decreased continuously during the unloading stage. The curves of ΔVpp and working stress for different diameters had similarities. When the working stress decreased gradually, one ΔVpp corresponded to two different stress levels of the rebar. Therefore, in the data analysis, $d\Delta Vpp$ could be chosen as the fit variable to characterize the variation of the magnetic properties of prestressed rebar with working stress, which was recorded as Method 2. D16-P90-T2, D18-P90-T2, and D20-P90-T2 were used as examples.

As shown in Figure 9, the working stress of rebar could be uniquely determined by the $d\Delta Vpp$. In the unloading stage, the trend between the $d\Delta Vpp$ and the working stress was basically the same. With the decrease of working stress, the $d\Delta Vpp$ decreased gradually and was linearly correlated. Therefore, to clarify the relationship between the two variables, a linear fit was made between the $d\Delta Vpp$ and working stress of the three specimens. The goodness of fit (R^2) was used to indicate the linear fit of the specimens. The R^2 corresponding to D16-P90-T2, D18-P90-T2, and D20-P90-T2 was 0.97112, 0.97041, and 0.91294, respectively. Therefore, it was preliminarily shown that there was a good linear relationship between the working stress and the $d\Delta Vpp$ curve. The R^2 of all specimens was calculated, and the results are shown in Figure 10.

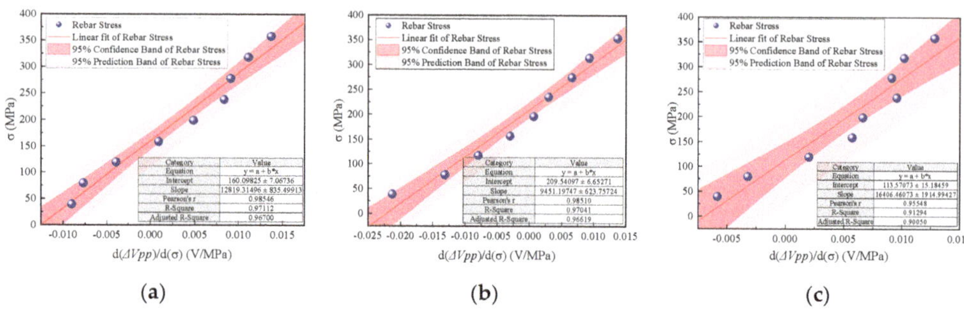

Figure 9. The overall linear fitting of the relationship between working stress and $d\Delta Vpp$ of three diameter rebars: (**a**) D16-P90-T2; (**b**) D18-P90-T2; (**c**) D20-P90-T2.

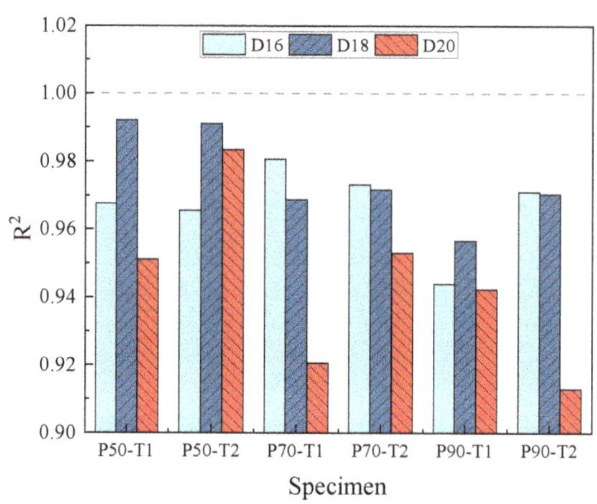

Figure 10. The overall linear fit R^2 of the relationship between working stress and the $d\Delta Vpp$ of each group of specimens.

It can be seen from Figure 10 that the R^2 of each specimen was more significant than 0.9, indicating an excellent linear fit between $d\Delta Vpp$-σ. Therefore, the working stress of rebar could be determined by the linear relationship of $d\Delta Vpp$-σ. Among them, the minimum R^2 was 0.91293 for D20-P90-T2, and the maximum R^2 was 0.99208 for D18-P50-T1. The R^2 for each of the three diameters was discussed by taking the average values of each specimen. The average values of R^2 for Group 1 and Group 2 were similar: 0.96699 and 0.97510, respectively. The average value of the R^2 of Group 3 was slightly lower, 0.94390. This was because the relative effective working area of the rebar decreased with increasing diameter due to the skin effect at a high alternating frequency.

4.2.3. Working Stress Monitoring Error Analysis

To propose a more reliable evaluation method for the working stress of vertical prestressed rebar, the errors of Method 1 and Method 2 proposed were compared. The calculation steps can be shown in Figure 11.

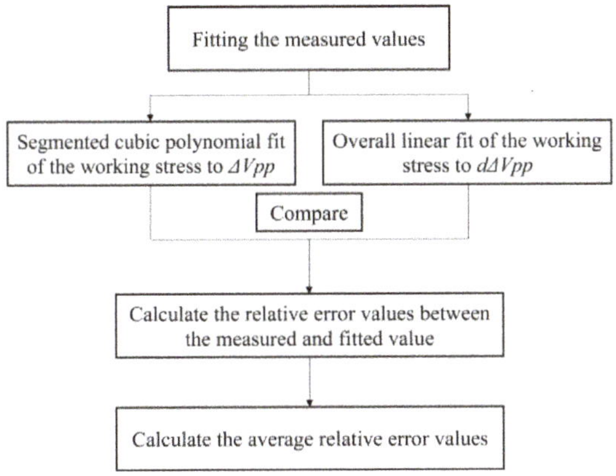

Figure 11. Calculation flow chart.

For Method 1, the curve fit degree was good; all groups' R^2 were greater than 0.96. This showed that working stress had an excellent functioning relationship with the ΔVpp. The fit relationship was generalized to Equation (16). The measured ΔVpp was substituted into the fit equation. The results were compared with the actual measured working stress. The relative error values of each specimen were calculated as shown in Figure 12.

$$F = AV^3 + BV^2 + CV + D \tag{16}$$

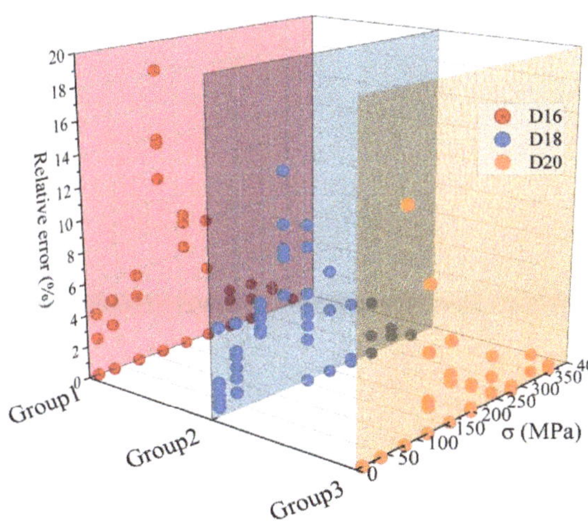

Figure 12. The relative error values of the fitted and measured values of Method 1.

From the error analysis of the fitted and measured values under the unloading stage, it was found that the relative error values did not exceed 20% under any working conditions. The relative error values were concentrated below 10% in the high stress section. The measured ΔVpp was substituted into the corresponding fit equation under different working conditions. The percentage of relative error at the turning point of D20-P90-T1 was the highest, 18.21%, and the maximum relative error between the fitted and measured value was 21.73 MPa. The high relative errors were concentrated near the turning point. Therefore, increasing the measurement points near the measured turning points during the calibration in the laboratory could significantly reduce the relative error. The relative error values of all specimens were normalized, and the average relative error values with robustness were used for comparative analysis. The maximum average relative error values for Group 1, Group 2, and Group 3 at different stress levels were 3.68%, 4.16%, and 2.79%, respectively. The average relative error values for all specimens were less than 5%, close to the results of the REME method for testing strand stresses [28].

For Method 2, the R^2 were greater than 0.90. This showed that working stress has a good linear correlation with the $d\Delta Vpp$. The measured induced voltage peak-to-peak values were substituted into Method 2. The results were compared with the actual working stress. To ensure the consistency of the $d\Delta Vpp$ loading step, only the prestress levels with a design stress above 20% of the yield strength ratio were analyzed. The results are shown in Figure 13.

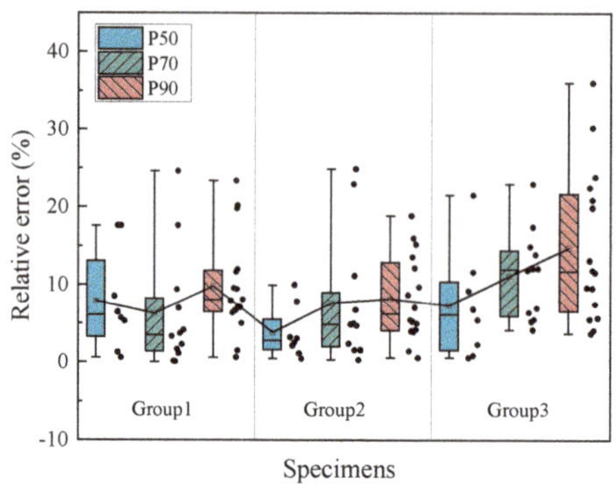

Figure 13. The relative error values of the fitted and measured values of Method 2.

More than 75% of the test points had relative error values below 20%, and the maximum relative error value was 23.85%. The errors of all specimens were normalized and the average relative error value with robustness was used for comparative analysis. The results are shown in Figure 12. In the error analysis, it was found that the maximum average relative error values of each group were 9.77%, 8.20%, and 14.53%, respectively. Therefore, under any working conditions, the average relative error values were less than 15%, better than the 25% average relative error value of the ultrasonic guided wave method [42]. This result showed that using Method 2 to monitor vertical prestressed rebar's working stress loss had good reliability. However, the error was greater than that of the traditional magnetoelastic method [22].

In summary, Method 1 could avoid high error by increasing the measurement points near the turning point. Therefore, the Method 1 test error value can be considered as low and could meet engineering needs. Method 2 avoided the uncertainty of the turning point in the laboratory calibration process, but its error was greater than the traditional magnetoelastic method. Therefore, the cubic polynomial segmental fit (Method 1) was selected to establish the mapping relationship between working stress and the ΔVpp. Then, the working stress monitoring method of prestressed rebar based on magnetic resonance was proposed.

5. Conclusions

In this paper, the relationship between the sensor induced voltage and the rebar stress was derived based on the electromagnetic induction law, magnetoelastic effect, and magnetic resonance theory. Working stress monitoring experiments with different design stress levels were carried out for rebars with diameters of 16 cm, 18 cm, and 20 cm. The induced voltage peak-to-peak values under working stress variations were collected with a magnetic resonance sensor. The main conclusions were as follows:

(1) The curves of the working stress and the induced voltage peak-to-peak values at different design stress levels showed nonlinear correlation. Due to the hysteresis effect, the induced voltage peak-to-peak values measured in the loading stage differed

from those in the unloading stage. Two characteristic indicators, the ΔVpp and $d\Delta Vpp$, were proposed for evaluating the working stress. The correlation between the two characteristic indicators and the working stress was analyzed. On this basis, the mapping relationships from the characteristic indicators to the working stress were obtained by nonlinear fitting and linear fitting, respectively.

(2) For the $d\Delta Vpp$ overall linear fit method, the R^2 was greater than 0.90. The average relative error values in different design conditions were less than 15%. This method ignored the influence of different turning points caused by external factors, but the measurement accuracy and stability needed further improvement. For the ΔVpp segmented polynomial fit method, the cubic polynomial fit was better than the quadratic polynomial and linear fit. The R^2 of the cubic polynomial fit was greater than 0.96, and the relative error values in the high stress section were all concentrated below 10%. The high errors were concentrated near the turning points, and the errors could be reduced by increasing the measurement points near the turning points. The average relative error values in different design conditions were less than 5%.

(3) According to the actual demand, the method of ΔVpp segmented polynomial fit was selected to monitor the working stress of the rebar. The magnetic resonance sensor has the advantages of small power supply, small size, light weight, and high accuracy, which is suitable for the internal monitoring of working stress of rebar. This paper verified the applicability of the induced voltage peak-to-peak value to characterize the rebar working stress.

This paper provided a new method for the working stress monitoring of vertical prestressed rebars.

Author Contributions: J.X.: methodology, validation, investigation, and writing—original draft; S.Z.: term and writing—review and editing; H.L.: validation and investigation; L.L.: software, formal analysis, and data curation; Y.S.: methodology and formal analysis. All authors have read and agreed to the published version of the manuscript.

Funding: This research was funded by the National Natural Science Foundation of China (52278146, 51978111), the Chongqing Natural Science Foundation of China (cstc2022ycjh-bgzxm0086), and the Innovation Fund Project of Graduate Education of Chongqing Jiaotong University (2023S0022).

Data Availability Statement: The data presented in this study are available on request from the corresponding author.

Conflicts of Interest: The authors declare no conflict of interest.

References

1. Abdel-Jaber, H.; Glisic, B. Monitoring of prestressing forces in prestressed concrete structures—An overview. *Struct. Control. Health Monit.* **2019**, *26*, e2374. [CrossRef]
2. Liu, Z.; Cui, Y.; Wang, W.; Cao, W.; Wang, X.; Xue, X. Strengthening Shear Resistance of Beams without Web Reinforcements Using Vertical Prestressed Steel Bars. *Adv. Mater. Sci. Eng.* **2022**, *2022*, 3869978. [CrossRef]
3. Ferreira, D.; Bairan, J.M.; Marí, A. Shear strengthening of reinforced concrete beams by means of vertical prestressed reinforcement. *Struct. Infrastruct. Eng.* **2015**, *12*, 394–410. [CrossRef]
4. Yin, H.J.; Li, Z.Q.; Hao, X.W.; Zhao, B. Analysis on stress state of box-girder web under prestressing effect. *ITM Web Conf.* **2018**, *17*, 03007. [CrossRef]
5. Li, S.; Xin, J.; Jiang, Y.; Wang, C.; Zhou, J.; Yang, X. Temperature-induced deflection separation based on bridge deflection data using the TVFEMD-PE-KLD method. *J. Civ. Struct. Health Monit.* **2023**, *13*, 781–797. [CrossRef]
6. Xin, J.; Jiang, Y.; Zhou, J.; Peng, L.; Liu, S.; Tang, Q. Bridge deformation prediction based on SHM data using improved VMD and conditional KDE. *Eng. Struct.* **2022**, *261*, 114285. [CrossRef]
7. Xin, J.; Zhou, C.; Jiang, Y.; Tang, Q.; Yang, X.; Zhou, J. A signal recovery method for bridge monitoring system using TVFEMD and encoder-decoder aided LSTM. *Measurement* **2023**, *214*, 112797. [CrossRef]
8. Bonopera, M.; Chang, K.; Lee, Z. State-of-the-Art Review on Determining Prestress Losses in Prestressed Concrete Girders. *Appl. Sci.* **2020**, *10*, 7527. [CrossRef]
9. Sawicki, B.; Brühwiler, E.; Bassil, A. Deformational behavior and damage mechanism of R-UHPFRC beam subjected to fatigue loading. *Mater. Struct.* **2021**, *54*, 158. [CrossRef]

10. Cui, H.; Li, X.; Zhang, B.; Yuan, F.; Chen, B. Research on Inductance–capacitance electromagnetic resonance measurement of prestressed concrete beam. *Measurement* **2022**, *189*, 110618. [CrossRef]
11. Wu, B.; Cui, W.; Sun, J.L.; Long, Y.Y. Research for Non-Destructive Detection of Vertical Prestressed Tensile Force in Box Girder. *Adv. Mater. Res.* **2013**, *838–841*, 2052–2057. [CrossRef]
12. Zhong, X.; Yang, T.; Shen, M.; Yu, J.; Xie, X. Method for Testing the Tension of Vertical Prestressing Bars in Webs of Concrete Box Girder Bridges. *J. Bridg. Eng.* **2011**, *16*, 438–444. [CrossRef]
13. Zhong, X.; Zhang, T.; Zhao, C.; Shu, X.; Shen, M.; Chen, Y.F. New non-destructive dynamic tensile testing of prestressing fine-rolled screw-threaded steel bars. *Eng. Struct.* **2018**, *182*, 153–163. [CrossRef]
14. Zhong, X.; Zhang, T.; Zhao, C.; Shu, X.; Shen, M.; Chen, Y.F. Tension Test for Axially Vibrated Prestressing Fine-Rolled Screw-Threaded Steel Bars Based on Thread-Contact Connection. *Adv. Civ. Eng.* **2019**, *2019*, 1806532. [CrossRef]
15. Chen, P.; He, X.; Wang, X. Ultrasonic Measurement of Axial Stress Using High-Frequency Cylindrical Guided Wave. *IEEE Sensors J.* **2021**, *21*, 6691–6697. [CrossRef]
16. Xiu, C.; Ren, L.; Li, H.; Jia, Z. Study on an innovative self-inductance tension eddy current sensor based on the inverse magnetostrictive effect. *Sens. Rev.* **2017**, *37*, 43–53. [CrossRef]
17. Xiu, C.; Ren, L.; Li, H. Investigation on Eddy Current Sensor in Tension Measurement at a Resonant Frequency. *Appl. Sci.* **2017**, *7*, 538. [CrossRef]
18. Alonso, M.S.G.; Hernando, A.; Vinolas, J.; García, M.A. Magnetic detection of high mechanical stress in iron-based materials using eddy currents and phase shift measurements. *J. Appl. Phys.* **2021**, *129*, 243901. [CrossRef]
19. Liang, K.; Angelopoulos, S.; Ktena, A.; Bi, X.; Hristoforou, E. Residual Stress Distribution Monitoring and Rehabilitation in Ferromagnetic Steel Rods. *Sensors* **2022**, *22*, 1491. [CrossRef] [PubMed]
20. Cappello, C.; Zonta, D.; Laasri, H.A.; Glisic, B.; Wang, M. Calibration of Elasto-Magnetic Sensors on In-Service Cable-Stayed Bridges for Stress Monitoring. *Sensors* **2018**, *18*, 466. [CrossRef]
21. Joh, C.; Lee, J.W.; Kwahk, I. Feasibility Study of Stress Measurement in Prestressing Tendons Using Villari Effect and Induced Magnetic Field. *Int. J. Distrib. Sens. Networks* **2013**, *9*. [CrossRef]
22. Deng, D.; Wu, X. Feasibility study of determining axial stress in ferromagnetic bars using reciprocal amplitude of initial differential susceptibility obtained from static magnetization by permanent magnets. *J. Magn. Magn. Mater.* **2018**, *449*, 243–256. [CrossRef]
23. Yan, S.; Wang, Y.; Li, P.; Gao, Z.; Wu, B.; Liu, X. Comprehensive Indicators for Evaluating and Seeking Elasto-Magnetic Parameters for High-Performance Cable Force Monitoring. *Sensors* **2022**, *22*, 7776. [CrossRef]
24. Duan, Y.-F.; Zhang, R.; Dong, C.-Z.; Luo, Y.-Z.; Or, S.W.; Zhao, Y.; Fan, K.-Q.; Or, D.S.W. Development of Elasto-Magneto-Electric (EME) Sensor for In-Service Cable Force Monitoring. *Int. J. Struct. Stab. Dyn.* **2016**, *16*, 1640016. [CrossRef]
25. Zhang, S.; Zhou, J.; Chen, J. Cable Tension Monitoring Based on the Elasto-Magnetic Effect and the Self-Induction Phenomenon. *Materials* **2019**, *12*, 2230. [CrossRef]
26. Kurs, A.; Karalis, A.; Moffatt, R.; Joannopoulos, J.D.; Fisher, P.; Soljačić, M. Wireless Power Transfer via Strongly Coupled Magnetic Resonances. *Science* **2007**, *317*, 83–86. [CrossRef]
27. Hughes, R.; Fan, Y.; Dixon, S. Investigating electrical resonance in eddy-current array probes. *AIP Conf. Proc.* **2016**, *1706*, 090001. [CrossRef]
28. Zhang, S.; Zhang, H.; Liu, H.; Zhou, J.; Yin, C.; Liao, L. Resonance enhanced magnetoelastic method with high sensitivity for steel stress measurement. *Measurement* **2021**, *186*, 110139. [CrossRef]
29. Zhang, K.C.; Chen, J.Z.; Liu, H.Q.; Zheng, Y. Research of Vertical Prestressed Losses during Construction Process of PC Continuous Box Girder Bridge. *Appl. Mech. Mater.* **2013**, *351–352*, 1142–1145. [CrossRef]
30. Liu, X.; Chen, Y.; Hu, H.; Feng, S.; Feng, Z. Measurement method of natural frequencies and tension forces for cables based on elasto-magnetic sensors calibrated by frequencies. *AIP Adv.* **2022**, *12*, 015301. [CrossRef]
31. Zhang, H.; Li, H.; Zhou, J.; Tong, K.; Xia, R. A multi-dimensional evaluation of wire breakage in bridge cable based on self-magnetic flux leakage signals. *J. Magn. Magn. Mater.* **2023**, *566*, 170321. [CrossRef]
32. Zhang, S.; Zhou, J.; Zhang, H.; Liao, L.; Liu, L. Influence of cable tension history on the monitoring of cable tension using magnetoelastic inductance method. *Struct. Health Monit.* **2021**, *20*, 3392–3405. [CrossRef]
33. Stefański, T.P.; Gulgowski, J. Electromagnetic-based derivation of fractional-order circuit theory. *Commun. Nonlinear Sci. Numer. Simul.* **2019**, *79*, 104897. [CrossRef]
34. De Angelis, G.; De Angelis, A.; Moschitta, A.; Carbone, P.; Pintelon, R. Online Identification of the LC Product in Coupled Resonant Circuits. *IEEE Trans. Instrum. Meas.* **2019**, *69*, 4592–4603. [CrossRef]
35. Burdin, D.; Chashin, D.; Fetisov, L.; Saveliev, D.; Ekonomov, N.; Vopson, M.; Fetisov, Y. Towards Fabrication of Planar Magneto-electric Devices: Coil-Free Excitation of Ferromagnet-Piezoelectric Heterostructures. *Actuators* **2021**, *10*, 294. [CrossRef]
36. Ru, G.; Gao, B.; Liu, D.; Ma, Q.; Li, H.; Woo, W.L. Structural Coupled Electromagnetic Sensing of Defects Diagnostic System. *IEEE Trans. Ind. Electron.* **2023**, *70*, 951–964. [CrossRef]
37. Weng, G.; Wang, J.; Liu, Y.; Zhu, X.; Dai, J. Magnetic Stress Sensing System for Nondestructive Stress Testing of Structural Steel and Steel Truss Components Based on Existing Magnetism. *Sensors* **2020**, *20*, 4043. [CrossRef]
38. Zhang, H.; Ma, X.; Jiang, H.; Tong, K.; Zheng, Y.; Zhou, J. Grading Evaluation of Overall Corrosion Degree of Corroded RC Beams via SMFL Technique. *Struct. Control. Health Monit.* **2023**, *2023*, 6672823. [CrossRef]

39. Tong, K.; Zhou, J.; Ma, X.; Ying, H.; Zhao, R. Investigation of the effect of initial magnetization state on the force-magnetic coupling effect of rebars. *J. Magn. Magn. Mater.* **2023**, *569*, 170382. [CrossRef]
40. Diogenes, A.G.; de Moura, E.P.; Machado, A.D.S.; Gonçalves, L.L. Corrosion evaluation of carbon steel bars by magnetic non-destructive method. *Nondestruct. Test. Evaluation* **2022**, *37*, 315–331. [CrossRef]
41. Zhang, R.; Duan, Y.; Or, S.W.; Zhao, Y. Smart Elasto-Magneto-Electric (EME) Sensors for Stress Monitoring of Steel Cables: Design Theory and Experimental Validation. *Sensors* **2014**, *14*, 13644–13660. [CrossRef] [PubMed]
42. Zhang, Z.; Tang, F.; Cao, Q.; Pan, H.; Wang, X.; Lin, Z. Deep Learning-Enriched Stress Level Identification of Pretensioned Rods via Guided Wave Approaches. *Buildings* **2022**, *12*, 1772. [CrossRef]

buildings

MDPI

Article

Theoretical and Numerical Examination of a Novel Method for Identifying Bridge Moving Force Using an Instrumented Vehicle

Dexin Liu [1], Bo Liu [1], Xingui Li [1,2,*] and Kang Shi [3]

1 CCCC First Highway Engineering Group Co., Ltd., Beijing 100024, China
2 School of Civil Engineering, Guizhou University, Guiyang 550025, China
3 School of Civil Engineering, Chongqing University, Chongqing 400044, China
* Correspondence: li18610690238@163.com

Abstract: This study presents a novel and highly efficient technique to identify moving forces by utilising the acceleration response of an instrumented moving vehicle. The complex task of determining moving forces is transformed into one of solving linear equations, made easier through the Newmark-β method. To overcome the ill-conditioned problem and improve the recognition accuracy, Tikhonov regularisation is employed. Uniquely, this approach necessitates the placement sensors on the vehicle alone to record the responses of the vehicle–bridge coupled system as it traverses the bridge. Once the coupled responses are acquired, the moving force is quickly identified using the proposed method. The reliability of this method is numerically verified using a case study of a single degree-of-freedom vehicle crossing a simply supported beam. The performance of the proposed method is further demonstrated by examining several typical external factors. The results indicate that the method presents a high recognition accuracy, demonstrating good robustness and reliability even amidst substantial environmental noise interference. This proposed method offers a new perspective for identifying the moving force of small- to medium-span bridges.

Keywords: moving force identification; vehicle–bridge interaction; Newmark-β method; Tikhonov regularisation

Citation: Liu, D.; Liu, B.; Li, X.; Shi, K. Theoretical and Numerical Examination of a Novel Method for Identifying Bridge Moving Force Using an Instrumented Vehicle. *Buildings* **2023**, *13*, 1481. https://doi.org/10.3390/buildings13061481

Academic Editors: Humberto Varum and Rajai Zuheir Al-Rousan

Received: 7 April 2023
Revised: 28 May 2023
Accepted: 1 June 2023
Published: 7 June 2023

1. Introduction

The installation of structural health monitoring systems (SHMSs) has emerged as an effective approach with which to gauge bridge responses, detect potential damage, and evaluate the overall condition of bridges [1,2]. To date, numerous SHMSs comprising a multitude of sensors have been set up on a range of modern bridges, notably the Qingma bridge and the Hong Kong–Zhuhai–Macao bridge [1]. Such systems accumulate extensive data during the operational phase of these structures. The data amassed are analysed using various data-driven assessment algorithms, including machine learning, deep learning, and other intelligence approaches, which are proposed based on the dataset recorded via the SHMS [3]. These data-driven algorithms aid in the extraction of crucial features' parameters to evaluate the health condition of bridges. Despite the SHMS's capability in detecting anomalies in loading and responses, offering real-time information for timely safety assessments after disasters and extreme events, its application is largely limited to long-span or keynote bridges. This limitation is due to the high costs associated with its instalment and maintenance. Additionally, anomalies within the data are inevitable due to potential sensor faults, transmission failures, and the aging of sensor components [4–6]. These challenges necessitate manual inspection as the primary method for monitoring the health status of small- to medium-span bridges.

Moving force identification (MFI) plays a crucial role in the health state monitoring of bridges and can provide valuable references for bridges' management and maintenance during their service period [7–9]. Given recent reports of bridge collapse accidents due to overload, it is crucial for bridge managers to have timely knowledge of moving forces as vehicles cross bridges. While an installed bridge weigh-in-motion device can directly measure moving forces, these devices need to be installed during the bridge construction stage, and their functionality can be affected by the pavement status [10,11]. Moreover, the installation and maintenance of such devices are generally expensive, making it impractical to deploy these devices on every bridge, especially considering the large quantity of small- and medium-span bridges.

To this end, there has been growing interest in the MFI technique over the past few decades. This technique identifies moving forces using bridge responses captured by installing sensors such as strain gauges, displacement sensors, and accelerometers [12–21]. Law and Chan proposed a time domain method to identify moving loads by combining the use of bending moments and acceleration responses [12], while Zhu and Law developed a time domain method for a multi-span continuous bridge using strain and displacement measurements [13]. Liu et al. proposed a load identification method utilising a new global kernel function matrix to reduce ill-posedness and improve the identification accuracy under the displacement response [14,15]. The literature [16] also shows that the identification of the moving load can be determined with the bending moment and accelerations. A notable contribution was made by Yu and Chan, who proposed a frequency–time domain method for identifying moving vehicle axle loads, relying on the bending moment responses [17]. Lage et al. also proposed a method using a concept of response transmissibility based on measuring the displacement. Their methods include two steps: initially identifying the number of forces and their respective locations, followed by reconstructing the load vector [18]. Aucejo and De proposed a space–force multiplicative regularisation method to avoid the preliminary definition of any regularisation parameter [19]. He et al. developed a frequency domain method for load bounds' identification in uncertain structures [20]. These studies showcase the major identification techniques categorized into two primary methods: time domain [12–16] and frequency domain methods [17–21].

For the time domain method, the motion equation is initially transformed into a second-order differential equation in modal coordinates, corresponding to the structural response and load. Subsequently, the differential equation is decoupled via the convolution integral within the time domain [12–16]. By contrast, the frequency domain method starts by transforming the time domain motion equation into its frequency domain counterpart using Fourier transformation. The subsequent stage involves solving the equation in the frequency domain. The dynamic spectrum is determined based on the relationship between the transfer function matrix and the system response spectrum [17–21].

However, owing to unknown initial conditions and state variables, the MFI presents a typical ill-posed problem, indicating that the inverse identification problems may not have unique solutions [22]. The discrete MFI equations are ill-conditioned, and the identified results are sensitive to external environmental noise. To overcome the ill-posed problem, effective techniques, including regularisation techniques, have been adopted [23]. For instance, Chen and Chan proposed various methods, such as the truncated generalized singular value decomposition method (TGSVD), piecewise polynomial truncated singular value decomposition (PP-TSVD) method, modified preconditioned conjugate gradient (M-PCG) method, and preconditioned least square QR factorization method, to solve the ill-posed problems in identifying the moving force from the response of the bridge deck [24–27].

In further research, Chen et al. compared these four methods to evaluate their overall performance through numerical simulations and laboratory verification [28]. Recently, Chen proposed a modified truncated singular value decomposition (MTSVD) method for moving force identification, aiming to overcome the ill-posed problems, and a comparative study was conducted with its conventional counterparts: the SVD and TSVD methods [29]. Pan and his co-authors proposed MFI methods to address the ill-posed problem [30–33]. In 2018, Pan et al. proposed a hybrid moving force identification method based on weighted L1-norm regularisation and redundant concatenated dictionary [30].

The dictionary, including trigonometric functions and rectangular functions, serves to match the principal features of unknown moving forces, while the weighted L1-norm regularisation method forms the MFI equation. Later, Pan et al. introduced a matrix regularisation-based method for solving large-scale inverse problems for force identification. Moreover, they proposed a constrained sparse regularisation-based method, taking into account unknown moving forces and initial conditions [31,32]. Recently, Pan et al. devised an equivalent load-based method for identifying the gross weight of a vehicle moving on a beam-like bridge [33]. Zhou et al. proposed an integral time domain method, effectively eliminating errors generated in the discrete unit impulse response function [34]. Zhang et al. proposed a fresh MFI method based on learning dictionary with double sparsity to a fixed dictionary for expressing moving forces with sufficient sparsity. The sparse K-singular value decomposition (K-SVD) algorithm was employed to realize the learning process [35]. He et al. developed a time domain MFI method that considered the uncertainty in the finite element model and verified the effectiveness and superiority of the method using numerical and experimental examples [36].

The aforementioned literature establishes that the proposed MFI methods can be transformed into a problem of solving linear equations, applicable to both time domain and frequency domain methods [24–27,30–33]. The Newmark-β method, valued for its stability, has been widely applied to forward structural dynamic problems [37]. In recent years, this method has been applied to the solution of inverse problems. For instance, Liu et al. firstly proposed a Newmark-based method to identify dynamic loads, demonstrating smaller cumulative errors and a superior recognition accuracy compared to the traditional state-space method [38]. Jiang et al. developed an inverse Newmark-based algorithm for estimating pavement roughness and proposed a dynamic load identification method for continuous systems based on the Newmark-β method [39,40]. Pourzeynali et al. introduced a moving load identification method based on the Newmark-β method and generalized Tikhonov regularisation method and verified the feasibility by numerical and experimental studies [41].

A review of the existing studies reveals that the majority of sensors are typically arranged in critical sections of the bridge to capture the bridge dynamic responses [24–27,30–33,37–41]. However, traditional identification methods exhibit lower efficiency in the actual bridge process, making it challenging to swiftly identify moving loads for numerous small- and medium- span bridges [42–44]. To address this issue, the vehicle scanning method (VSM) proposed by Yang has recently gained widespread attention [45–47]. The method necessitates the placement of sensors in a movable detection vehicle which traverses the bridge under testing, thus capturing vibration data to identify modal parameters indicative of the bridge's health status, such as the frequency, damping, and mode [46]. Otherwise, there are no related studies identifying the moving contact force using VSM. In this context, this study aims to propose a new MFI method inspired by the core idea of the VSM, with sensors installed on the moving vehicle rather than the bridge. Section 2 of this paper introduces the theoretical background of the method. The use of Tikhonov regularisation to address the ill-posed problem is discussed in Section 3. The finite element method for vehicle–bridge interaction analysis and a detailed identification procedure for the proposed method are outlined in Sections 4 and 5. Section 6 describes the employment of a single degree-of-freedom (Dof) vehicle crossing a simply supported bridge to verify the accuracy of the algorithm. Section 7 presents a parametric analysis of several typical external factors, while Section 8 concludes the study.

2. Theoretical Background of the Problem

The vehicle model is simplified to a single DOF moving mass with a support spring to better clarify the moving load force identification method proposed in this study. The bridge is set according to simple support boundary conditions. The mechanical model of the vehicle moving over the bridge is shown in Figure 1. The vehicle runs at a constant speed of v. As depicted in Figure 1, EI represents bridge bending stiffness, m_b represents unit length weight, L represents bridge span, $v(x,t)$ represents bridge vertical displacement, and u_c represents contact displacement at the current time step; for the vehicle model, m_v represents vehicle mass, c_v represents the vehicle damping coefficient, and k_v represents the vehicle stiffness coefficient.

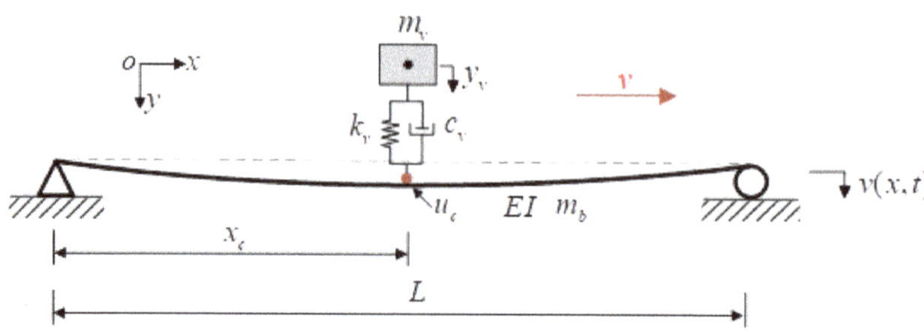

Figure 1. Mechanical model of the vehicle–bridge system.

Based on Dalembert's principle, the vehicle dynamics equation can be expressed as:

$$m_v\ddot{y}_v(t) + c_v\dot{y}_v(t) + k_v y_v(t) = P(t) \tag{1}$$

where $P(t)$ is the contact force between the vehicle and bridge; and $\ddot{y}_v(\mathrm{t})$, $\dot{y}_v(t)$, and $y_v(\mathrm{t})$ are the acceleration, velocity, and displacement of the vehicle, respectively.

Based on the Newmark algorithm, the interrelationship between the dynamic responses of moment t_i and moment t_{i+1} can be established as:

$$\dot{y}_v^{i+1} = \dot{y}_v^i + (1-\gamma)\Delta t\ddot{y}_v^i + \gamma\Delta t\ddot{y}_v^{i+1} \tag{2}$$

$$y_v^{i+1} = y_v^i + \Delta t\dot{y}_v^i + \left(\frac{1}{2}-\beta\right)\Delta t^2\ddot{y}_v^i + \beta\Delta t^2\ddot{y}_v^{i+1} \tag{3}$$

where $\gamma = 0.5$, $\beta = 0.25$; and the superscript "i" and "i + 1" denote the vehicle response at the t_i and t_{i+1} moments, respectively. For simplicity, the same, simplified notations will be adopted in the following equations.

By rewriting Equation (3), the acceleration at the t_{i+1} moment can be obtained as:

$$\ddot{y}_v^{i+1} = \frac{1}{\beta\Delta t^2}\left(y_v^{i+1} - y_v^i\right) - \frac{1}{\beta\Delta t}\dot{y}_v^i - \left(\frac{1}{2\beta}-1\right)\ddot{y}_v^i \tag{4}$$

Substituting Equation (4) into Equation (2) yields the velocity response at the t_{i+1} moment, illustrated as:

$$\dot{y}_v^{i+1} = \frac{\gamma}{\beta\Delta t}\left(y_v^{i+1} - y_v^i\right) + \left(1-\frac{\gamma}{\beta}\right)\dot{y}_v^i + \left(1-\frac{\gamma}{2\beta}\right)\Delta t\ddot{y}_v^i \tag{5}$$

At the t_{i+1} moment, the vehicle's motion equation can be expressed as:

$$m_v\ddot{y}_v^{i+1} + c_v\dot{y}_v^{i+1} + k_v y_v^{i+1} = P(t_{i+1}) \tag{6}$$

Subsequently, by substituting Equations (4) and (5) into Equation (6) and merging the similar items, Equation (6) can be simplified as:

$$\hat{K}y_v^{i+1} = \hat{P}_{i+1} \tag{7}$$

where $\hat{K}$ and $\hat{P}_{i+1}$ represent the equivalent stiffness and external load, respectively, as illustrated:

$$\hat{K} = k_v + \frac{1}{\beta \Delta t^2} m_v + \frac{\gamma}{\beta \Delta t} c_v \tag{8}$$

$$\begin{aligned} \hat{P}_{i+1} = P(t_{i+1}) + \left[\tfrac{1}{\beta \Delta t^2} m_v + \tfrac{\gamma}{\beta \Delta t} c_v\right] y_v^i + \left[\tfrac{1}{\beta \Delta t} m_v + \left(\tfrac{\gamma}{\beta} - 1\right) c_v\right] \dot{y}_v^i \\ + \left[\left(\tfrac{1}{2\beta} - 1\right) m_v + \tfrac{\Delta t}{2}\left(\tfrac{\gamma}{\beta} - 1\right) c_v\right] \ddot{y}_v^i \end{aligned} \tag{9}$$

For Equation (7), the displacement response y_v^{i+1} at the end of the time step can be obtained as:

$$y_v^{i+1} = A_0 P(t_{i+1}) + A_d y_v^i + A_v \dot{y}_v^i + A_a \ddot{y}_v^i \tag{10}$$

with

$$A_0 = \hat{K}^{-1}, \tag{11a}$$

$$A_d = \hat{K}^{-1}\left[\frac{1}{\beta \Delta t^2} m_v + \frac{\gamma}{\beta \Delta t} c_v\right], \tag{11b}$$

$$A_v = \hat{K}^{-1}\left[\frac{1}{\beta \Delta t} m_v + \left(\frac{\gamma}{\beta} - 1\right) c_v\right], \tag{11c}$$

$$A_a = \hat{K}^{-1}\left[\left(\frac{1}{2\beta} - 1\right) m_v + \frac{\Delta t}{2}\left(\frac{\gamma}{\beta} - 1\right) c_v\right] \tag{11d}$$

After obtaining the displacement response y_v^{i+1}, the velocity response $\dot{y}_v^{i+1}$ at this moment can easily be calculated using Equation (5). Additionally, the constant coefficients matrices in Equation (5) are converted into unit ones in terms of $I = \hat{K}^{-1}\hat{K}$. After combining the same terms, Equation (5) can be unified into an expression similar to Equation (10), expressed as:

$$\ddot{y}_v^{i+1} = B_0 P(t_{i+1}) + B_d y_v^i + B_v \dot{y}_v^i + B_a \ddot{y}_v^i \tag{12}$$

with

$$B_0 = \frac{\gamma}{\beta \Delta t} \hat{K}^{-1}, \tag{13a}$$

$$B_d = -\frac{\gamma}{\beta \Delta t} \hat{K}^{-1} K, \tag{13b}$$

$$B_v = \frac{\gamma}{\beta \Delta t} \hat{K}^{-1}\left[\left(\frac{\beta \Delta t}{\gamma} - \Delta t\right) k_v + \frac{1}{\gamma \Delta t} m_v\right], \tag{13c}$$

$$B_a = \frac{\gamma}{\beta \Delta t} \hat{K}^{-1}\left[\left(\frac{\beta \Delta t^2}{\gamma} - \frac{\Delta t^2}{2}\right) k_v + \left(\frac{1}{\gamma} - 1\right) m_v\right] \tag{13d}$$

Similarly, the acceleration response $\ddot{y}_v^{i+1}$ can be simplified to the following expression by substituting Equation (10) into Equation (4), expressed as:

$$\ddot{y}_v^{i+1} = C_0 P(t_{i+1}) + C_d y_v^i + C_v \dot{y}_v^i + C_a \ddot{y}_v^i \tag{14}$$

with

$$C_0 = \frac{1}{\beta \Delta t^2} \hat{K}^{-1}, \tag{15a}$$

$$C_d = -\frac{1}{\beta \Delta t^2} \hat{K}^{-1} K, \tag{15b}$$

$$C_v = -\frac{1}{\beta \Delta t^2} \hat{K}^{-1}(c_v + \Delta t k_v), \tag{15c}$$

$$C_a = \frac{1}{\beta \Delta t^2} \hat{K}^{-1} \left[(\gamma - 1)\Delta t c_v - \beta \Delta t^2 \left(\frac{1}{2\beta} - 1 \right) k_v \right] \tag{15d}$$

Combing Equations (10), (12) and (14), one yields:

$$\begin{bmatrix} y_v^{i+1} \\ \dot{y}_v^{i+1} \\ \ddot{y}_v^{i+1} \end{bmatrix} = \begin{bmatrix} A_0 \\ B_0 \\ C_0 \end{bmatrix} P(t_{i+1}) + \begin{bmatrix} A_d & A_v & A_a \\ B_d & B_v & B_a \\ C_d & C_v & C_a \end{bmatrix} \begin{bmatrix} y_v^i \\ \dot{y}_v^i \\ \ddot{y}_v^i \end{bmatrix} \tag{16}$$

Equation (16) clearly indicates that a recursion formula for the vehicle's response between the t_i and t_{i+1} moments is established. On this basis, the displacement, velocity, and acceleration responses of the vehicle at the t_i moment are written using the above recursion formula as:

$$\begin{bmatrix} y_v^i \\ \dot{y}_v^i \\ \ddot{y}_v^i \end{bmatrix} = \sum_{j=0}^{i-1} \begin{bmatrix} A_d & A_v & A_a \\ B_d & B_v & B_a \\ C_d & C_v & C_a \end{bmatrix}^j \begin{bmatrix} A_0 \\ B_0 \\ C_0 \end{bmatrix} P_i + \begin{bmatrix} A_d & A_v & A_a \\ B_d & B_v & B_a \\ C_d & C_v & C_a \end{bmatrix}^i \begin{bmatrix} y_v^0 \\ \dot{y}_v^0 \\ \ddot{y}_v^0 \end{bmatrix} \tag{17}$$

where y_v^0, $\dot{y}_v^0$, and $\ddot{y}_v^0$ are the initial displacement, velocity, and acceleration responses, respectively.

In the context of this study, the acceleration transducers are installed on the test vehicle. The vehicle's acceleration response is recorded first, followed by the application of the proposed algorithm to retrieve the moving loads through the recorded data. Supposing the output response of the test vehicle is represented by vector $\mathbf{y}(t) \in \mathbf{R}^{m\times 1}$, and m is the number of measured responses,

$$Y = R_a \ddot{y}_v + R_v \dot{y}_v + R_d y = \begin{bmatrix} R_d & & \\ & R_v & \\ & & R_a \end{bmatrix} \begin{bmatrix} y_v \\ \dot{y}_v \\ \ddot{y}_v \end{bmatrix} = R \begin{bmatrix} y_v \\ \dot{y}_v \\ \ddot{y}_v \end{bmatrix} \tag{18}$$

where $\mathbf{R} = diag\begin{bmatrix} R_d & R_v & R_a \end{bmatrix}$. Given the installation of a single acceleration transducer on the test vehicle, $R_a = 1$, $R_d = R_v = 0$.

Substituting Equation (17) into Equation (18), one yields the following equation:

$$Y(t_i) = \sum_{j=0}^{i-1} \mathbf{R} \begin{bmatrix} A_d & A_v & A_a \\ B_d & B_v & B_a \\ C_d & C_v & C_a \end{bmatrix}^j \begin{bmatrix} A_0 \\ B_0 \\ C_0 \end{bmatrix} P_i + \mathbf{R} \begin{bmatrix} A_d & A_v & A_a \\ B_d & B_v & B_a \\ C_d & C_v & C_a \end{bmatrix}^i \begin{bmatrix} y_v^0 \\ \dot{y}_v^0 \\ \ddot{y}_v^0 \end{bmatrix} \tag{19}$$

Assuming the null initial conditions, i.e., $y_v^0 = \dot{y}_v^0 = \ddot{y}_v^0 = 0$, a new matrix $\mathbf{H}_k$ is defined as:

$$\mathbf{H}_k = \mathbf{R} \begin{bmatrix} A_d & A_v & A_a \\ B_d & B_v & B_a \\ C_d & C_v & C_a \end{bmatrix}^k \begin{bmatrix} A_0 \\ B_0 \\ C_0 \end{bmatrix} \tag{20}$$

Using the matrix $\mathbf{H}_k$, Equation (19) can be written in matrix convolution form within the time interval $[t_1 \ t_n]$, expressed as:

$$\mathbf{Y} = \mathbf{H}_L \mathbf{P} \tag{21}$$

where

$$\mathbf{Y} = \begin{Bmatrix} \mathbf{y}(t_1) \\ \mathbf{y}(t_2) \\ \cdots \\ \mathbf{y}(t_n) \end{Bmatrix}, \tag{22a}$$

$$\mathbf{H}_L = \begin{bmatrix} \mathbf{H}_0 & 0 & \ldots & 0 \\ \mathbf{H}_1 & \mathbf{H}_0 & \ldots & 0 \\ \ldots & \ldots & \ldots & \ldots \\ \mathbf{H}_{n-1} & \mathbf{H}_{n-2} & \ldots & \mathbf{H}_0 \end{bmatrix}, \quad (22b)$$

$$\mathbf{P} = \begin{Bmatrix} \mathbf{P}(t_1) \\ \mathbf{P}(t_2) \\ \ldots \\ \mathbf{P}(t_n) \end{Bmatrix} \quad (22c)$$

Equation (21) can be directly solved to acquire the load vector **P** as:

$$\mathbf{P} = \left(\mathbf{H}_L^T\mathbf{H}_L\right)^{-1}\mathbf{H}_L^T\mathbf{Y} \quad (23)$$

Given that the linear correlation between the columns of matrix $\mathbf{H}_L$ is large, the determinant of $\mathbf{H}_L^T\mathbf{H}_L$ is close to zero, implying that the direct solution for Equation (23) becomes an unsettled problem. In other words, the load vector cannot be calculated using Equation (23). To overcome this obstacle, the Tikhonov regularisation approach is utilised in the subsequent section.

3. Tikhonov Regularisation

In practice, the equation used for identifying a moving load often manifests ill-posedness due to the coefficient matrix. The ill-posedness problem is exhibited as a loss of rank and the presence of pathology; thus, the paramount task lies in solving these issues for the force identification. When data collected at a high sampling frequency from measurement points exceed the number of moving loads, the problem of rank loss no longer exists. Consequently, the critical aspect for identifying the moving load centres on solving the pathological problem. To this end, this study employs Tikhonov regularisation to solve the instability encountered during the process of moving load identification.

As depicted in Equation (23), the solution for identifying the moving load is $\mathbf{P} = (\mathbf{H}_L^T\mathbf{H}_L)^{-1}\mathbf{H}_L^T\mathbf{Y}$. Nevertheless, the inversion process for $\mathbf{H}_L^T\mathbf{H}_L$ could become unstable due to the significant impacts of the numerous coefficient conditions and external noise in the measured data. As a result of such instabilities, the solution of the load vector **P** deviates greatly from the actual vehicle axle load. In addressing the pathological problem, the Tikhonov regularisation method applies a weak smoothness constraint and selects the approximate solutions of the equation. The actual process for Tikhonov regularisation thus involves solving the minimum value of the following equation:

$$min = \left\{\|\mathbf{H}_L\mathbf{P} - \mathbf{Y}\|_2 + \lambda^2\|\mathbf{P}\|_2\right\} \ (\lambda \geq 0) \quad (24)$$

where $\|\cdot\|_2$ signifies the two-dimensional norm of the vector, and λ signifies a regularisation parameter. From Equation (24), it can be observed that the Tikhonov regularisation method actually serves to find the minimum of the residual two-dimensional norm $\|\mathbf{H}_L\mathbf{P} - \mathbf{Y}\|_2$ and the two-dimensional norm of the solution.

The L-curve principle is employed to obtain the optimal solution and to identify the optimal regular parameter λ. As a raph-based parameter selection method, the L-curve principle involves plotting two quantities, the solution norm $\log\|\mathbf{P}_\lambda\|_2$ (y axis) and the residual norm $\log\|\mathbf{H}_L\mathbf{P}_\lambda - \mathbf{Y}\|_2$ (x axis), on the logarithmic coordinate scale to find the equilibrium point, illustrated in Figure 2. At the corner point of the L-curve, both the solution norm and the residual norm are maintained in a balanced dynamic equilibrium, indicating that the point represents the optimal regularisation parameter value.

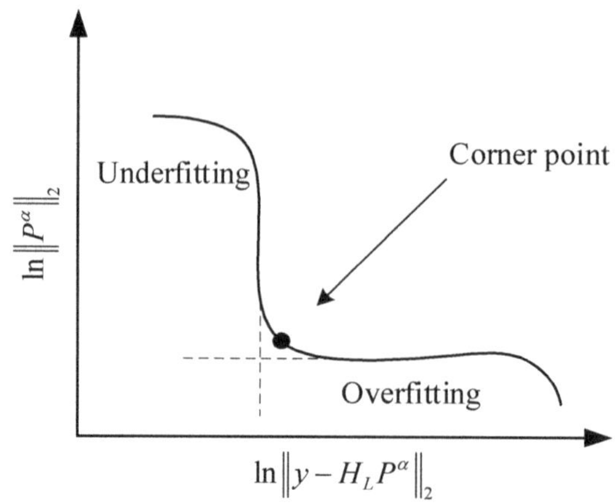

Figure 2. L-curve.

4. Finite Element Method for Vehicle–Bridge Interaction Analysis

A vehicle–bridge interaction (VBI) element with road surface roughness r was used, illustrated in Figure 3. The test vehicle is modelled as a sprung mass m_v supported by a spring of stiffness k_v and a dashpot of damping coefficient c_v. The coordinate x_c represents the position of the contact point in the element local coordinate. The equation of motion for the VBI element can be expressed as:

$$\begin{bmatrix} m_v & 0 \\ \mathbf{0} & [m_b] \end{bmatrix} \begin{Bmatrix} \ddot{y}_v \\ \{\ddot{u}_b\} \end{Bmatrix} + \begin{bmatrix} c_v & -c_v\{N\}_c^T \\ -c_v\{N\}_c & [c_b] + c_v\{N\}_c\{N\}_c^T \end{bmatrix} \begin{Bmatrix} \dot{y}_v \\ \{\dot{u}_b\} \end{Bmatrix} + \begin{bmatrix} k_v & -k_v\{N\}_c^T - vc_v\{N'\}_c^T \\ -k_v\{N\}_c & [k_b] + vc_v\{N\}_c\{N'\}_c^T + k_v\{N\}_c\{N\}_c^T \end{bmatrix} \begin{Bmatrix} y_v \\ \{u_b\} \end{Bmatrix} = \begin{Bmatrix} c_v v r_c' + k_v r_c \\ -(c_v v r_c' + k_v r_c + m_v g)\{N\}_c \end{Bmatrix} \tag{25}$$

where y_v and $\{u_b\}$ represent the displacement of the vehicle and that of the bridge element, respectively, and a dot represents the derivative with respect to time t; $[m_b]$, $[c_b]$, and $[k_b]$ denote the mass, damping, and stiffness matrices of the bridge element, respectively; r_c represents the road surface roughness at the contact point x_c; and $\{N\}_c$ represents the cubic Hermitian function evaluated at the contact point x_c, whose detailed expression is based on the literature.

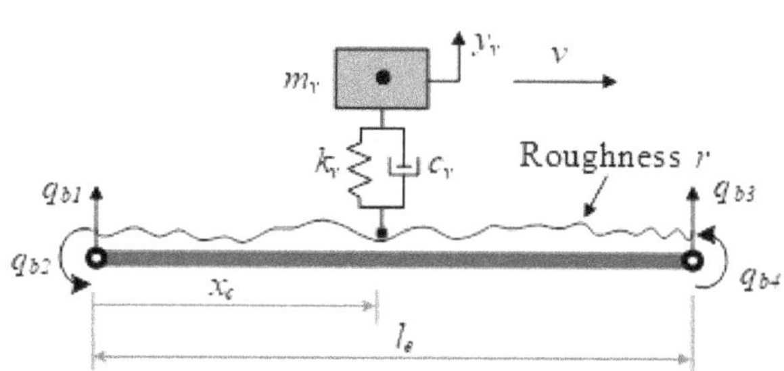

Figure 3. VBI element.

The global equation of motion for the entire bridge system is formulated by assembling all the bridge elements (with and without the vehicle) and imposing the boundary conditions, i.e., zero vertical displacement and moments, at the two ends. Subsequently, the Newmark-β method, with β = 0.25 and γ = 0.5, is adopted for the response to the coupled dynamic equation.

5. Identifying Procedures for the Proposed Method

The proposed identification method primarily includes the following three steps:

Step 1: The motion equation of the vehicle–bridge coupling system under a moving test vehicle is established. The vehicle's displacement, velocity, and acceleration responses are obtained via numerical calculation.

Step 2: A_i, B_i, and C_i ($i = 0, d, v,$ and a) are calculated based on the properties of the test vehicle. According to the sensor type and its layout position on the vehicle, the matrix is determined, and the vector $\mathbf{Y}$ and matrix $\mathbf{H}_k$ are calculated using the vehicle response obtained in Step 1.

Step 3: Substituting the vector $\mathbf{Y}$ and matrix $\mathbf{H}_k$ obtained in Step 2 into Equation (23) and using the Tikhonov regularisation method to solve the ill-conditioned problem, the moving force as the test vehicle traverses on the bridge can finally be identified.

6. Numerical Verifications

In order to authenticate the reliability of the proposed method, a numerical verification is conducted within this section. This is achieved by employing the identification procedure delineated in Section 5. The contact response, obtained through numerical calculation, is taken as the *real* value, whereas the contact response identified in reverse is taken as the *predicted* one. The accuracy of the proposed method is verified by comparing these two values. The properties of the vehicle and bridge are listed in Table 1. The vehicle traverses the bridge at a constant speed of 10 m/s. A random road roughness sample with Grade A is generated based on the ISO 8606 specification, with the simulated road roughness sample illustrated in Figure 4. It is assumed that the vehicle maintains a zero initial state, with a step size of 0.001 s adopted for the numerical calculation. Employing the vehicle and bridge parameters as listed in Table 1, the vehicle acceleration response is demonstrated in Figure 5.

Table 1. Properties of the vehicle and bridge.

		Vehicle	
Mass	m_v	kg	1200
Spring stiffness	k_v	N/m	5.0×10^4
		Bridge	
Young's modulus	E	N/m^2	2.75×10^{10}
Moment of inertia	I	m^4	0.20
Mass per unit length	m	kg/m	2400
Length	L	m	30

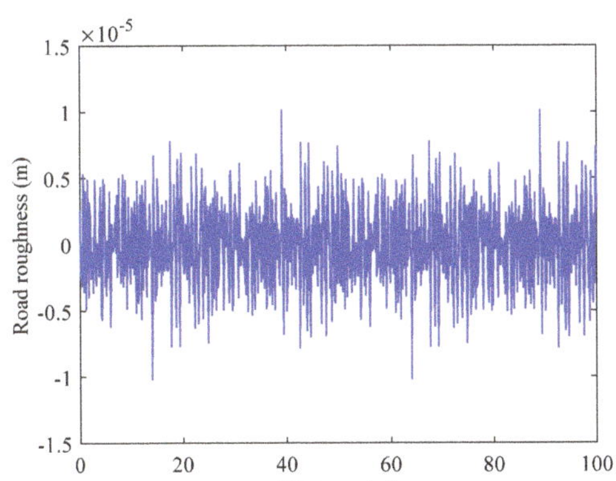

Figure 4. Random road roughness sample.

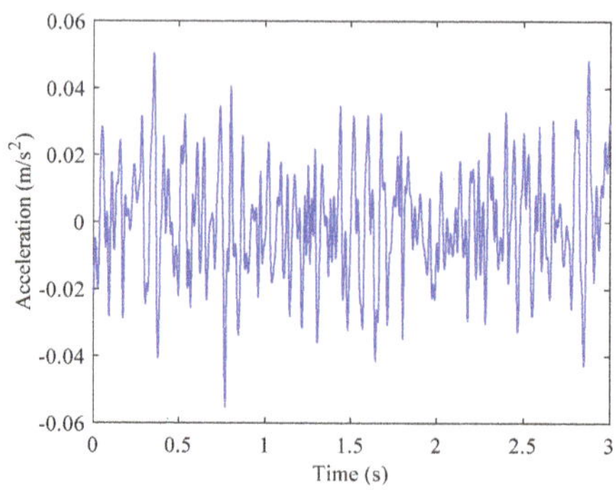

Figure 5. Acceleration response of the test vehicle.

To evaluate the predicted accuracy of the proposed method, the root mean square error (RMSE) between the predicted value and the real one is used herein, calculated as:

$$\mathrm{RMSE}(y,\hat{y}) = \sqrt{\frac{1}{n}\sum_{i=1}^{n}(y_i - \hat{y}_i)^2} \tag{26}$$

where y_i and $\hat{y}_i$ denote the predicted and real contact forces at time step i, respectively, and n stands for the total number of time steps. As derived from the definition of the RMSE, the value is smaller, while the predicted accuracy is higher for the proposed method.

Figure 6 presents a comparison of the moving load identification results between the real and predicted values. It can be observed from the figure that the results for the predicated moving load align well with the real ones, despite the influence of road roughness. The RMSE value is 0.1379, being close to null for the case considered, implying that the proposed method has a high accuracy in identifying the moving contact force. Given the aforementioned comparison results, the reliability of the proposed method is verified.

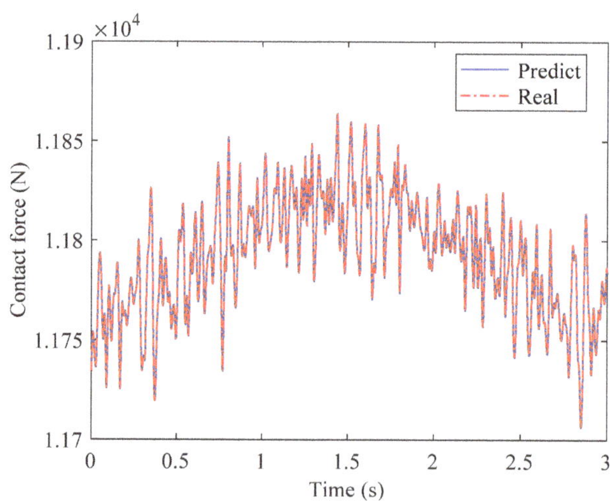

Figure 6. Comparison results of the contact force for the vehicle–bridge system.

Furthermore, the conventional time domain method (CTM) is utilised to further validate the feasibility of the proposed method. Differing from the method presented here, the CTM involves installing sensors on the bridge to record the bridge acceleration responses as the vehicle passes over it. Two different sensor deployments are considered for the CTM, referred to as Case 1 and Case 2. The specific deployments for both cases are listed in Table 2.

Table 2. Simulation cases for the conventional time domain (CTM) method.

No. Case	Locations for the Sensors	Quantity Sensors
Case 1	1/4 span, 1/2 span, and 3/4 span	Three
Case 2	1/6 span, 1/3 span, 1/2 span, 2/3 span, and 5/6 span	Five

Figure 7 depicts the comparison results of the moving contact force for various identified methods. It can be observed that the contact force identified using the proposed method coincides closely with Case 2 using the CTM and the real values, whereas a distinct disparity is observed with the prediction result of Case 1. In comparison with the CTM, the proposed method utilises fewer sensors to obtain the moving contact force with high precision. However, a greater number of sensors must be adopted to ensure the identification accuracy of the CTM. Consequently, the comparison results further validate the accuracy and advantages of the proposed method.

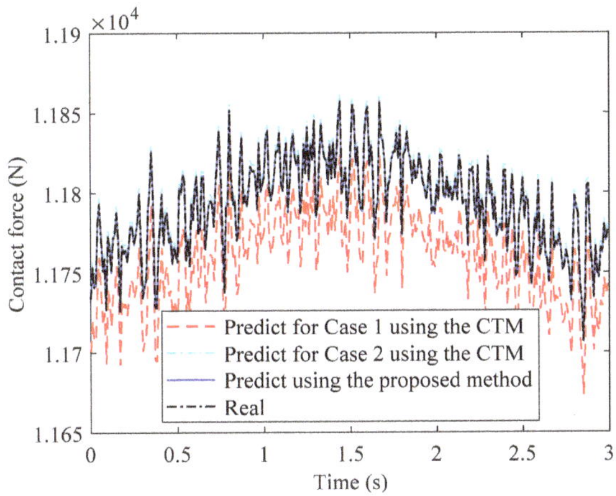

Figure 7. Comparison results of the contact force for various identified methods.

7. Parametric Analyses of Various Influencing Factors

The accuracy of the proposed method was verified in the preceding section. This section further studies the robustness of the method, performing a sensitivity analysis of various external factors, i.e., road roughness, running speed, and noise level, that may affect the recognition accuracy.

7.1. Effect of Road Roughness

Using the simulation method outlined in the ISO 8606 specification, random road roughness samples for Grades B and C were generated, as illustrated in Figure 8. It becomes evident that pavement classes B and C possess greater amplitudes and higher frequencies in comparison to Class A. Figure 9 showcases a comparison of the moving force identification results at various pavement roughness levels. It can be observed that the vehicle–bridge contact force gradually rises with the increasing road surface grade. Notwithstanding the heightened roughness of classes B and C, the moving load identification employing the proposed method remains consistent with the real one. The RMSE values for Grades B and C stand at 0.1380 and 0.1384, respectively. Comparatively, the identified accuracy for these cases is slightly reduced compared to the result for Grade A, as depicted in Section 6, yet continues to present high precision. Consequently, the proposed method exhibits minimal susceptibility to road roughness in identifying the moving load. Furthermore, higher pavement grades lead to a more significant acceleration response, favouring the identification of the moving contact force.

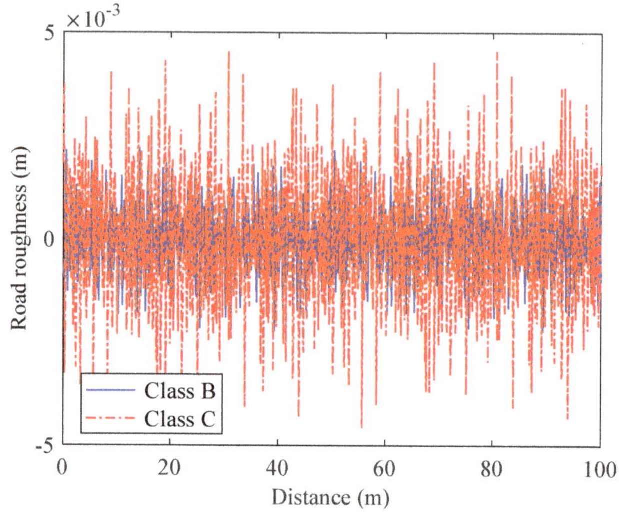

Figure 8. Road roughness samples for various class levels.

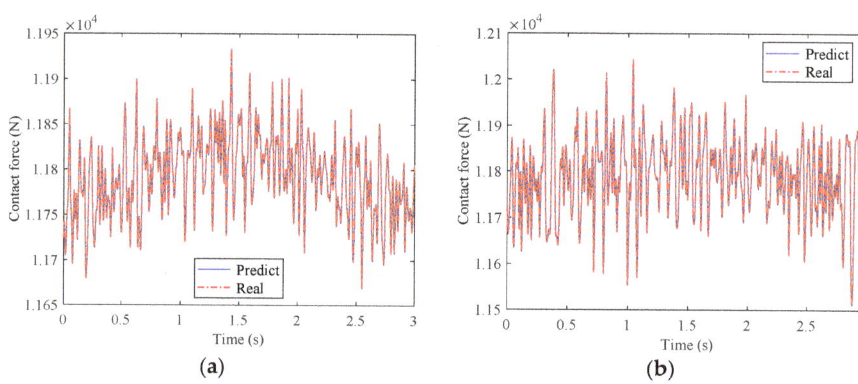

Figure 9. Comparison results of contact force for the vehicle–bridge system at various class levels: (**a**) Class B. (**b**) Class C.

7.2. Effect of Running Speed

The efficiency of identification using the proposed method is directly determined by the vehicle running speed in real-world projects. Therefore, this section considers two speeds, namely, $v = 5$ and 15 m/s. The pavement grade adheres to Class C, and the other parameters align with those used in Section 3. Figure 10 compares the contact force identification results at different vehicle operating speeds. It can be observed that the contact force rises with the increasing vehicle speed. The moving force remains well identified at different vehicle running speeds using the proposed method, implying that the running speed of the vehicle exerts little effect on it. From the perspective of identification accuracy, the RMSE values for the running speeds of 5 and 15 m/s are 0.0687 and 0.2087, respectively. The results indicate that the identified precision may undergo limited reductions, while the proposed method can maintain a high precision even at a rapid running speed. To ensure the test precision and efficiency, however, an intermediate speed is recommended for field applications employing the proposed method. The primary reason for this lies in the fact that a smaller volume of data, obtained at the same sampling frequency for a limited length of the bridge, will result in an inferior identification accuracy. Thus, for the case under consideration, a test speed of 10 m/s will be optimal.

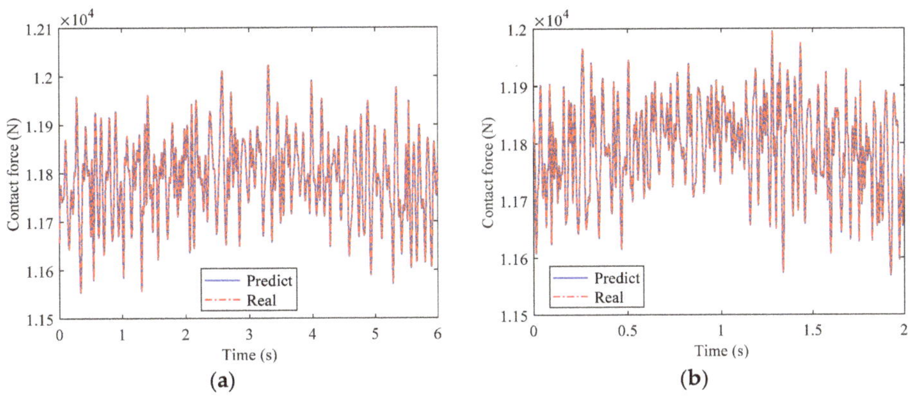

Figure 10. Comparison results of the contact force for the vehicle–bridge system at various running speeds: (**a**) 5 m/s. (**b**) 15 m/s.

7.3. Effect of the Environmental Noise Level

In practice, measured data inevitably incur contamination from environmental noise. To investigate the feasibility of the proposed method, four environmental noise levels of 2%, 5%, 10%, and 20% are applied to the measured data. The expression is defined as:

$$\ddot{y}_{vp} = \ddot{y}_v + E_p N_s \sigma_{\ddot{y}_v} \tag{27}$$

where $\ddot{y}_v$ denotes the original vehicle acceleration response. E_p denotes the noise level. N_s and $\sigma_{\ddot{y}_v}$ represent the mean and standard deviation of the vehicle's acceleration response.

Figure 11 demonstrates a comparison of the moving force identification results at varying environmental noise levels. The RMSE values for E_p = 2%, 5%, 10%, and 20% stand at 3.1940, 4.9653, 14.5237, and 36.3568, respectively. It can be seen that the recognition results are worse than those without noise, but the recognition accuracy is generally maintained at a higher level. Even at a 20% environmental noise level, the recognition results remain highly accurate. Therefore, the proposed method for identifying moving contact force presents a certain robustness to environmental noise. However, it is suggested that the environmental noise level should be controlled within 10% to ensure test precision when applying the proposed method in field tests.

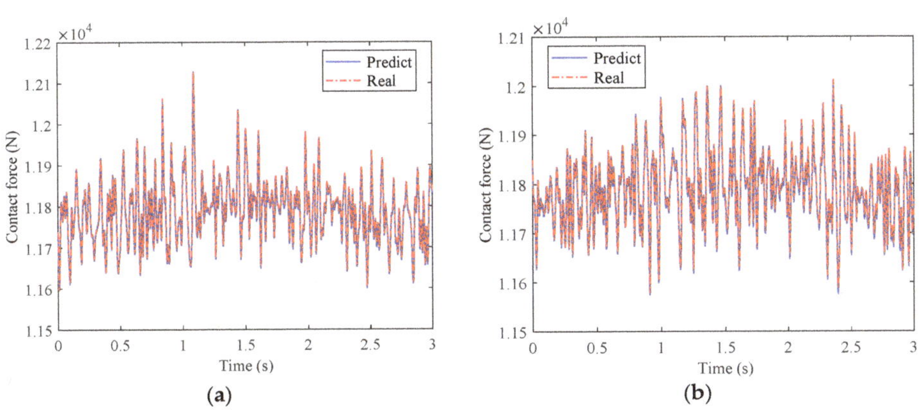

Figure 11. *Cont.*

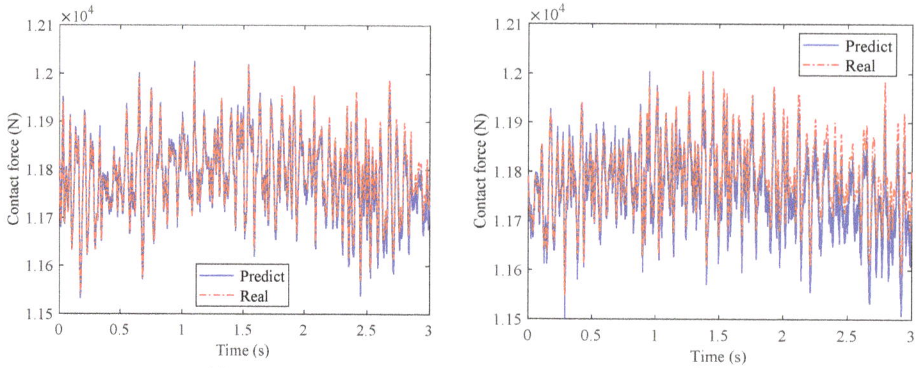

(c) (d)

Figure 11. Comparison results of contact force for the vehicle–bridge system under various environmental noise levels: (**a**) $E_p = 2\%$. (**b**) $E_p = 5\%$. (**c**) $E_p = 10\%$. (**d**) $E_p = 20\%$.

8. Conclusions

In this study, we introduced a novel method for moving load identification on bridges, using a vehicle–bridge coupled vibration theory and the Newmark-β algorithm. This approach requires sensor arrangement solely on the vehicle, recording its responses as it traverses the bridge. Based on the obtained responses, the moving force can be swiftly identified via the proposed method. Tikhonov regularisation is employed to overcome the ill-conditioned nature of the inverse problem so as to enhance the recognition accuracy. A single DoF vehicle traversing a simply supported beam is used as an example to numerically verify the proposed method's reliability by comparing the predicted values with the real ones. Additionally, we examined several typical external factors to discuss the robustness of the proposed method. Based on the theoretical derivations and the data adopted in our analysis, the following conclusions were drawn:

(1) Unlike the existing identification methods, our approach does not necessitate the installation of sensors on the bridge. Instead, the moving contact force can be identified readily through sensors installed on the moving vehicle. This provides a new idea for moving load identification on small- and medium-span bridges.
(2) The proposed method demonstrates high precision in identifying the moving contact force by comparing with the real values and those identified using the CTM method.
(3) The moving load can be effectively identified even under conditions of poor road roughness in Class C.
(4) The effect of moving speed is limited for the proposed method; however, an intermediate speed of 10 m/s is recommended for field testing.
(5) Environmental noise can affect the identification accuracy to an acceptable extent. Control of the environment noise level within 10% for the proposed method is recommended in actual applications.

Though the proposed method is based on a single-DOF vehicle model moving over a simply supported bridge, it is adaptable to multi-DOF vehicle models and other bridge types, i.e., elastic, clamped, fixed, and general support boundary conditions. Consequently, our future studies will focus on identifying the moving force using multi-DOF vehicle models and conduct further experimental validations of the proposed method' accuracy.

Author Contributions: Conceptualization, D.L.; methodology, B.L.; software, K.S.; validation, X.L.; formal analysis, K.S.; writing—original draft preparation, K.S.; writing—review and editing, X.L.; visualization, K.S.; supervision, B.L.; project administration, D.L.; funding acquisition, D.L. All authors have read and agreed to the published version of the manuscript.

Funding: This research was funded by the National Natural Science Foundation of China (grant number 52008060), Chongqing Municipal Natural Science Foundation (grant No. CSTB2022NSCQ-MSX1448) and Guizhou Provincial Science and Technology Projects (Grant No. Qiankehe Support [2022] General 026).

Data Availability Statement: Not applicable.

Conflicts of Interest: The authors declare no conflict of interest.

References

1. He, X.H.; Shi, K.; Wu, T. An integrated structural health monitoring system for the Xijiang high-speed railway arch bridge. *Smart Struct. Syst.* **2018**, *21*, 611–621.
2. He, Z.; Li, W.; Salehi, H.; Zhang, H.; Zhou, H.; Jiao, P. Integrated structural health monitoring in bridge engineering. *Autom. Constr.* **2022**, *136*, 104168. [CrossRef]
3. Tian, Y.; Chen, C.; Sagoe-Crentsil, K.; Zhang, J.; Duan, W. Intelligent robotic systems for structural health monitoring: Applications and future trends. *Autom. Constr.* **2022**, *139*, 104273. [CrossRef]
4. Niu, J.; Li, Z.; Zhuo, Y.; Di, H.; Wei, J.; Wang, X.; Guo, Y.; Li, S. Robust correlation mapping of train-induced stresses for high-speed railway bridge using convolutional denoising autoencoder. *Struct. Health Monit.* **2023**, *22*, 832–845. [CrossRef]
5. Lei, X.; Xia, Y.; Wang, A.; Jian, X.; Zhong, H.; Sun, L. Mutual information based anomaly detection of monitoring data with attention mechanism and residual learning. *Mech. Syst. Signal Process.* **2023**, *182*, 109607. [CrossRef]
6. Tang, Z.; Chen, Z.; Bao, Y.; Li, H. Convolutional neural network-based data anomaly detection method using multiple information for structural health monitoring. *Struct. Control Health Monit.* **2019**, *26*, e2296. [CrossRef]
7. Ouyang, H. Moving-load dynamic problems: A tutorial (with a brief overview). *Mech. Syst. Signal Process.* **2011**, *25*, 2039–2060. [CrossRef]
8. Zhu, X.Q.; Law, S.S. Recent developments in inverse problems of vehicle–bridge interaction dynamics. *J. Civ. Struct. Health Monit.* **2016**, *6*, 107–128. [CrossRef]
9. Liu, R.; Dobriban, E.; Hou, Z.; Qian, K. Dynamic load identification for mechanical systems: A review. *Arch. Comput. Methods Eng.* **2022**, *29*, 831–863. [CrossRef]
10. Yu, Y.; Cai, C.S.; Deng, L. State-of-the-art review on bridge weigh-in-motion technology. *Adv. Struct. Eng.* **2016**, *19*, 1514–1530. [CrossRef]
11. Paul, D.; Roy, K. Application of bridge weigh-in-motion system in bridge health monitoring: A state-of-the-art review. *Struct. Health Monit.* **2023**, 14759217231154431. [CrossRef]
12. Law, S.S.; Chan, T.H.; Zeng, Q.H. Moving force identification: A time domain method. *J. Sound Vib.* **1997**, *201*, 1–22. [CrossRef]
13. Zhu, X.Q.; Law, S.S. Moving forces identification on a multi-span continuous bridge. *J. Sound Vib.* **1999**, *228*, 377–396. [CrossRef]
14. Liu, J.; Meng, X.; Jiang, C.; Han, X.; Zhang, D. Time-domain Galerkin method for dynamic load identification. *Int. J. Numer. Methods Eng.* **2016**, *105*, 620–640. [CrossRef]
15. Liu, J.; Meng, X.; Zhang, D.; Jiang, C.; Han, X. An efficient method to reduce ill-posedness for structural dynamic load identification. *Mech. Syst. Signal Process.* **2017**, *95*, 273–285. [CrossRef]
16. Chen, Z.; Qin, L.; Chan, T.H.; Yu, L. A novel preconditioned range restricted GMRES algorithm for moving force identification and its experimental validation. *Mech. Syst. Signal Process.* **2021**, *155*, 107635. [CrossRef]
17. Yu, L.; Chan, T.H. Moving force identification based on the frequency–time domain method. *J. Sound Vib.* **2003**, *261*, 329–349. [CrossRef]
18. Lage, Y.E.; Maia, N.M.M.; Neves, M.M.; Ribeiro, A.M.R. Force identification using the concept of displacement transmissibility. *J. Sound Vib.* **2013**, *332*, 1674–1686. [CrossRef]
19. Aucejo, M.; De Smet, O. A space-frequency multiplicative regularization for force reconstruction problems. *Mech. Syst. Signal Process.* **2018**, *104*, 1–18. [CrossRef]
20. He, Z.C.; Lin, X.Y.; Li, E. A novel method for load bounds identification for uncertain structures in frequency domain. *Int. J. Comput. Methods* **2018**, *15*, 1850051. [CrossRef]
21. Yan, W.J.; Yuen, K.V. A new probabilistic frequency-domain approach for influence line extraction from static transmissibility measurements under unknown moving loads. *Eng. Struct.* **2020**, *216*, 110625. [CrossRef]
22. Sanchez, J.; Benaroya, H. Review of force reconstruction techniques. *J. Sound Vib.* **2014**, *333*, 2999–3018. [CrossRef]
23. Tran, H.; Inoue, H. Development of wavelet deconvolution technique for impact force reconstruction: Application to reconstruction of impact force acting on a load-cell. *Int. J. Impact Eng.* **2018**, *122*, 137–147. [CrossRef]
24. Chen, Z.; Chan, T.H. A truncated generalized singular value decomposition algorithm for moving force identification with ill-posed problems. *J. Sound Vib.* **2017**, *401*, 297–310. [CrossRef]
25. Chen, Z.; Chan, T.H.; Nguyen, A. Moving force identification based on modified preconditioned conjugate gradient method. *J. Sound Vib.* **2018**, *423*, 100–117. [CrossRef]
26. Chen, Z.; Chan, T.H.; Nguyen, A.; Yu, L. Identification of vehicle axle loads from bridge responses using preconditioned least square QR-factorization algorithm. *Mech. Syst. Signal Process.* **2019**, *128*, 479–496. [CrossRef]

27. Chen, Z.; Qin, L.; Zhao, S.; Chan, T.H.; Nguyen, A. Toward efficacy of piecewise polynomial truncated singular value decomposition algorithm in moving force identification. *Adv. Struct. Eng.* **2019**, *22*, 2687–2698. [CrossRef]
28. Chen, Z.; Chan, T.H.; Yu, L. Comparison of regularization methods for moving force identification with ill-posed problems. *J. Sound Vib.* **2020**, *478*, 115349. [CrossRef]
29. Chen, Z.; Deng, L.; Kong, X. Modified truncated singular value decomposition method for moving force identification. *Adv. Struct. Eng.* **2022**, *25*, 2609–2623. [CrossRef]
30. Pan, C.D.; Yu, L.; Liu, H.L.; Chen, Z.P.; Luo, W.F. Moving force identification based on redundant concatenated dictionary and weighted l1-norm regularization. *Mech. Syst. Signal Process.* **2018**, *98*, 32–49. [CrossRef]
31. Pan, C.; Ye, X.; Zhou, J.; Sun, Z. Matrix regularization-based method for large-scale inverse problem of force identification. *Mech. Syst. Signal Process.* **2020**, *140*, 106698. [CrossRef]
32. Pan, C.; Huang, Z.; You, J.; Li, Y.; Yang, L. Moving force identification based on sparse regularization combined with moving average constraint. *J. Sound Vib.* **2021**, *515*, 116496. [CrossRef]
33. Chudong, P.; Liwen, Z.; Xijun, Y.; Zhuo, S. Vehicle weight identification based on equivalent loads reconstructed from responses of beam-like bridge. *J. Sound Vib.* **2022**, *534*, 117072. [CrossRef]
34. Zhou, H.C.; Li, H.N.; Yang, D.H.; Yi, T.H. Moving force identification of simply supported bridges through the integral time domain method. *J. Sound Vib.* **2022**, *534*, 117046. [CrossRef]
35. Zhang, Z.H.; He, W.Y.; Ren, W.X. Moving force identification based on learning dictionary with double sparsity. *Mech. Syst. Signal Process.* **2022**, *170*, 108811. [CrossRef]
36. He, W.Y.; Li, Y.L.; Yi, J.X.; Ren, W.X. Time-domain identification of moving load on beam type bridges considering interval uncertainty in finite element model. *Mech. Syst. Signal Process.* **2023**, *191*, 110168. [CrossRef]
37. Newmark, N.M. A method of computation for structural dynamics. *J. Eng. Mech. Div.* **1959**, *85*, 67–94. [CrossRef]
38. Liu, K.; Law, S.S.; Zhu, X.Q.; Xia, Y. Explicit form of an implicit method for inverse force identification. *J. Sound Vib.* **2014**, *333*, 730–744. [CrossRef]
39. Jiang, J.; Seaid, M.; Mohamed, M.S.; Li, H. Inverse algorithm for real-time road roughness estimation for autonomous vehicles. *Arch. Appl. Mech.* **2020**, *90*, 1333–1348. [CrossRef]
40. Jiang, J.; Ding, M.; Li, J. A novel time-domain dynamic load identification numerical algorithm for continuous systems. *Mech. Syst. Signal Process.* **2021**, *160*, 107881. [CrossRef]
41. Pourzeynali, S.; Zhu, X.; Ghari Zadeh, A.; Rashidi, M.; Samali, B. Comprehensive study of moving load identification on bridge structures using the explicit form of Newmark-β method: Numerical and experimental studies. *Remote Sens.* **2021**, *13*, 2291. [CrossRef]
42. Xin, J.; Jiang, Y.; Zhou, J.; Peng, L.; Liu, S.; Tang, Q. Bridge deformation prediction based on SHM data using improved VMD and conditional KDE. *Eng. Struct.* **2022**, *261*, 114285. [CrossRef]
43. Jiang, Y.; Hui, Y.; Wang, Y.; Peng, L.; Huang, G.; Liu, S. A novel eigenvalue-based iterative simulation method for multi-dimensional homogeneous non-Gaussian stochastic vector fields. *Struct. Saf.* **2023**, *100*, 102290. [CrossRef]
44. Li, S.; Xin, J.; Jiang, Y.; Wang, C.; Zhou, J.; Yang, X. Temperature-induced deflection separation based on bridge deflection data using the TVFEMD-PE-KLD method. *J. Civ. Struct. Health Monit.* **2023**, *13*, 781–797. [CrossRef]
45. Yang, Y.B.; Lin, C.W.; Yau, J.D. Extracting bridge frequencies from the dynamic response of a passing vehicle. *J. Sound Vib.* **2004**, *272*, 471–493. [CrossRef]
46. Yang, Y.B.; Yang, J.P.; Wu, Y.; Zhang, B. *Vehicle Scanning Method for Bridges*; John Wiley & Sons: Hoboken, NJ, USA, 2019.
47. Yang, Y.B.; Wang, Z.L.; Shi, K.; Xu, H.; Wu, Y.T. State-of-the-art of vehicle-based methods for detecting various properties of highway bridges and railway tracks. *Int. J. Struct. Stab. Dyn.* **2020**, *20*, 2041004. [CrossRef]

buildings

MDPI

Article

Research on the Diffusion Model of Cable Corrosion Factors Based on Optimized BP Neural Network Algorithm

Shiya Li [1,2], Guowen Yao [1,2], Wei Wang [3,*], Xuanrui Yu [1,2], Xuanbo He [1,2], Chongyang Ran [1,2] and Hong Long [1,2]

1. State Key Laboratory of Mountain Bridge and Tunnel Engineering, Chongqing Jiaotong University, Chongqing 400074, China; lshiya0507@163.com (S.L.); 990020050526@cqjtu.edu.cn (G.Y.); 611190080010@mails.cqjtu.edu.cn (X.Y.); 611220080009@mails.cqjtu.edu.cn (X.H.); 13658308605@163.com (C.R.); 17323410209@163.com (H.L.)
2. School of Civil Engineering, Chongqing Jiaotong University, Chongqing 400074, China
3. Wanzhou District Urban Management Bureau, Chongqing 404199, China

* Correspondence: 622220080005@mails.cqjtu.edu.cn

Abstract: Corrosion factors enter the cable via diffusion and penetration from the defect position of the cable or the connection position between the anchoring system and the cable section, seriously affecting the cable's durability. Exploring the transmission mechanism of corrosion factors in the cable structure is essential to reveal the durability and the long-term performance of the cable structure and to judge the corrosion damage of steel wires in the cable structure. Based on the machine learning (ML) method and the analytical solution of Fick's second law, the laws between different temperatures, humidity, cable inclinations, cable defect areas, etc., and the diffusion coefficient of corrosion factors and the concentration of surface corrosion factors are obtained, also a spatial diffusion model of corrosion factors is established. According to the research, the optimum simulation result is achieved by employing the optimized back propagation (BP) neural network algorithm, which has a faster convergence speed and better robustness. Although ambient temperature, humidity, and corrosion time all impact the diffusion rate of corrosion factors, the tilt angle of the cable and the size of cable defects are the main factors influencing the diffusion coefficient of corrosion factors and the concentration of surface corrosion factors. The error between the concentration of corrosion factors calculated by the model in this article and the measured values at each spatial point of the cable is controlled within 15%, allowing for the spatial diffusion of corrosion factors to be effectively predicted and evaluated in practical engineering.

Keywords: bridge engineering; stay cable; corrosion factors; neural networks; spatial diffusion model

Citation: Li, S.; Yao, G.; Wang, W.; Yu, X.; He, X.; Ran, C.; Long, H. Research on the Diffusion Model of Cable Corrosion Factors Based on Optimized BP Neural Network Algorithm. *Buildings* **2023**, *13*, 1485. https://doi.org/10.3390/buildings13061485

Academic Editor: Fabrizio Greco

Received: 8 May 2023
Revised: 29 May 2023
Accepted: 7 June 2023
Published: 8 June 2023

1. Introduction

Stay cables, as significant components of cable-stayed bridges, carry and transmit loads, and their service conditions directly impact the safety of bridge operation. The combined impact of load and environment can accelerate the corrosion degradation of cable systems, resulting in failure, jeopardizing the safety of bridge operation, and incurring considerable economic losses and safety mishaps [1–3].

Scholars have conducted relevant experimental research to uncover the corrosion mechanism of cables in corrosive settings, delay the corrosion rate of cables, and increase the durability of cables. In terms of cable damage and failure, the cable tension plays an increasingly important role in the cable-stayed bridge system and affects the feasibility of optimizing and updating this type of bridge [4]. Greco et al. [5] conducted a nonlinear analysis of bridge cables based on the characteristic parameters of the bridge structure, taking into account the dynamic amplification effect and failure mechanism of the cables under dynamic loads. Ammendolea et al. [6] reproduced the damage law of cables under the coupling effect of the bridge and dynamic load based on the theory of continuous damage mechanics. Mozos [7,8] analyzed 10 cable-stayed bridges and investigated the effects of cable layout, single and double cable planes, and main beam cross-section on the ultimate failure state of the cables, laying the foundation for the evaluation of the safe service state and long-term performance of the cables. In terms of corrosion fatigue, Yan et al. [9] examined the corrosion of steel wires in service. They discovered that lower anchoring section steel wires were more susceptible to corrosion than higher ones. Stewart et al. [10] conducted tests to investigate the corrosion process of steel wires, and the results revealed that the corrosion form of steel wires is connected to their environment. Through accelerated corrosion testing on parallel steel wires, Rou [11] presented a positive and negative electrode chemical reaction formula for hydrogen evolution and oxygen absorption corrosion. Changqing [12] conducted a corresponding exploration of the law of waste steel wire corrosion based on actual engineering. Betti [13] investigated the corrosion damage mechanism under different environmental corrosion by simulating salt spray corrosion experiments of high-strength galvanized steel wires in acid rain. Furuya et al. [14] conducted atmospheric exposure studies on cable segments in natural environments. The study revealed that humidity and temperature are the primary causes of the deterioration of the cable interior environment. Suzumura [15,16] conducted accelerated corrosion tests on high-strength galvanized steel wires in different environments and comparatively analyzed the effects of temperature, relative humidity, and NaCl solution concentration on corrosion rate. Furthermore, the galvanized layer has variable degrees of effect on corrosion parameters such as corrosion potential and polarization resistance of steel wire [17–20]. Heying [21] discovered that corroded steel wires' elongation and fatigue strength dramatically decrease after corrosion, with the decline of elongation occurring primarily in the latter stages of corrosion [22]. Hamilton [23,24] used seawater as the corrosion media in accelerated corrosion studies on defective cable-stayed cables under static tension. On this basis, researchers investigated the damaging effect of corrosion on steel wires under environmental load coupling using experiments and numerical simulations, as well as analyzed the law of mechanical properties degradation and damage evolution of corroded steel wires [25–29]. Similarly, Rosso et al. [30] compared the model with the half-joint in the actual project and analyzed the degradation mechanism of the mechanical and physical properties of the half-joint of the bridge under different corrosion levels by simulating the corrosion development process.

With the development of computer technology, the machine learning (ML) method, as a relatively advanced data processing approach, has become widely applied in the practical engineering of related bridges. Xin et al. [31–33] used the machine learning method to identify and process the deformation monitoring data of the bridge more efficiently, and reliably laid the foundation for the early warning of bridge deformation. Kim et al. [34] investigated the diffusion of the Cl^- in concrete structures using a neural network model and examined the time-varying law of the Cl^- diffusion coefficient utilizing measured data from 30 concrete specimens as input to the database. Gupta [35] used adaptive artificial neural network (ANN), and ANN approaches to model the permeation law of Cl^- in concrete structures, considering the effect of environmental temperature on the Cl^- permeability coefficient. Yong et al. [36] investigated the shear strength of recycled concrete beams (RAC) based on the ANN and random forest (RF) models. Bukhsh [37] and Pengyong [38] designed a bridge state-level prediction approach based on the ML model to solve bridge degradation influenced by various unpredictable factors, boosting forecast accuracy. Boyu [39] promoted the idea of constructing a decision tree (DT) based on density, which lowered the size of the DT and, to some extent, prevented overfitting. The intelligent assessment model of bridge safety risk created using the ML algorithm solved the problem of insufficient use of previous assessment data [40–42]. Shuheng [43] utilized the sparrow search algorithm (SSA) to optimize and alter the initial weights and thresholds of the back propagation (BP) neural network, demonstrating that the SSA has excellent accuracy and can optimize the BP neural network.

The current study mainly focuses on the overall distribution, macroscopic performance, and mechanical changes of cable corrosion. There has been little research on the non-stationary dispersion of internal corrosion factors in cables. The diffusion of corrosion factors and external elements have a complicated nonlinear relationship that is difficult to obtain using one-time conversion or simple linear regression (LR) approaches. Therefore, this article establishes a data-driven model based on the ML method to reveal the effects of cable inclination angle, environmental temperature, humidity, and cable defect size on the concentration and diffusion coefficient of corrosion factors on the cable's surface. Based on multiple nonlinear regression analysis methods, appropriate empirical formulas are provided and validated, providing a reference for theoretical analysis of the long-term performance of stay cables in practical engineering.

2. Diffusion Mechanism and the Test Method of Corrosion Factors

2.1. Diffusion Mechanism

The corrosion factors diffuse from the outside to the interior of the cable due to the impact of the concentration gradient difference of corrosion factors within and outside the cable. The diffusion forms of corrosion factors in the cable can be divided into two types: (1) The corrosion factors diffuse radially along the gap between steel wires until the concentration of corrosion factors reaches saturation in each layer of steel wire on the section. (2) Corrosion factors will diffuse upwards in the longitudinal and circumferential directions of the cable due to the gaps between the layers of steel wires, eventually causing three-dimensional damage to the cable. The diffusion mechanism is shown in Figure 1.

Figure 1. Transfer form of corrosion factors in the cable: (**a**) radial transmission; (**b**) circular and axial transmission.

The diffusion mechanism of corrosion factors is mainly described by Fick's law. Fick's law is made up of the first and second laws of Fick. Fick's first law is proposed based on the concentration gradient difference of diffusing substances, which believes that the greater the concentration gradient difference of the diffusing substance, the greater the flow of the substance per unit of time through the unit cross-sectional area perpendicular to the direction of diffusion, proportional to the concentration gradient at the cross-section. However, Fick's first law is only applicable to the analysis of steady-state diffusion processes, in which each volume element at any moment has an equal quantity of incoming and exiting material and a constant concentration throughout the process. This circumstance is not typical, in actuality. The diffusion of substances will be influenced by both environmental factors and the properties of the materials themselves. The diffusion rate is likely to alter as diffusion time and depth change. Based on Fick's first law, the second law to describe the non-stationary diffusion of substances was proposed to explain the diffusion law of substances in actual processes more accurately. The law of variation of concentration of corrosion factors with diffusion time and depth was obtained. Fick's second law can be expressed as the following equation:

$$\frac{\partial C}{\partial t} + \nabla \times (-D \times \nabla c) = 0 \quad (1)$$

where C is the concentration of corrosion factors, D is the diffusion coefficient of corrosion factors, c is the gradient difference of concentration of corrosion factors, and t is the corrosion time.

The more significant the gradient difference in concentration of corrosion factors, the faster the diffusion rate of corrosion factors, which is more likely to cause damage to parallel steel wires in the cable. For spatial diffusion issues, the concentration gradient difference of corrosion factors and the Hamiltonian operator in Equation (1) can be expressed as follows [44].

$$\left.\begin{array}{l} \nabla c = \frac{\partial C}{\partial x} + \frac{\partial C}{\partial y} + \frac{\partial C}{\partial z} \\ \nabla = \frac{\partial}{\partial x} + \frac{\partial}{\partial y} + \frac{\partial}{\partial z} \end{array}\right\} \quad (2)$$

By substituting Equation (2) into Equation (1), the analytical solution model of Fick's second law's error function can be produced, as shown below:

$$C(x,y,z) = C_0 + (C_S - C_0)\left(1 - erf\frac{x}{2\sqrt{D \times t}} \times erf\frac{y}{2\sqrt{D \times t}} \times erf\frac{z}{2\sqrt{D \times t}}\right) \quad (3)$$

where $C(x,y,z)$ represents the concentration at any spatial point within the component, C_s is the concentration of corrosion factors on the element's surface, C_0 is the concentration of corrosion factors within the structure, and *erf* is the error function.

The calculation method is shown the following equation:

$$erf(x) = \frac{2}{\sqrt{\pi}} \times \int_0^z \exp\left(\beta^2\right)d\beta \quad (4)$$

The concentration of initial corrosion factors within the structure is relatively low in general. The effect of the concentration of corrosion factors within the structure on the predicted results is neglected to make the solution easier. Equation (3) is further simplified to obtain the following equation:

$$C(x,y,z) = C_S \times \left(1 - erf\frac{x}{2\sqrt{D \times t}} \times erf\frac{y}{2\sqrt{D \times t}} \times erf\frac{z}{2\sqrt{D \times t}}\right) \quad (5)$$

The key to constructing a spatial diffusion model of corrosion factors, according to Equation (5), is getting the concentration and the diffusion coefficient of corrosion factors on the surface of the cable. The concentration and the diffusion coefficient of corrosion factors on the surface of the cable can be calculated using the procedures below.

The initial defect size of a square cable is shown as an example for demonstration. Create a three-dimensional coordinate system using the cable's defect position as the center point, then characterize the cable's damage range. Any point within the damage range can be represented as follows. Define the distance from the coordinate origin at the center position of different sampling points as R_i, as shown in Figure 2. The concentration of corrosion factors at each site is measured using electrochemical measuring equipment. After completing the test of the concentration of corrosion factors at each location of each working condition, the concentration and the diffusion coefficient of corrosion factors on the surface of the cable under various operational circumstances can be obtained by substituting results into Equation (5).

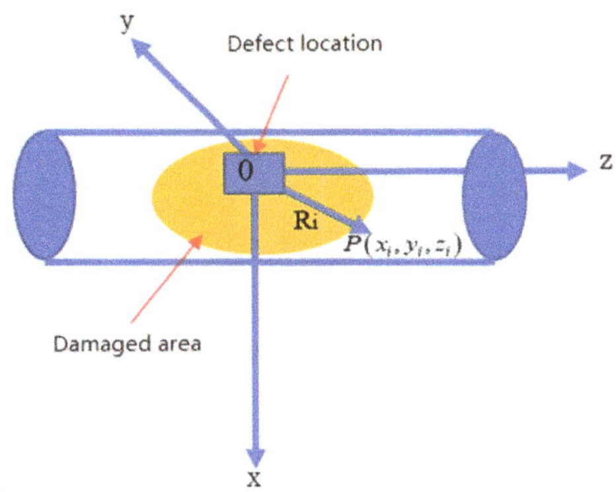

Figure 2. Schematic diagram of sampling points for corrosion factors.

The following are the specific procedures for building the spatial diffusion model of corrosion factors using the ML method: (1) Using the testing procedures mentioned above, determine the concentration of corrosion factors at each location in the cable space along the path above. (2) Using ambient temperature, humidity, cable inclination angle, and cable defect area as input variables, the prediction model for diffusion coefficient and concentration of surface corrosion factors is created based on the ML approach. (3) According to Fick's second law, the spatial diffusion model of corrosion factors is established by comprehensively considering the influence of environmental and material factors on the diffusion rate of corrosion factors. (4) Combine the results of the salt spray corrosion test and the prototype cable corrosion test to verify the accuracy of the prediction model.

2.2. Test Method

Due to the influence of the gradient difference of the concentration of corrosion factors inside and outside the cable, the corrosion factors diffuse from the outside to the inside. The diffusion forms of corrosion factors in the cable can be divided into two types: (1) The corrosion factors diffuse along the steel wire and the gap between the steel wires in the radial direction until the concentration of corrosion factors in each layer of the steel wire on this section reaches saturation. (2) Due to the gap between the layers of steel wires, corrosion factors will spread upward along the longitudinal and ring of the cable, resulting in three-dimensional damage to the cable. In order to clarify the diffusion law of corrosion factors in the cable, the concentration of corrosion factors on the surface of each layer of steel wire in the cable was measured by the BION-1881 chemical analyzer under different periods, defect areas, ambient temperature, and humidity. The test procedure is as follows: Firstly, the coordinate system is established with the cable defect position as the origin. Secondly, after the cable was corroded for a corresponding time, it was dissected and cut into sections of 5 cm, and its spatial position in the cable was sorted out and recorded, respectively. Finally, an electrochemical analyzer was used to measure the concentration of corrosion factors on the surface of each cut steel wire. The test procedure of the concentration of corrosion factors is as follows: (1) Each layer of corroded steel wire is processed and cut into a 5 cm section. After the processing is completed, it is numbered; (2) Wash the cut steel wire with water and repeatedly brush it with a brush to completely dissolve the corrosion factors on the surface of the steel wire in water; (3) Completely immerse one electrode rod of the electrochemical analyzer into a beaker, test the concentration of corrosion factors in the water, and take readings after stabilization to obtain the concentration of corrosion factors on the surface of the corroded steel wire. The testing process of corrosion factor concentration is shown in Figure 3. According to the test methods above, the different distribution laws of corrosion factors along the cable in radial, circumferential, and axial directions and the distribution forms of corrosion factors in three directions can be obtained, respectively.

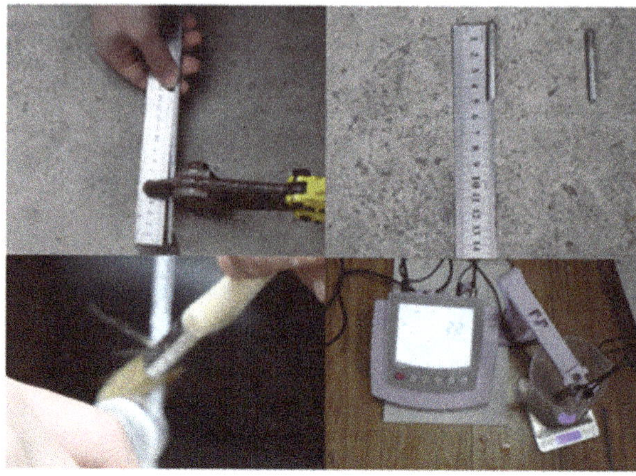

Figure 3. Testing process of corrosion factor concentration.

3. Optimization of BP Neural Network Model

3.1. Data-Driven Model

The Scikit-learn software tool package is used to create five data-driven models: LR, RF, ridge regression (RR), DT, and BP neural network. The cross-validation analysis approach is used to verify the model's accuracy to acquire the best output results. A total of 60% of the data samples (154 total) were categorized as training subsets and 40% as testing subsets (102 total) using random functions to select the training and testing sets. The following are the specific steps: To begin, split the data into groups and number them using a random function. The training and testing sets are then chosen at random. Finally, statistical performance indicators such as the decisive correlation coefficient (R^2), root mean square error (RMSE), and relative root mean square error (MAE) were used to evaluate the model's computational accuracy. R^2 shows the degree of fitting of the proposed model to the experimental data. R^2 lies between 0 and 1, and the closer its value is to the upper limit, the better the model's fit to the experimental data. Otherwise, the worse. RMSE, commonly known as the cost function, plays a positive role in the learning process of ML algorithms. The more significant the RMSE and MAE values, the worse the model's fitting effect on the experimental data and the lower the model's accuracy. Table 1 displays the test and verification results of various models.

Table 1. Calculation results of each data-driven model.

Mechanical Algorithm Model	R^2	RMSE	MAE
BP neural network model	0.868	0.845	0.327
DT model	0.748	1.279	0.523
RF model	0.834	0.924	0.375
LR model	0.638	1.338	0.813
RR model	0.764	1.132	0.601

As shown in Table 1, the BP neural network model has the highest R^2 value (0.868) from the training set, while the MAE (0.845) and RMSE (0.327) values are the lowest, indicating that the BP neural network model is most appropriate for simulating the diffusion coefficient of corrosion factors and concentration of surface corrosion factors. Similarly, the RF model's R^2, RMSE, and MAE values for the testing set are 0.834, 0.924, and 0.375, respectively. The results indicate a significant correlation between the measured and predicted diffusion coefficient values of corrosion factors and the concentration of surface corrosion factors. The LR model performs the poorest, with the lowest R^2 (0.638) and the highest MAE (0.813). The simulation results of the diffusion coefficient of corrosion factors and the concentration of surface corrosion factors using the BP neural network are better, with the highest R^2 and the lowest MAE and RMSE values, as shown by the analysis findings above. The drawbacks of the BP neural network model include that it is prone to collapsing into local minima and has poor stability, which slows down the convergence speed and ultimately leads to a drop in model accuracy. As a result, updating the BP neural network model is required to increase its accuracy.

3.2. Algorithm Optimization Based on Sparrow Search Algorithm

3.2.1. Sparrow Search Algorithm

Introducing the SSA with stronger search ability and faster convergence speed to update the BP neural network to improve the accuracy of the model. The SSA is a swarm intelligence optimization algorithm that is based on the foraging and anti-predation behavior of the sparrow. The sparrow with the highest fitness is prioritized in the SSA. They are separated into discoverers and joiners during the foraging process of sparrows. Discoverers are responsible for discovering food in the population and offering foraging places and instructions for the entire sparrow population, whereas joiners rely on discoverers for food. Sparrows can usually forage as discoverers and joiners to obtain food. Individuals in the population monitor the behavior of others in the community, and attackers in this population will fight for food supplies with high-intake peers to raise their predation rate. Furthermore, when the sparrow population detects a threat, it will engage in anti-predatory behavior. During each iteration, the description of the discoverer's location update is shown in the following equation [43]:

$$X_{i,j}{}^{q+1} = \begin{cases} X_{i,j}{}^{q} \exp\left(\frac{-i}{\alpha \times iter_{\max}}\right), if\ R_2 \leq ST \\ X_{i,j}{}^{q} + Q \times L, if\ R_2 > ST \end{cases} \tag{6}$$

where q is the current number of iterations, $iter_{\max}$ is the largest iteration algebra, X_{ij} is the position information of the i-th sparrow in the j-th dimension, α is the random number($\alpha \in (0, 1]$), R_2 is the warning value ($R_2 \in [0, 1]$), ST is the safety value ($ST \in [0.5, 1]$), Q is the random number subject to normal distribution, and L is a matrix of $1 \times$ d, where each element within the matrix is all 1.

When $R_2 < ST$, it indicates that there are no predators in the foraging environment, allowing for the discoverers to conduct lengthy search activities. If $R_2 \geq ST$, it means that some sparrows in the population have detected predators and alerted other sparrows. At this time, the sparrows must fly immediately to other safe areas for feeding. The description of the joiner's location update is shown in the following equation:

$$X_{i,j}{}^{q+1} = \begin{cases} X_{best}{}^{q} + \beta \times \left|X_{i,j}{}^{q} - X_{best}{}^{q}\right|, if\ f_i > f_g \\ X_{i,j}{}^{q+1} + K \times \left(\frac{X_{i,j}{}^{q} - X_{worst}{}^{q}}{(f_i - f_w) + \varepsilon}\right), if\ \ f_i = f_g \end{cases} \tag{7}$$

where X_{best} is the optimal position occupied by the discoverer; X_{worst} is the current global position; β is the control parameter of step size, which follows the normal distribution random number with the mean of 0 and variance of 1; $K \in [-1, 1]$ represents the random number; f_i is the fitness value of current sparrow individuals; f_g is the fitness value of the sparrow in the current global optimal position; f_w is the fitness value of the sparrow in the current global worst position; and ε is a constant.

Among these, $f_i > f_g$ indicates that sparrows are on the periphery of the population and highly vulnerable to predators. When $f_i = f$, it demonstrates that sparrows in the middle of the population know the danger and should approach other sparrows to reduce the predation risk. K controls the direction of the sparrow's movement and step size.

3.2.2. Optimization of BP Neural Network Algorithm

During the calculation process, the selection of the number of neurons in the hidden layer significantly impacts the model's calculation accuracy and speed. The more neurons in the hidden layer, the higher the model's accuracy, but the slower the computation speed. On the contrary, the smaller the number of neurons, the lower the computational accuracy of the model, but the faster the model operates. As a result, it is critical to reasonably determine the number of neurons in the hidden layer, which not only preserves a certain accuracy in the model but also increases its computational speed. According to the relevant literature [32], the number of neurons in the model can be calculated using the following formula:

$$Q = \sqrt{M + N} + A \tag{8}$$

where Q is the number of neurons in the hidden layer, M is the number of neurons in the input layer, N is the number of neurons in the output layer, and A is an integer between 1 and 10.

Considering the influence of overfitting and underfitting, after repeated debugging, Q is 10, and the test accuracy of the model is the highest. The calculation process of the entire model is shown in Figure 4.

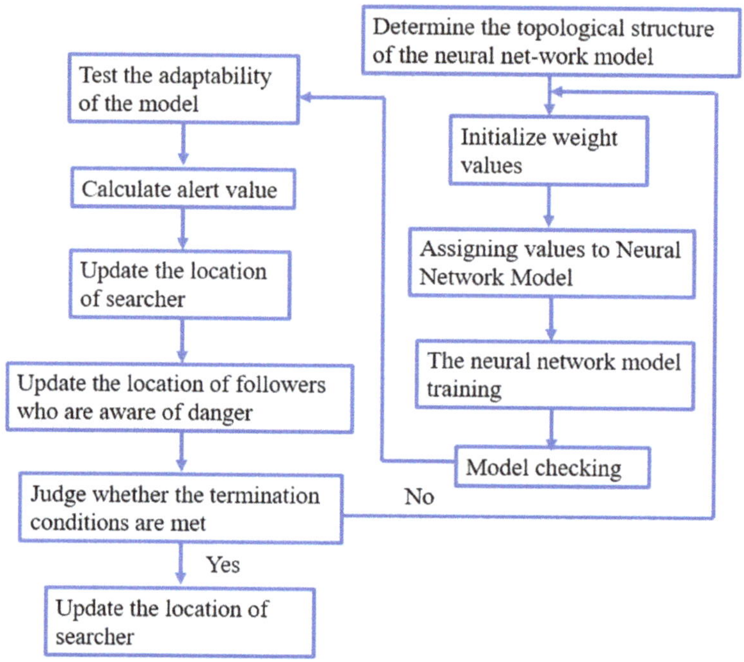

Figure 4. Process of the SSA-BP model.

Figures 5 and 6 demonstrate the prediction results of the SSA-BP neural network and the BP neural network. Figures 7 and 8 show that the SSA-BP model has a high R^2 (0.93) and the lowest RMSE (0.0378), which indicates that the updated model can accurately forecast the concentration of surface corrosion factors and the diffusion coefficient of corrosion factors. In addition, BP requires 13 iterations to achieve convergence, whereas the SSA-BP model only requires 9 iterations, indicating a significant improvement in the convergence speed of the SSA-BP model. The model was trained and simulated 20 times to compare the robustness of the SSA-BP model with the BP model, and the results are shown in Figure 9. The coefficient of variation of the SSA-BP model was found to be between 5% and 18%, whereas the coefficient of variation of the BP model was between 7% and 35%.

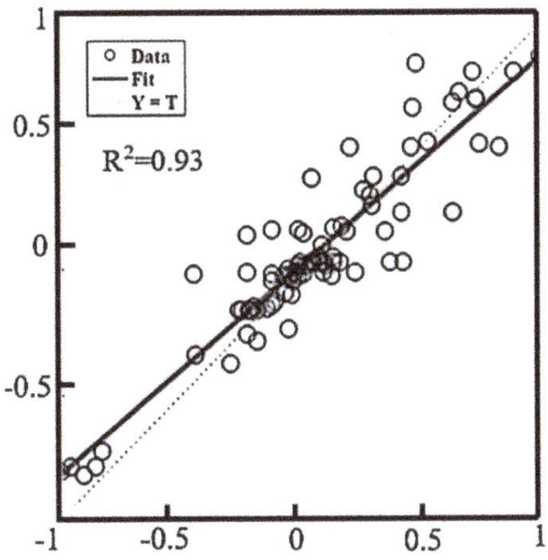

Figure 5. Prediction results of the SSA-BP model.

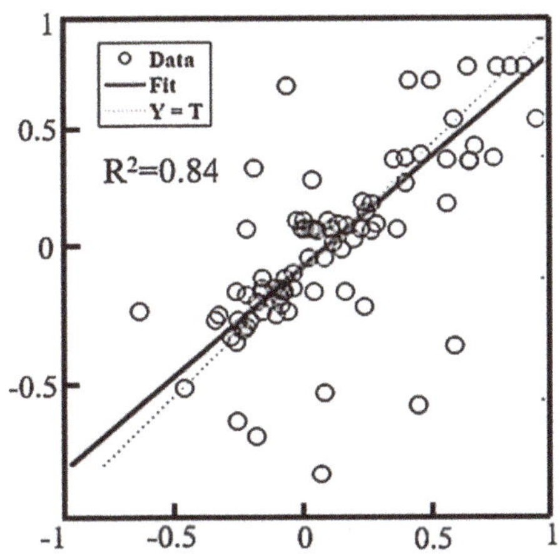

Figure 6. Prediction results of the BP model.

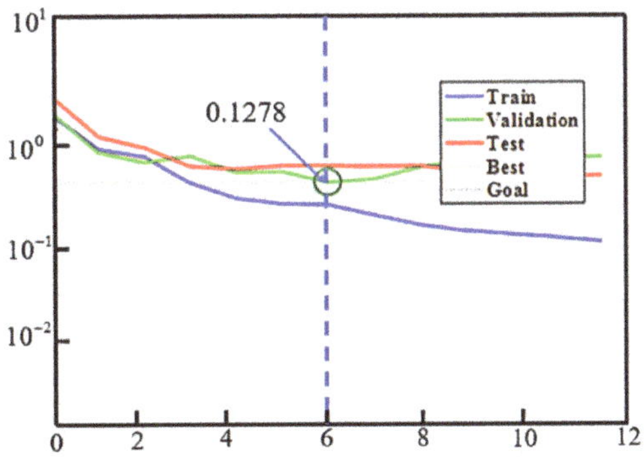

Figure 7. Convergence process of the BP model.

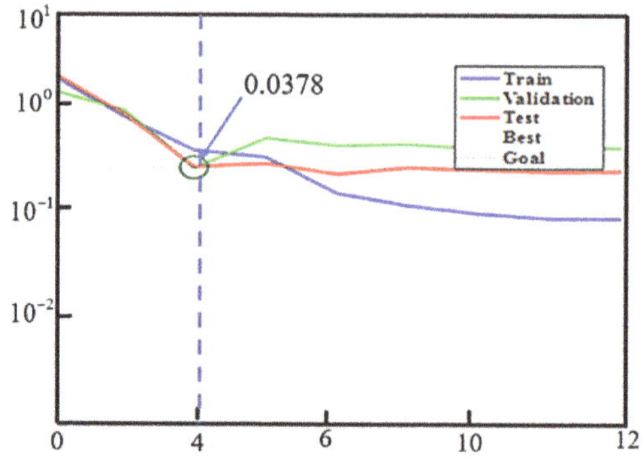

Figure 8. Convergence process of the SSA-BP model.

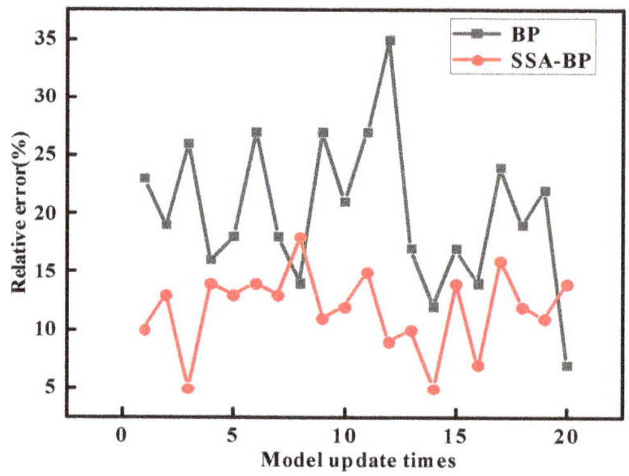

Figure 9. Analysis results of robustness of the SSA-BP and BP model.

4. Analysis of Parameter Sensitivity

The SSA-BP neural network model was used to calculate the weights between the input and hidden layer to analyze the impact of various parameters on the diffusion coefficient D of corrosion factors and the concentration Cs of surface corrosion factors. The results are shown in Tables 2 and 3, respectively.

Table 2. Results of connection weights for the input and hidden layer.

Hidden Layer	Defect Size	Dip Angle	Temperature	Humidity	Corrosion Time	Deviation
B1	−0.06	−0.14	−0.38	−0.15	0.12	0.05
B2	−0.35	0.92	0.27	0.31	0.87	0.08
B3	−0.72	−0.23	−0.01	−0.34	−0.15	−0.65
B4	0.25	0.14	0.52	−0.22	−0.05	−0.53
B5	0.52	−0.25	0.12	0.64	−0.05	−0.27
B6	−0.23	0.08	0.38	−0.59	0.07	0.42
B7	−0.36	0.17	−0.27	−0.19	−0.34	0.28
B8	0.54	−0.09	−0.32	0.65	−0.58	−0.47
B9	0.47	−0.51	−0.33	−1.04	0.85	−0.02
B10	−0.72	0.23	−0.47	0.58	0.09	−0.41

Table 3. Weight values of the hidden and output layer.

B1	B2	B3	B4	B5	B6	B7	B8	B9	B10	Bias
−0.53	0.41	0.36	0.54	−0.47	−0.38	−1.22	1.08	0.17	0.41	0.62

The Garson formula was used to calculate the significant impacts of input parameters on the transfer coefficient of corrosion factors and the concentration of surface corrosion factors, as indicated in the following equation:

$$Z_{ik} = \frac{\sum_{j=1}^{L}\left(\frac{w_{ij}}{\sum_{r=1}^{N} w_{rj}} v_{jk}\right)}{\sum_{i=1}^{N}\left(\sum_{j=1}^{L}\left(\frac{w_{ij}}{\sum_{r=1}^{N} w_{rj}} v_{jk}\right)\right)} \tag{9}$$

where w_{rj} is the connection weight value between the input neuron and the hidden layer neuron j, and v_{jk} is the connection weight value between the hidden layer neuron j and the output neuron K.

The impact of input parameters on the concentration of surface corrosion factors and diffusion coefficient of corrosion factors can be obtained using the steps above based on the connection weights value and biases between the input, hidden, and output layers. The results are shown in Figure 10.

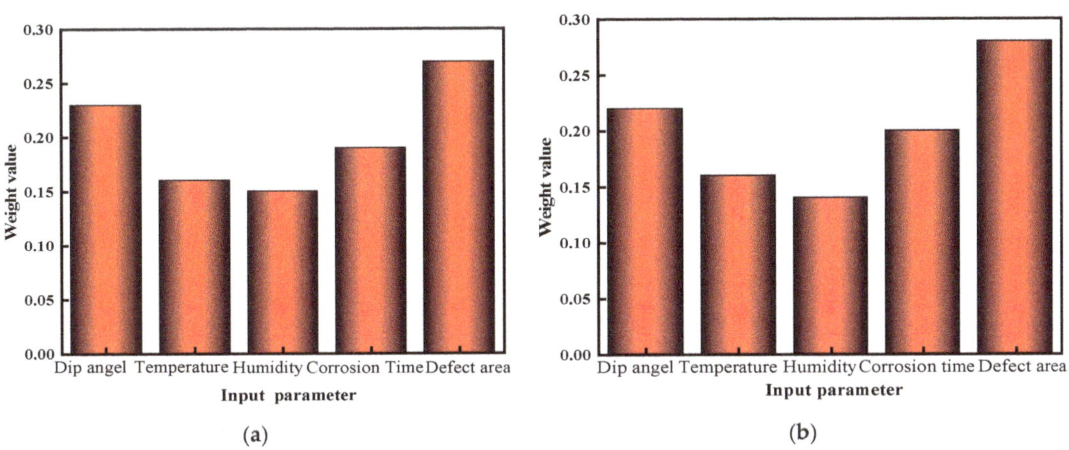

Figure 10. Weight analysis results of each input parameter on the corrosion factor of cable. (**a**) Concentration of surface corrosion factors; (**b**) Transfer coefficient of corrosion factors.

Figure 10 depicts the impact of various parameters on the concentration of surface corrosion factors and the transfer coefficient of corrosion factors. The results demonstrate that the dip angle and the defect area of the cable have the most significant effect on both, with weight values all greater than 0.25. Although environmental temperature, humidity, and corrosion time all impact the diffusion rate of corrosion factors, their weight values are all within 0.2. According to the results of the weight analysis of the SSA-BP neural network model, the concentration of surface corrosion factors and diffusion coefficient of corrosion factors of the cable above the defect location (A1 segment cable) and the cable below the defect location (B1 segment cable) were fitted using the tilt angle of cables by the distinguishing criterion, and were based on multiple nonlinear regression criteria. The fitting equation for the concentration of surface corrosion factors and the diffusion coefficient of corrosion factors on the A1 segment cable is shown in the following equation:

$$\left.\begin{array}{l} C_s = \alpha\cdot(t)^{A_1}\cdot(h)^{A_2}\cdot(T_1)^{A_3}(W)^{A_4}\cdot(\cos\theta)^{A_5} + H \\ D = \left[\beta\cdot(t)^{B_1}\cdot(h)^{B_2}\cdot(T_1)^{B_3}(W)^{B_4}\cdot(\cos\theta)^{B_5} + Q\right] \times 10^{-10} \end{array}\right\} \tag{10}$$

Referring to Equation (10), according to the difference between the concentration of surface corrosion factors and the diffusion coefficient of corrosion factors on the A1 and B1 cable sections, the concentration of surface corrosion factors and the diffusion coefficient of corrosion factors on the B1 cable section can be expressed as the following equation:

$$\left.\begin{aligned} C_s &= \eta_1\left[\alpha\cdot(t)^{A_1}\cdot(h)^{A_2}\cdot(T_1)^{A_3}(W)^{A_4}\cdot(\cos\theta)^{A_5}+H\right] \\ D &= \eta\left[\beta\cdot(t)^{B_1}\cdot(h)^{B_2}\cdot(T_1)^{B_3}(W)^{B_4}\cdot(\cos\theta)^{B_5}+Q\right]\times 10^{-10} \end{aligned}\right\} \quad (11)$$

where t is the corrosion time of the cable, h is the relative humidity of the environment, T_1 is the ambient temperature, W is the defect size area, θ is the dip angle of the cable, and A_1–A_5, B_1–B_5, H and Q are undetermined parameters.

MATLAB software programming was used to calculate Equation (10) and bring the known parameters into Equation (10) based on the nonlinear regression function to calculate the concentration of surface corrosion factors and diffusion coefficient of corrosion factors on the cable under different working conditions. The expression of the concentration of surface corrosion factors and the diffusion coefficient of corrosion factors on the A1 segment cable is as follows:

$$\left.\begin{aligned} C_s &= 0.387\cdot(t)^{0.32}\cdot(h)^{0.08}\cdot(T_1)^{0.12}(W)^{0.52}\cdot(\cos\theta)^{0.43}+0.216 \\ D &= \left[11.42\cdot(t)^{0.23}\cdot(h)^{0.05}\cdot(T_1)^{0.11}(W)^{0.54}\cdot(\cos\theta)^{0.41}+1.231\right]\times 10^{-10} \end{aligned}\right\} \quad (12)$$

The diffusion rate of corrosion factors differs between the A1 and B1 segment cables due to the inclination of the cable. The diffusion rate is faster in the B1 section of the cable but slower in the A1 section. According to Equation (12), the expression of the concentration of surface corrosion factors and the diffusion coefficient of corrosion factors on the B1 segment cable is as follows:

$$\left.\begin{aligned} C_s &= \eta_1\left[0.387\cdot(t)^{0.32}\cdot(h)^{0.08}\cdot(T_1)^{0.12}(W)^{0.52}\cdot(\cos\theta)^{0.43}+0.216\right] \\ D &= \eta\left[11.42\cdot(t)^{0.23}\cdot(h)^{0.05}\cdot(T_1)^{0.11}(W)^{0.54}\cdot(\cos\theta)^{0.41}+1.231\right]\times 10^{-10} \end{aligned}\right\} \quad (13)$$

The values of η_1 and η under different tilt angles based on the differences in the concentrations of corrosion factors between different layers inside the A1 and B1 cable segments are shown in Table 4.

Table 4. The range of values of η_1 and η under different tilt angles.

Tilt Angle	η_1	η
0	1.02	1.01
30	0.92	1.12
45	0.85	1.18
60	0.72	1.26

Table 4 shows that when the inclination of the cable increases, the concentration of surface corrosion factors on the B1 cable section gradually decreases compared to the A1 cable section. However, the diffusion coefficient of corrosion factors follows the reverse law, with the diffusion coefficient of corrosion factors on the B1 cable segment being significantly greater than that on the A1 cable section, which indicates that the diffusion rate of corrosion factors downward along the cable's inclination angle is more significant than that upward along the cable's inclination angle. This phenomenon is caused by the following: as the cable's inclination increases, substances such as water, corrosion factors, and oxygen travel lower along the cable, and the corrosion factors continually penetrate from the surface to the interior. As a result, the concentration of surface corrosion factors in section B1 is low, and the diffusion coefficient of corrosion factors is large. The gravity effect will obstruct the upward diffusion of corrosion factors and other substances along the A1 cable segment, resulting in the aggregation of corrosion factors on the cable surface and a decrease in the diffusion coefficient of corrosion factors.

The above depicts the process of solving the concentration of surface corrosion factors and diffusion coefficient of corrosion factors on square-hole defect cable. The diffusion form of corrosion factors in the annular hole cable is studied using the analysis method described above. The spatial diffusion model of corrosion factors in the annular hole defect cable is built to obtain the expression of the concentration of surface corrosion factors and the diffusion coefficient of corrosion factors. It can still be separated into two segments for convenience of representation: the C1 segment above the defect location and the D1 segment below the defect location. The expression is as follows.

(1) The expression of concentration of surface corrosion factors and diffusion coefficient of corrosion factors in the C1 cable segment:

$$\left.\begin{array}{l} C_s = 0.393\cdot(t)^{0.24}\cdot(h)^{0.12}\cdot(T_1)^{0.16}(W)^{0.57}\cdot(\cos\theta)^{0.46} + 0.413 \\ D = \left[13.32\cdot(t)^{0.26}\cdot(h)^{0.06}\cdot(T_1)^{0.13}(W)^{0.63}\cdot(\cos\theta)^{0.47} + 1.572\right] \times 10^{-10} \end{array}\right\} \quad (14)$$

According to the expression method of the concentration of surface corrosion factors and diffusion coefficient of corrosion factors on the C1 cable section, the concentration of surface corrosion factors and diffusion coefficient of corrosion factors on the D1 cable section can be expressed as follows.

$$\left.\begin{array}{l} C_s = \eta_2\left[0.393\cdot(t)^{0.24}\cdot(h)^{0.12}\cdot(T_1)^{0.16}(W)^{0.57}\cdot(\cos\theta)^{0.46} + 0.413\right] \\ D = \eta_3\left[13.32\cdot(t)^{0.26}\cdot(h)^{0.06}\cdot(T_1)^{0.13}(W)^{0.63}\cdot(\cos\theta)^{0.47} + 1.572\right] \times 10^{-10} \end{array}\right\} \quad (15)$$

Table 5 shows the values of η_2 and η_3 at different tilt angles. The amplification coefficients of the concentration of surface corrosion factors and diffusion coefficients of corrosion factors on the D1 segment cable under tilt angles are shown in Table 5. Compared with the results shown in Table 4, the tilt angle has a more significant impact on the concentration of surface corrosion factors and diffusion coefficient of corrosion factors on the ring defect cable.

Table 5. The range of values of η_2 and η_3 under different tilt angles.

Tilt Angle	η_2	η_3
0	1.04	1.02
30	0.89	1.17
45	0.81	1.22
60	0.68	1.35

5. Spatial Diffusion Model of Corrosion Factors

Based on the analysis results above, the spatial diffusion model of corrosion factors in the cable can be established under various tilt angles, temperatures, humidity, corrosion periods, and defect areas, divided into upper and lower cable sections for display. Firstly, the spatial diffusion model of corrosion factors in the square hole defective cable is exhibited. The results are shown below.

(1) The spatial diffusion model of corrosion factors in the A1 cable segment.

$$\left.\begin{array}{l} C_s = 0.387\cdot(t)^{0.32}\cdot(h)^{0.08}\cdot(T_1)^{0.12}(W)^{0.52}\cdot(\cos\theta)^{0.43} + 0.216 \\ D = \left[11.42\cdot(t)^{0.23}\cdot(h)^{0.05}\cdot(T_1)^{0.11}(W)^{0.54}\cdot(\cos\theta)^{0.41} + 1.231\right] \times 10^{-10} \\ C(x,y,z) = C_S \times \left(1 - erf\frac{x}{2\sqrt{D\times t}} \times erf\frac{y}{2\sqrt{D\times t}} \times erf\frac{z}{2\sqrt{D\times t}}\right) \end{array}\right\} \quad (16)$$

(2) The spatial diffusion model of corrosion factors in the B1 cable segment.

$$\left.\begin{array}{l} C_s = \eta_1\left[0.387\cdot(t)^{0.32}\cdot(h)^{0.08}\cdot(T_1)^{0.12}(W)^{0.52}\cdot(\cos\theta)^{0.43}+0.216\right] \\ D = \eta\left[11.42\cdot(t)^{0.23}\cdot(h)^{0.05}\cdot(T_1)^{0.11}(W)^{0.54}\cdot(\cos\theta)^{0.41}+1.231\right]\times 10^{-10} \\ C(x,y,z) = C_S\times\left(1-erf\frac{x}{2\sqrt{D\times t}}\times erf\frac{y}{2\sqrt{D\times t}}\times erf\frac{z}{2\sqrt{D\times t}}\right) \end{array}\right\} \quad (17)$$

Among them, the values of η_1 and η under different tilt angles are shown in Table 5. The spatial diffusion model of corrosion factors within the annular defect cable can be depicted as follows using the approach described above.

(1) The spatial diffusion model of corrosion factors in the C1 cable segment.

$$\left.\begin{array}{l} C_s = 0.393\cdot(t)^{0.24}\cdot(h)^{0.12}\cdot(T_1)^{0.16}(W)^{0.57}\cdot(\cos\theta)^{0.46}+0.413 \\ D = \left[13.32\cdot(t)^{0.26}\cdot(h)^{0.06}\cdot(T_1)^{0.13}(W)^{0.63}\cdot(\cos\theta)^{0.47}+1.572\right]\times 10^{-10} \\ C(x,y,z) = C_S\times\left(1-erf\frac{x}{2\sqrt{D\times t}}\times erf\frac{y}{2\sqrt{D\times t}}\times erf\frac{z}{2\sqrt{D\times t}}\right) \end{array}\right\} \quad (18)$$

(2) The spatial diffusion model of corrosion factors in the D1 cable segment.

$$\left.\begin{array}{l} C_s = \eta_2\left[0.393\cdot(t)^{0.24}\cdot(h)^{0.12}\cdot(T_1)^{0.16}(W)^{0.57}\cdot(\cos\theta)^{0.46}+0.413\right] \\ D = \eta_3\left[13.32\cdot(t)^{0.26}\cdot(h)^{0.06}\cdot(T_1)^{0.13}(W)^{0.63}\cdot(\cos\theta)^{0.47}+1.572\right]\times 10^{-10} \\ C(x,y,z) = C_S\times\left(1-erf\frac{x}{2\sqrt{D\times t}}\times erf\frac{y}{2\sqrt{D\times t}}\times erf\frac{z}{2\sqrt{D\times t}}\right) \end{array}\right\} \quad (19)$$

Among them, Table 5 shows the values of η_2 and η_3 under various tilt angles. The concentration of corrosion factors at each spatial position of the cable was calculated using this model to verify the accuracy of the model above. The calculated results were compared with the measured results, as shown in Figures 11–14. The findings demonstrate that the prediction results of the model are lower than the experimental test results, and the relative error between the two is within 15%, illustrating that the prediction model proposed in this article has a certain accuracy.

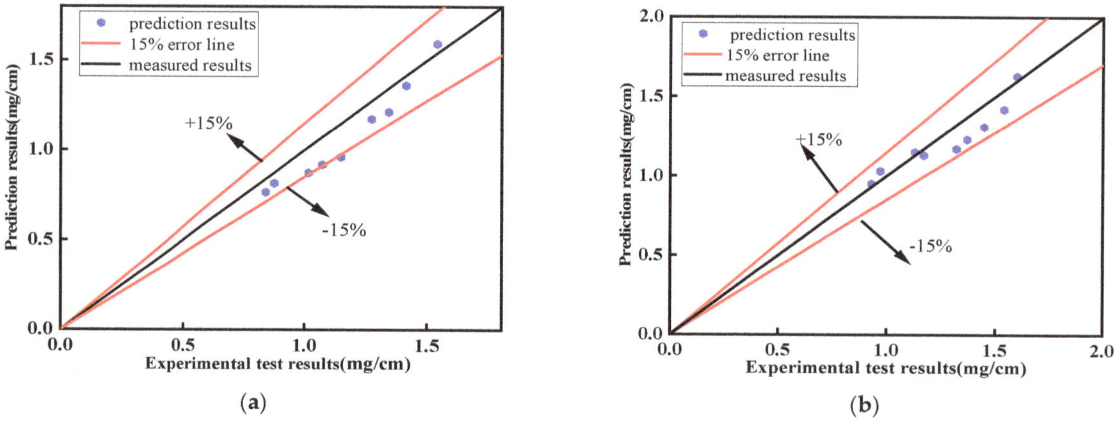

Figure 11. Comparison results of the A1 segment cable (defect size is 2 cm × 2 cm): (**a**) tilt angle 30°; (**b**) tilt angle 45°.

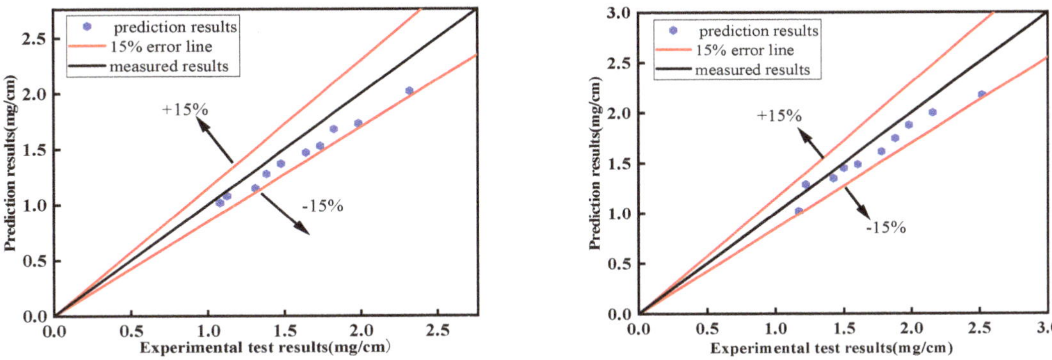

(**a**) (**b**)

Figure 12. Comparison results of the B1 segment cable (defect size is 2 cm × 2 cm): (**a**) tilt angle 30°; (**b**) tilt angle 45°.

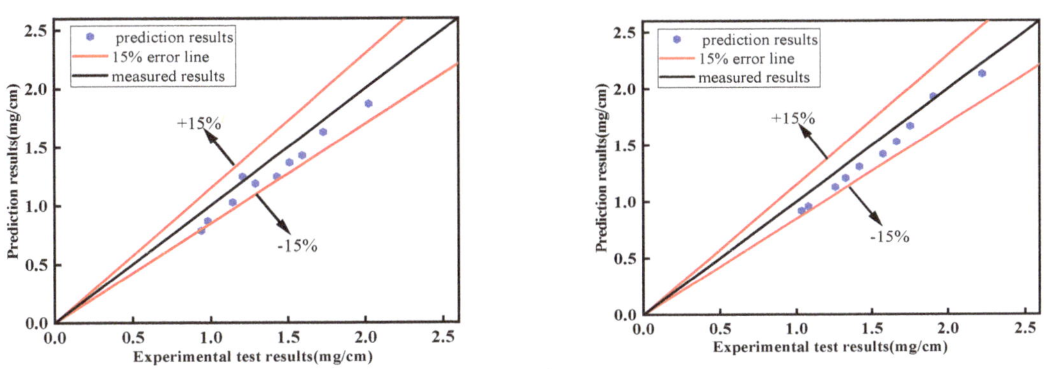

(**a**) (**b**)

Figure 13. Comparison results of the C1 segment cable (defect size width is 3 cm): (**a**) tilt angle 0°; (**b**) tilt angle 30°.

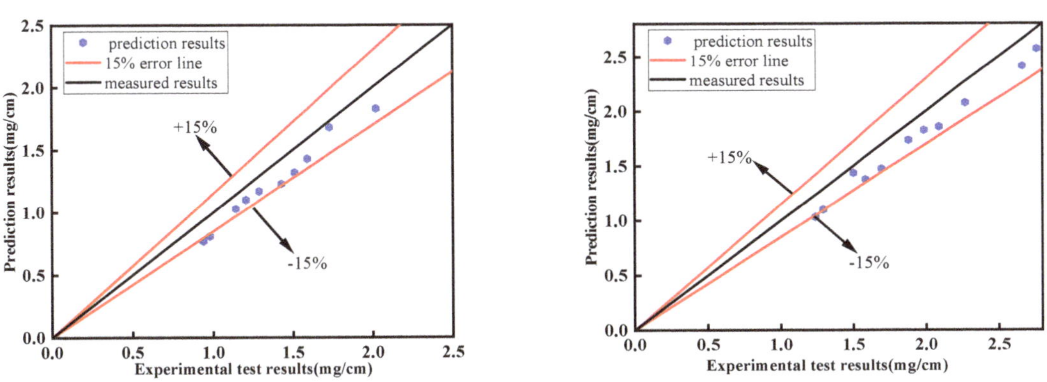

(**a**) (**b**)

Figure 14. Comparison results of the D1 segment cable (defect size width is 3 cm): (**a**) tilt angle 0°; (**b**) tilt angle 30°.

6. Conclusions

The spatial diffusion model of corrosion factors was built using the ML method based on the test results of concentrations of corrosion factors at various positions in the space of the cable system and Fick's second law. The relevant empirical formulae were presented, and the accuracy of the theoretical model was verified using multiple nonlinear regression analysis methods, taking into account the effect of various parameters on the diffusion coefficient of corrosion factors and the concentration of surface corrosion factors. The specific conclusions are as follows:

(1) Five ML algorithms were used to investigate the impacts of environmental temperature, humidity, cable inclination angle, and cable defect size on the diffusion coefficient of corrosion factors and the concentration of surface corrosion factors. According to the simulation findings, the optimized BP neural network method has the best simulation effect, with fast convergence speed and good robustness.

(2) The inclination angle and the defect size of the cable are the primary parameters influencing the diffusion coefficient of corrosion factors and the concentration of surface corrosion factors—their weight values are all above 0.25. Although environmental temperature, humidity, and corrosion time all affect the diffusion rate of corrosion factors, they are more limited, and their weight values are all within 0.2.

(3) Due to the increase in the inclination of the cable, substances such as water, corrosion factors, and oxygen will move downwards along the cable, and corrosion factors will continually penetrate from the surface to the interior of the cable. The concentration of surface corrosion factors on the cable above the defect is low, and the diffusion coefficient of corrosion factors is high. The gravity effect will impede the upward diffusion of corrosion factors and other substances along the cable, resulting in the aggregation of corrosion factors on the surface of the cable below the defect and a drop in the diffusion coefficient of corrosion factors.

(4) The cross-sectional loss or corrosion weight loss of the steel wire inside the cable mainly occurs where corrosion factors gather. The area with the highest concentration of corrosion factors is the area where corrosion occurs most severely. The diffusion path of corrosion factors is the area where corrosion is most likely to occur during the service process of the cable. Studying the diffusion trend of corrosion factors can provide guidance for the corrosion protection of steel wires in different parts. The concentration of corrosion factors at each spatial position of the cable was calculated using the model in this article. The predicted results of the spatial diffusion model of cable were lower than the measured results of the experiment, with a relative error of 15%, showing a good level of agreement, which can effectively forecast and assess the spatial diffusion status of cable's corrosion factors in practical engineering.

Author Contributions: Conceptualization, S.L., G.Y., W.W. and X.Y.; methodology, G.Y.,W.W. and X.Y.; software, S.L., X.Y. and X.H.; validation, S.L., X.H. and C.R.; formal analysis, S.L., G.Y., W.W., X.Y., C.R. and H.L.; investigation, G.Y., W.W. and X.Y.; resources, X.H., C.R. and H.L.; data curation, S.L., X.H., C.R. and H.L.; writing—original draft preparation, S.L., G.Y., W.W., X.Y., X.H. and C.R.; writing—review and editing, S.L. and G.Y.; visualization, X.Y., X.H., C.R. and H.L.; supervision, S.L. and W.W. All authors have read and agreed to the published version of the manuscript.

Funding: The National Natural Science Foundation of China (Grant No. 52178273), the Natural Science Foundation of Chongqing (Grant No. cstc2021jcyj-msxmX1159), the Chongqing Talent Plan Project (Grant No. cstc2022ycjh-bgzxm0124), the Open Fund of State Key Laboratory of Mountain Bridge and Tunnel Engineering (Grant No. SKLBT-YF2105), the Team Building Project for Graduate Tutors in Chongqing (Grant No. JDDSTD2022003), the Joint Training Base Construction Project for Graduate Students in Chongqing (Grant No. JDLHPYJD2020004) and Research and Innovation Program for Graduate Students in Chongqing (Grant No. CYS22392 and 2022S0007).

Data Availability Statement: The data presented in this study are available on request from the corresponding author.

Conflicts of Interest: The authors declare no conflict of interest.

References

1. Liu, Z.; Guo, T.; Huang, L.; Pan, Z. Fatigue life evaluation on short suspenders of long-span suspension bridge with central clamps. *J. Bridge Eng.* **2017**, *22*, 04017074. [CrossRef]
2. Morcous, G.; Lounis, Z.; Cho, Y. An integrated system for bridge management using probabilistic and mechanistic deterioration models: Application to bridge decks. *KSCE J. Civ. Eng.* **2010**, *14*, 527–537. [CrossRef]
3. Zheng, G.; Tang, W.; Wang, P. *Bridge Cable Structures*; China Communications Press: Beijing, China, 2015.
4. Aloisio, A.; Pasca, D.P.; Rosso, M.M.; Briseghella, B. Role of cable forces in the model updating of cable-stayed bridges. *J. Bridge Eng.* **2023**, *28*, 05023002. [CrossRef]
5. Greco, F.; Lonetti, P.; Pascuzzo, A. Dynamic Analysis of Cable-Stayed Bridges Affected by Accidental Failure Mechanisms under Moving Loads. *Math. Probl. Eng.* **2013**, *2013*, 302706. [CrossRef]
6. Ammendolea, D.; Bruno, D.; Greco, F.; Lonetti, P.; Pascuzzo, A. An investigation on the structural integrity of network arch bridges subjected to cable loss under the action of moving loads. *Procedia Struct. Integr.* **2020**, *25*, 305–315. [CrossRef]
7. Mozos, C.M.; Aparicio, A.C. Parametric study on the dynamic response of cable stayed bridges to the sudden failure of a stay, Part I: Bending moment acting on the deck. *Eng. Struct.* **2010**, *32*, 3288–3300. [CrossRef]
8. Mozos, C.M.; Aparicio, A.C. Parametric study on the dynamic response of cable stayed bridges to the sudden failure of a stay, Part II: Bending moment acting on the pylons and stress on the stays. *Eng. Struct.* **2010**, *32*, 3301–3312. [CrossRef]
9. Qiao, Y.; Miao, C.; Sun, C. Evaluation of corrosion fatigue life for corroded wire for cable-supported bridge. *J. Civ. Environ. Eng.* **2017**, *39*, 115–121.
10. Stewart, M.G.; Al-Harthy, A. Pitting corrosion and structural reliability of corroding RC structures: Experimental data and probabilistic analysis. *Reliab. Eng. Syst. Saf.* **2006**, *93*, 373–382. [CrossRef]
11. Li, R.; Miao, C.; Wei, T. Experimental study on corrosion behavior of galvanized steel wires under stress. *Corros. Eng. Sci. Technol.* **2020**, *55*, 622–633. [CrossRef]
12. Miao, C.; Yu, J.; Mei, M. Distribution law of corrosion pits on steel suspension wires for a tied arch bridge. *Anti-Corros. Methods Mater.* **2016**, *63*, 166–170. [CrossRef]
13. Betti, R.; West, A.C.; Vermaas, G.; Cao, Y. Corrosion and embrittlement in high-strength wires of suspension bridge cables. *J. Bridge Eng.* **2005**, *10*, 151–162. [CrossRef]
14. Furuya, K.; Kitagawa, M.; Nakamura, S.I.; Suzumura, K. Corrosion mechanism and protection methods for suspension bridge cables. *Struct. Eng. Int.* **2000**, *10*, 189–193. [CrossRef]
15. Suzumura, K.; Nakamura, S.I. Environmental factors affecting corrosion of galvanized steel wires. *J. Mater. Civ. Eng.* **2004**, *16*, 1–7. [CrossRef]
16. Nakamura, S.I.; Suzumura, K. Hydrogen embrittlement and corrosion fatigue of corroded bridge wires. *J. Constr. Steel Res.* **2008**, *65*, 269–277. [CrossRef]
17. Sloane, M.J.D.; Betti, R.; Marconi, G.; Hong, A.L.; Khazem, D. Experimental analysis of a nondestructive corrosion monitoring system for main cables of suspension bridges. *J. Bridge Eng.* **2013**, *18*, 653–662. [CrossRef]
18. Zin, I.M.; Howard, R.L.; Badger, S.J.; Scantlebury, J.D.; Lyon, S.B. The mode of action of chromate inhibitor in epoxy primer on galvanized steel. *Prog. Org. Coat.* **1998**, *33*, 203–210. [CrossRef]
19. Yao, G.; Yang, S.; Zhang, J.; Leng, Y. Analysis of corrosion-fatigue damage and fracture mechanism of in-service bridge cables/hangers. *Adv. Civ. Eng.* **2021**, *2021*, 6633706.
20. Yang, S.; Yao, G.; Zhang, J.; Shi, K. The corrosion fatigue characteristic of steel strand experiencing an artificial accelerated salt fog aging. *Mater. Rep.* **2018**, *32*, 1988–1993.
21. Qin, H.; Shen, Q.; Ou, J.; Zhu, W. Long-term monitoring reliability and life prediction of fiber bragg grating-based self-sensing steel strands. *Adv. Civ. Eng.* **2020**, *2020*, 7687039. [CrossRef]
22. Furuya, T.; Kawafuku, J.; Satoh, H.; Shimogori, K.; Aoshima, A.; Takeda, S. A corrosion testing method for titanium in nitric acid environments. *ISIJ Int.* **1991**, *31*, 189–193. [CrossRef]
23. Hamilton, H.R., III. Bridge stay cable corrosion protection. I: Grout injection and load testing. *J. Bridge Eng.* **1998**, *3*, 64–71. [CrossRef]
24. Hamilton, H.R., III; Breen, J.E.; Frank, K.H. Bridge stay cable corrosion protection. II: Accelerated corrosion tests. *J. Bridge Eng.* **1998**, *3*, 72–81. [CrossRef]
25. Matsumoto, M.; Yagi, T.; Shigemura, Y.; Tsushima, D. Vortex-induced cable vibration of cable-stayed bridges at high reduced wind velocity. *J. Wind. Eng. Ind. Aerodyn.* **2001**, *89*, 633–647. [CrossRef]
26. Wu, Z.; Ding, Z.; Sun, C.; Zhang, L. Finite element analysis of section stress and failure mode of steel strand. *China Sci.* **2018**, *13*, 2623–2628.
27. Wang, X.; Wang, J.; Jin, M.; Yang, C. Analysis of coupling injury caused by corrosion and fatigue of fable stayed cables. *J. Taiyuan Univ. Sci. Technol.* **2019**, *40*, 472–476.
28. Yang, S.; Zhang, J.; Yao, G. Analysis on corrosion-fatigue damage and fracture mechanism of cables /hangers in service bridges. *J. Highw. Transp. Res. Dev.* **2019**, *36*, 80–86.

29. Guo, Z.; Li, L.; Yao, G. Corrosion behavior analysis of wire-steel in cables and its prediction under combined effect of cycling loading and eroded environment. *J. Chongqing Univ.* **2018**, *41*, 48–57.
30. Rosso, M.M.; Asso, R.; Aloisio, A.; Di, B.M.; Cucuzza, R.; Greco, R. Corrosion effects on the capacity and ductility of concrete half-joint bridges. *Constr. Build. Mater.* **2022**, *360*, 129555. [CrossRef]
31. Li, S.; Xin, J.; Jiang, Y.; Wang, C.; Zhou, J.; Yang, X. Temperature-induced deflection separation based on bridge deflection data using the TVFEMD-PE-KLD method. *J. Civ. Struct. Health Monit.* **2023**, *13*, 781–797. [CrossRef]
32. Xin, J.; Jiang, Y.; Zhou, J.; Peng, L.; Liu, S.; Tang, Q. Bridge deformation prediction based on SHM data using improved VMD and conditional KDE. *Eng. Struct.* **2022**, *261*, 114285. [CrossRef]
33. Xin, J.; Zhou, C.; Jiang, Y.; Tang, Q.; Yang, X.; Zhou, J. A signal recovery method for bridge monitoring system using TVFEMD and encoder-decoder aided LSTM. *Measurement* **2023**, *214*, 112797. [CrossRef]
34. Kim, D.S.; Lee, H.S.; Lee, S.M.; Wang, X.Y. A study on the evaluation of probabilistic durability life for RC structures deteriorated by chloride ion. *Key Eng. Mater.* **2007**, *76*, 417–420. [CrossRef]
35. Gupta, T.; Patel, K.A.; Siddique, S.; Sharma, R.K.; Chaudhary, S. Prediction of mechanical properties of rubberized concrete exposed to elevated temperature using ANN. *Measurement* **2019**, *147*, 106870. [CrossRef]
36. Yu, Y.; Zhao, X.; Xu, J.; Wang, S.; Xie, T. Evaluation of shear capacity of steel fiber reinforced concrete beams without stirrups using artificial intelligence models. *Materials* **2022**, *15*, 2407. [CrossRef]
37. Bukhsh, Z.A.; Stipanovic, I.; Saeed, A.; Doree, A.G. Maintenance intervention predictions using entity-embedding neural networks. *Autom. Constr.* **2020**, *116*, 103202. [CrossRef]
38. Miao, P. Prediction-based maintenance of existing bridges using neural network and sensitivity analysis. *Adv. Civ. Eng.* **2021**, *2021*, 4598337. [CrossRef]
39. Cao, B. An Improved Decision Tree Algorithm Based on Density. Master's Thesis, Dalian University of Technology, Dalian, China, 2016.
40. Qiao, P.; Liang, Z.; Xu, K.; Zhong, C.; Qin, F. Evaluation of technical condition of medium and small span bridge based on machine learning. *J. Chang. Univ. (Nat. Sci. Ed.)* **2021**, *41*, 39–52.
41. Wu, D.; Liu, L.; Miao, R. Neural network method in bridge condition assessment by B-TBU model. *J. Jiangsu Univ. (Nat. Sci. Ed.)* **2017**, *38*, 466–471.
42. Li, Y.; Zou, Z.; Xu, L.; Wang, Y. Risk assessment study of decision tree analysis technology during bridge construction. *J. China Foreign Highw.* **2019**, *39*, 297–302.
43. Yang, S. Research on maximum power point tracking of photovoltaic system based on sparrow search algorithm to optimize BP neural network. *Sci. Technol. Innov.* **2022**, *16*, 62–63, 66.
44. Ju, X.; Wu, L.; Liu, M.; Zhang, H.; Li, T. Service life prediction for reinforced concrete wharf considering the influence of chloride erosion dimension. *Mater. Rep.* **2021**, *35*, 24075–24080, 24087.

buildings

MDPI

Article

Research on Loading Scheme for Large-Scale Model Tests of Super-Long-Span Arch Bridge

Yonghui Fan, Jianting Zhou, Chao Luo, Jun Yang *, Jingzhou Xin and Shaorui Wang

State Key Laboratory of Mountain Bridge and Tunnel Engineering, Chongqing Jiaotong University, Chongqing 400074, China; fyh1995@mails.cqjtu.edu.cn (Y.F.); jtzhou@cqjtu.edu.cn (J.Z.); luochao@mails.cqjtu.edu.cn (C.L.); xinjz@cqjtu.edu.cn (J.X.); ruiruiplace@163.com (S.W.)
* Correspondence: yangjun@mails.cqjtu.edu.cn; Tel.: +86-152-1503-5160

Abstract: A reasonable and efficient loading scheme is needed to guarantee the success of large-scale bridge tests. In this study, an array-type, self-balancing pulley-group loading system was designed based on the world's largest spanning arch bridge using a 1:10 scale model test. Automatic statistics of the required load at each loading point were realized using ANSYS, and a load optimization algorithm for loading points at different construction stages was proposed. Tests were carried out separately for the loading system using a single set of pulley groups and an array-type pulley group. Finite element models of the model bridge and the original bridge were established separately using ANSYS, and the stress results of different components during different construction stages of the main arch ring were compared. The research results show the following: (1) The load magnification factor of the single-pulley-group loading device is approximately 6.6 times, with a mechanical efficiency of 94.26%. (2) In the array-type loading system, the actual load at each loading point can reach the design value. The self-balancing characteristic of this system can eliminate the impact of vertical deformation of the structure on loading accuracy, verifying the reliability of the system. (3) The simulation results of the original bridge and the model bridge coincide well, and the stress of each component during the construction process has the same trend. At key construction stages, the maximum relative errors of the stress results of the rigid steel frame and the concrete inside the pipe of the two bridges are 8.33% and 9.34%, respectively, and the maximum absolute error of the bottom plate concrete is 0.66 MPa, verifying the correctness of the counterweight-optimization method. The loading scheme proposed in this paper can provide a reference for the design of loading systems with the same type of scale model test.

Keywords: rigid-frame arch bridge; scale model test; test design; loading system; weight calculation method

Citation: Fan, Y.; Zhou, J.; Luo, C.; Yang, J.; Xin, J.; Wang, S. Research on Loading Scheme for Large-Scale Model Tests of Super-Long-Span Arch Bridge. *Buildings* **2023**, *13*, 1639. https://doi.org/10.3390/buildings13071639

Academic Editor: Andrea Benedetti

Received: 15 May 2023
Revised: 22 June 2023
Accepted: 27 June 2023
Published: 27 June 2023

1. Introduction

According to available data, all concrete arch bridges featuring a span exceeding 300 m worldwide utilize rigid-frame arch bridge structures [1]. The Tian'e Longtan Grand Bridge, which is presently under construction, boasts a main span reaching an impressive 600 m. With an increase in span for rigid-frame arch bridges, structural stress becomes increasingly complex [2], necessitating advanced requirements for bridge design principles and construction control methodologies. Scaled model experimentation serves as an efficacious approach to ascertain the genuine structural status of arch bridges under load-bearing circumstances [3], providing benefits such as elevated safety, robust control over test conditions, and abbreviated test durations.

Researchers have conducted studies on many large bridges around the world [4–7] focusing on the design of scaled bridge tests. The efficiency and accuracy of the loading system, as an essential component of scaled bridge tests, are some of the critical factors affecting the success of these experiments. Currently, research on loading systems in scaled bridge tests conducted by scholars both domestically and internationally can be roughly divided into three categories:

1. The first category mainly uses hydraulic jacks for loading. The base of the hydraulic jack is fixed to the ground or another rigid structure, and the load is applied to the structure through the jack head. This loading method is suitable for cases with fewer loading points and shorter loading cycles [8]. For example, Liu et al. [9] conducted a 1:16 scale test of the Wasiwo Bridge which has a clear span of 95 m, studying the failure modes of catenary arches under loading at the crown and L/4 span. Wang et al. [10] proposed a double-layer corrugated steel–concrete composite arch structure and studied the bearing capacity of this type of arch using a single-point loading test at the mid-span.
2. The second category is the multi-point suspended-weight method, which applies the self-weight of counterweights as the load on the structure. For example, Liu et al. [11] used 32 tons of iron bars to load a model test to study the mechanical behavior of a steel truss arch bridge. This method results in large counterweight volumes, occupying a significant amount of experimental space and affecting the normal use of other test equipment. Li et al. [12] investigated the dynamic response of a long-span, irregular steel tube-reinforced concrete arch bridge. They conducted 1:16 scale model tests of the Yitong River Bridge with a main span of 158 m. The test was designed to simulate vehicle loads using iron bricks to simulate the self-weight of the bridge. However, due to space limitations, it was not possible to achieve an equivalent weight loading of the self-weight.
3. To solve the problems of large counterweight space and high counterweight cost, some scholars have designed a third category of loading systems, the lever method. This method uses the principle of moment balance to multiply the self-weight load of the counterweight acting on the structure to be loaded. For example, Zhang et al. [13] conducted a 1:7.5 scale model test of the main arch ring of the 445-m main span of Beipan River Bridge, which is a rigid-frame arch bridge. Based on the lever principle, 48 sets of loading devices were used to achieve the equivalent counterweight of the arch rib. Tang et al. [14] conducted a 1:10 scale model test of the main arch ring of the 420-m main span of Wanxian Yangtze River Bridge, studying the feasibility of the main arch ring's formation sequence and the rationality of the internal force and deformation during the construction stage. The test achieved rib weighting at 27 loading points using the lever principle, but this method increased the workload of the weighting process as the vertical deflection of a loading point would change the load of adjacent loading points in real time, thus making it challenging to ensure the accuracy of the rib weighting at each loading point, which significantly affected the accuracy of the model test results in terms of inferring the bridge's response from the model test [15].

In summary, traditional model test loading systems have limitations, such as limited loading point numbers, large counterweight volumes, and low loading accuracy [16–19]. To address these issues, a 1:10 scale model loading system for the main arch ring of the 600-m main span of Tian'e Longtan Grand Bridge was designed. An innovative array-type, self-balancing pulley-group loading system and counterweight optimization algorithm were proposed. This loading system offers advantages, including small counterweight volume, long-term stability, elimination of repeated adjustments during multi-point loading, and high loading accuracy. Our aim is to provide a reference for the design of reduced-scale model tests for similar bridges in the field of engineering structures, overcoming the aforementioned limitations in existing loading systems.

2. Test Overview

2.1. Overview of the Original Bridge Project

The Tian'e Longtan Grand Bridge is an upper-bearing, steel tube-reinforced concrete rigid-frame arch bridge, with a main span of 600 m, a rise of 125 m, and a rise-to-span ratio of 1/4.8 [20], as shown in Figure 1. The main arch curve is in the form of a catenary, with an arch axis coefficient of m = 1.9. The arch ring adopts a concrete box section with constant width and variable height, with a box width of 6.5 m, a box height of 12 m at the arch foot, and a box height of 8 m at the arch crown. The transverse center distance of the arch rib is 16.5 m, and the thickness of the arch box web gradually changes from 0.45 m at the arch crown to 0.95 m at the arch foot. The top slab thickness is 0.65 m, and the bottom slab thickness gradually changes from 0.65 m at the arch crown to 1.3 m at the arch foot. The rigid steel skeleton primarily uses Q420 steel, while the internal concrete employs C80 self-compacting micro-expansion concrete, and the external concrete utilizes C60 concrete. Notably, the external concrete is cast using a segmented ring method, dividing the external concrete horizontally into three rings, including the bottom slab, the web, and the top slab, and adopting a symmetrical casting approach with eight working faces longitudinally.

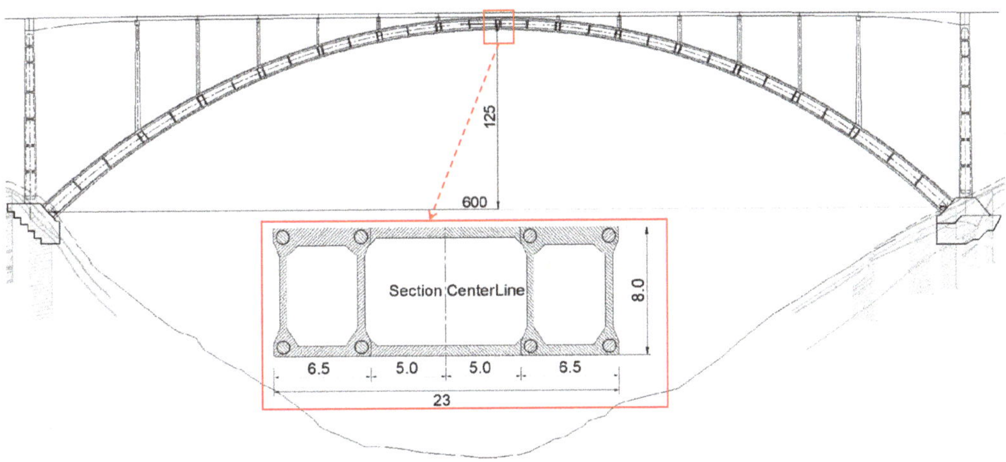

Figure 1. Elevation view of Tian'e Longtan Grand Bridge (unit: m).

2.2. 1:10 Scale Model Test of the Main Arch Ring

A 1:10 scale model experiment was conducted on the main arch ring of the original bridge, with stress equivalence as the design principle [21,22]. To minimize the influence of the scale effect on the structural response [23], the overall geometric parameters of the model bridge, such as span and rise, were strictly converted according to the scale ratio. Local designs adhered to the principle of achieving rigidity similarity and construction requirements in order to replicate the actual project conditions to the greatest extent and to strive for consistency of stress conditions between the model bridge and the original bridge. Additionally, by conducting nearly one hundred sets of experiments on the concrete mix ratio, we ensured that the concrete material properties of the model bridge closely resembled those of the original bridge [24], thereby reducing the potential impact of the scale effect on the material side.

It is important to note that the scale effect primarily affects discontinuous regions where beam theories [25], such as the Euler–Bernoulli beam theory, are not applicable. However, in the context of the present experiment, the rigidity of the skeleton structure remains continuous throughout the investigation process. When integrated with the results from Section 4.3, it becomes evident that the scale effect has not significantly impacted the precision of our experimental outcomes.

2.2.1. Scaled Model Test Scheme

The model test was carried out in the State Key Laboratory of Mountain Bridges and Tunnel Engineering at Chongqing Jiaotong University, with a main span of 60 m, as shown in Figure 2. This span exceeds the largest scaled test ever conducted for similar bridge types [14,26], making it the largest scaled model test for an arch bridge in the world to date.

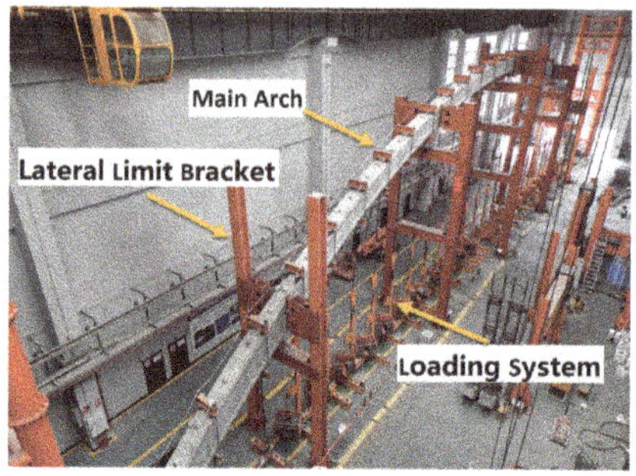

(a)

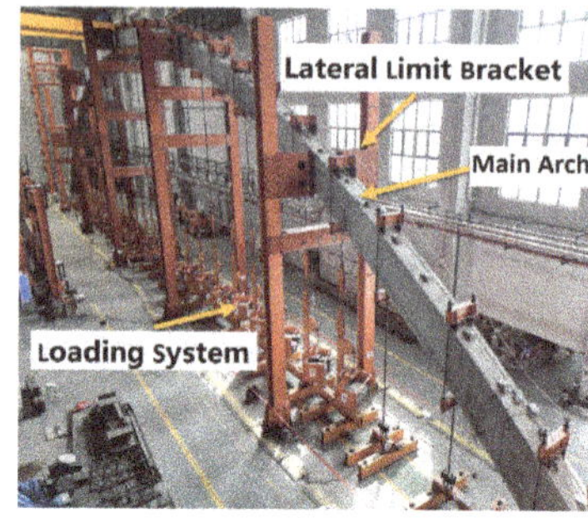

(b)

Figure 2. Picture of model test. (**a**) Figure of the arch bridge model test under viewpoint 1; (**b**) Figure of the arch bridge model test under viewpoint 2.

This structure's arch axis adopts a catenary, with an arch axis coefficient (m) of 1.9, a calculated rise of 12.5 m, and a span-to-rise ratio of 1/4.8. The arch rib adopts a concrete box-type section with uniform width and variable height, as shown in Figure 3. The box is 0.65 m wide, the box height at the foot of the arch is 1.23 m, and the box height at the apex of the arch is 0.83 m. The thickness of the box web plate gradually varies from 55 mm at the apex to 95 mm at the foot of the arch. The top plate is 65 mm thick, while the bottom plate thickness gradually varies from 65 mm at the apex to 130 mm at the foot of the arch. To enhance the local stability of the main arch ring, transverse partitions with a thickness of 50 mm are arranged every 2 m along the length of the bridge.

The model test's main arch ring mainly includes three parts: a rigid steel framework, C80 internal concrete in the pipe, and C60 external concrete. The rigid steel framework was assembled using the large-section lifting method, with a total weight of approximately 5.0 tons. The internal concrete was pumped from both sides of the arch base section symmetrically toward the arch crown using the pumping and jacking method, with a total volume of approximately 1.5 m^3.

The casting of the external concrete is consistent with the original bridge, dividing the external concrete into three rings in the transverse direction: the bottom plate, the web, and the top plate. In the longitudinal direction, each ring is divided into eight working faces, with each working face being divided into six segments for casting. The model test's segmented positions and the casting length of each segment are consistent with the original bridge, which has been scaled proportionally, and the age difference between the concrete segments is basically the same as the original bridge.

The main test content of this experiment included the strain, displacement, environmental temperature fields, and structural temperature field throughout the construction process [27]. The strain testing primarily involved the establishment of test cross-sections at 16 longitudinal points of the bridge (a total of 17 test sections are shown in Figure 4a), with fiber-optic strain sensors being arranged at each cross-section on the main steel tube, bottom plate, web, and top plate, as shown in Figure 4a. The displacement testing was mainly conducted at 8 longitudinal points of the bridge (a total of 7 test sections are shown in Figure 4b), with four measurement points set near the upper and lower chord steel tubes at each cross-section. The related testing was completed using wire-drawing displacement sensors, as shown in Figure 4b. The environmental temperature field and structural temperature field were monitored using PT100 sensors. Given that the temperature distribution of the lab environment and the structure was relatively uniform, an environmental temperature sensor and a structural temperature sensor were placed along the bridge longitudinally at intervals of 10 m.

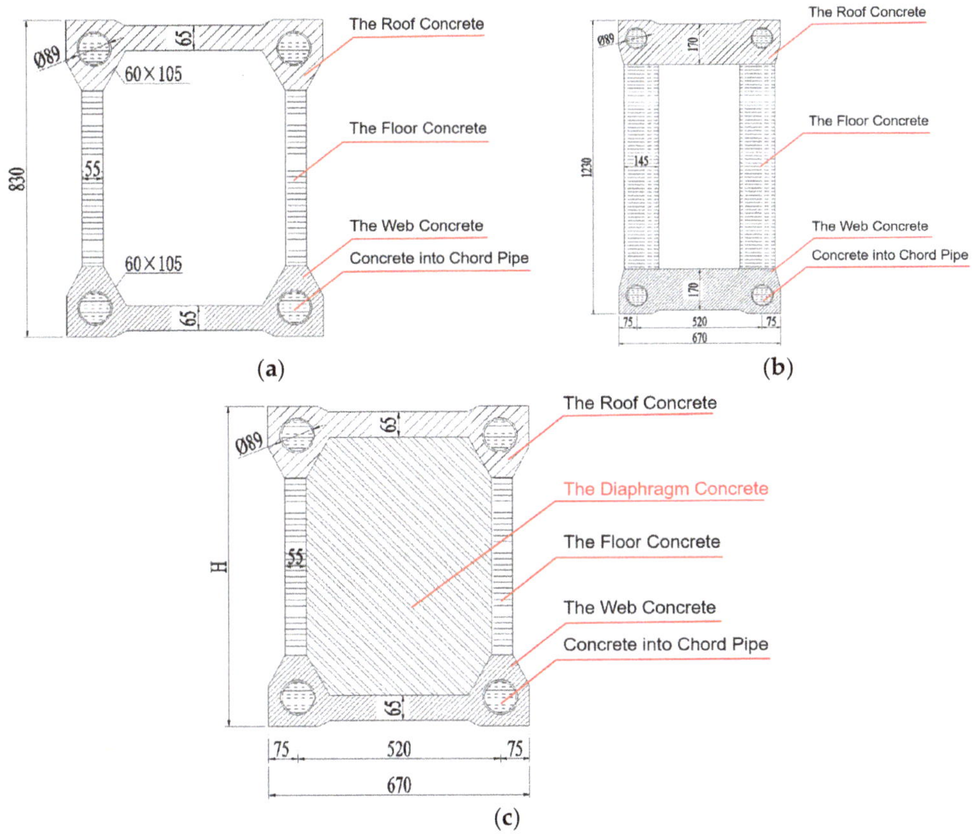

Figure 3. Cross-section of the main arch ring of the model bridge (unit: mm). (**a**) Dimensional drawing of mid-span section; (**b**) Dimensional drawing of arch foot section; (**c**) Dimensional drawing of the cross-section of the mid-span diaphragm.

2.2.2. Model Test Construction Stages

By referring to the original bridge's construction process, the construction of the main arch ring is divided into 20 stages, with the specific content of each stage shown in Table 1.

Table 1. Division of the construction stages of the main arch ring.

Serial Number	Construction Content
1	Closure of rigid steel frame
2	Pouring concrete into the pipe
3~8	Pouring concrete for base slab
9~14	Pouring web concrete
15~20	Pouring roof concrete

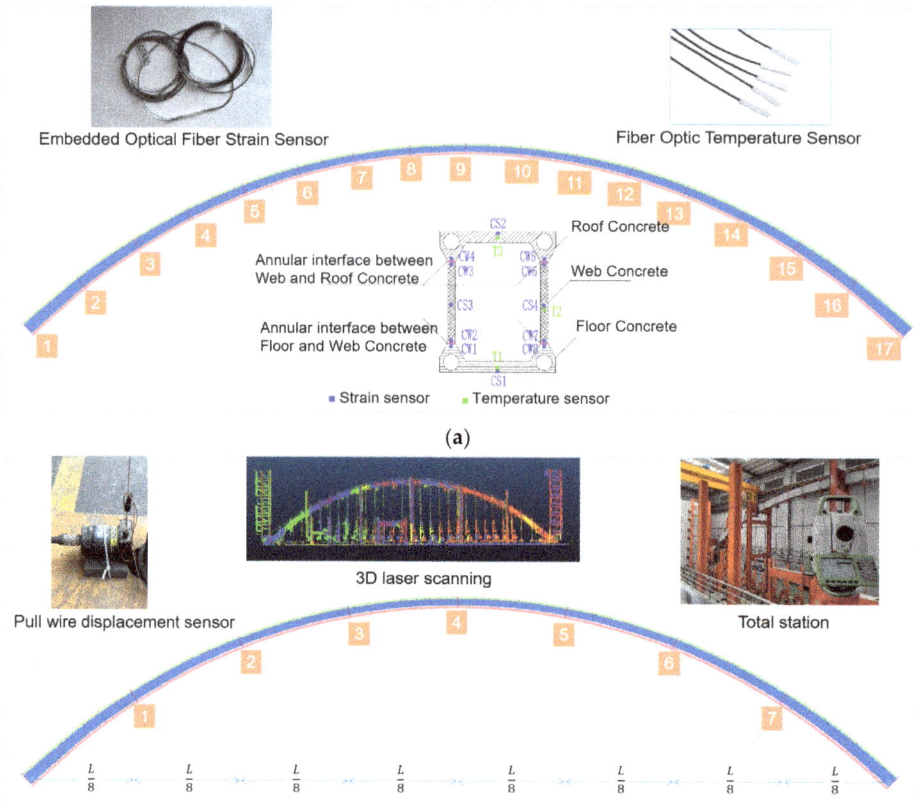

Figure 4. Schematic diagrams of test cross-sections and test methods. (**a**) Cross-section of strain test and sensor display diagram; (**b**) Displacement testing cross-section and display diagram of testing equipment.

3. Loading System Design

The model test was designed based on the principle of stress equivalence. Under a scale factor of 1:10, a nine-time self-weight load should be applied to the model arch. The self-weight load is approximately uniformly distributed along the longitudinal direction of the main arch ring. Theoretically, the nine-time self-weight load should also be applied as a uniformly distributed load. Considering the available space in the laboratory, operational feasibility, and finite element calculation results, a loading point was arranged every 2 m along the bridge's longitudinal direction, using 29 concentrated forces to replace the uniformly distributed load. To prevent local buckling or distortion buckling of the structure near the loading points during the construction stage, a transverse partition was set up at each loading point section.

3.1. Loading Scheme

3.1.1. Working Principle of Single-Point Loading Device

Reasonable and accurate counterweights directly affect the accuracy of the test data, and an efficient and feasible counterweight method is an important part of the test design. To solve the drawbacks of traditional loading methods, a pulley-group loading scheme was designed [28–30], and its working principle is shown in Figure 5. In the figure, pulleys X1, X2, and X3 form a fixed pulley group, which is fixed on the ground, and pulleys S1, S2, and

S3 form a movable pulley group, which is connected to the upper structure using elastic supports to simulate the vertical displacement of the main arch ring. The pulley groups are connected by a wire, starting from the axis of pulley S3, winding counterclockwise around pulleys X1, S3, X2, S2, X3, and S1 in sequence, and a counterweight block is suspended at the end of the wire.

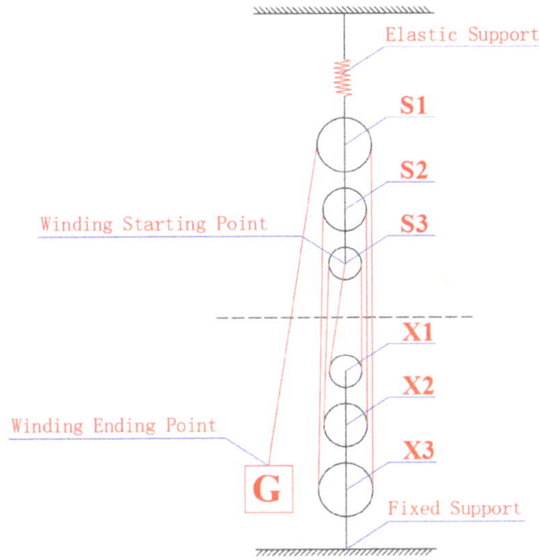

Figure 5. Diagram of the working principle of a loading system with a single set of pulleys.

In the aforementioned loading scheme, the ratio of the actual load (F) borne by the loading point to the self-weight load (G) of the counterweight block is denoted as the load magnification factor, represented by F_S. The schematic calculation diagram is shown in Figure 6. In Figure 6, taking pulley S2 as the research object, the load magnification factor $F_2 = F_6$ can be determined according to the principle of moment balance. By analogy, $F_i = G(i = 1, 2, \cdots, 7)$ can be obtained; thus, the theoretical value of the load magnification factor for the single-point loading device under this design scheme can be derived as follows:

$$F_S = \frac{F}{G} = \frac{\sum_{i=1}^{7} F_i}{G} = 7 \quad (1)$$

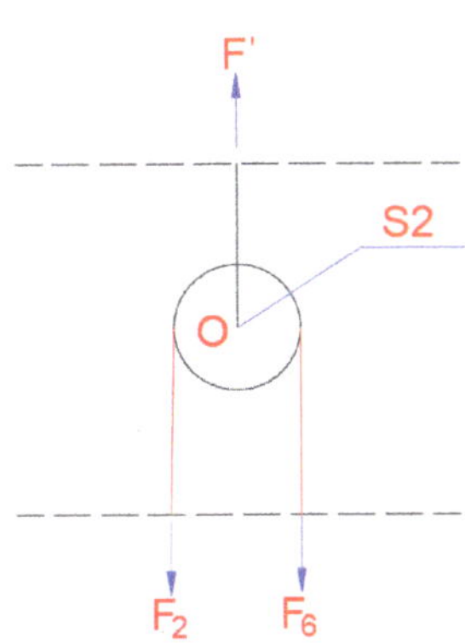

Figure 6. Simple diagram of single pulley force.

For the moving pulley system, the load magnification factor, F_S, alters along with the changes in the number of pulleys, and the relationship between the two is expressed as follows:

$$F_S = 2X_m + 1, (X_m \in N_+) \quad (2)$$

where X_m represents the number of moving pulleys.

It is worth noting that as the number and the diameter of the pulleys increase, the total frictional force between the pulleys and the rope will also escalate, causing a reduction in the mechanical efficiency of the single pulley system's loading apparatus. Simultaneously, a rise in frictional force extends the time required for the tension of the outer wire rope to reach the inner wire rope. Therefore, when designing an experiment, comprehensive consideration should be given to various factors, such as the load magnification factor, mechanical efficiency, and the time required for the system to stabilize.

3.1.2. Design Scheme of Loading Device with a Single Set of Pulley Group

Following the design principle discussed in the previous section, the design of the single-pulley-group loading device was completed, as shown in Figure 7. In this apparatus, the movable pulley group is connected to the main arch ring through the main arch ring's fixture and connecting rod. The stationary pulley group is connected to the ground using the lower jig and the lower jig base. Both pulley groups consist of three steel pulleys, with diameters of 310 mm, 230 mm, and 130 mm, respectively. The steel wire rope has a diameter of 8.7 mm. A counterweight box is used to house the counterweights, allowing sequential loading based on experimental requirements. The tension sensor can display and record the load values at the loading points on the main arch ring in real time.

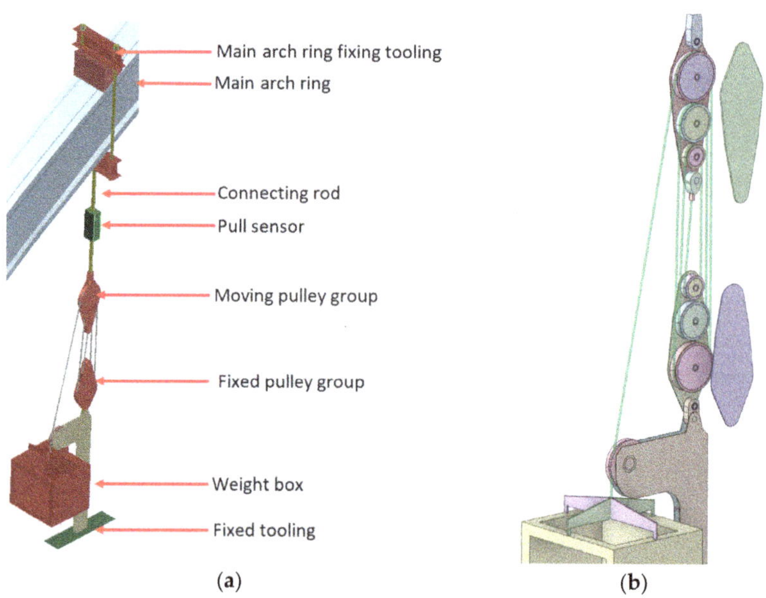

Figure 7. Schematic diagrams of a single-pulley-group loading system. (**a**) Overall diagram of a loading system with a single set of pulley group; (**b**) Partial view of a loading system with a single set of pulley group.

3.1.3. Loading System's Overall Layout Scheme

In accordance with the experimental requirements, the loading devices described in the previous section were sequentially arranged along the longitudinal direction of the bridge, forming an array-based pulley-group loading system, as shown in Figure 8. In this system, even if the counterweight near the loading point affects the vertical displacement of the current loading point, due to the equal tension on both sides of the same pulley's steel wire rope, the load amplification factor of the current loading point will gradually approach seven times over time, exhibiting a self-balancing characteristic. This system overcomes the shortcomings of the hydraulic jack-loading method, in which the load values are difficult to maintain stable over a long period; the lever-loading method, where load values are easily affected by adjacent loading points; and the direct-counterweight method, where the counterweight blocks are excessively large in volume.

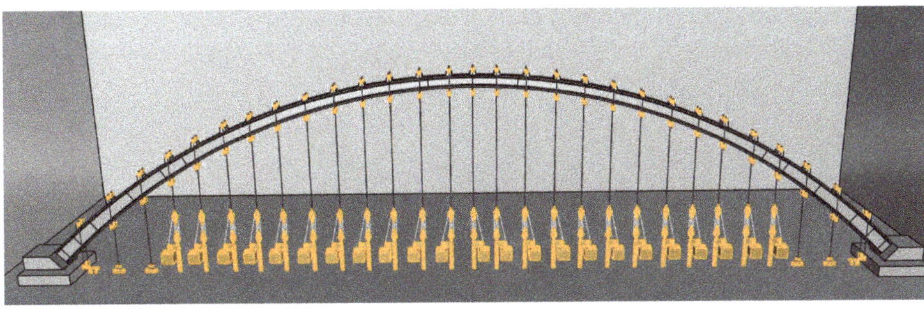

Figure 8. Overall layout plan of the model test's loading system.

3.2. *Array-Based Loading System's Load Calculation Method*

3.2.1. Load Point Counterweight Calculation Intervals

The main arch ring has a cross-section with constant width and varying height, and the thickness of the external concrete at different longitudinal positions is not the same. It would be unreasonable to simply distribute the weight of the main arch ring evenly among the various loading points. Instead, the counterweight should be calculated and allocated based on the actual size and weight of the structure near each loading point. The specific calculation method is as follows:

Let the longitudinal coordinates of the centroid of the arch springing plane on the left and right arch feet be 0 and 60, respectively. Denote the longitudinal coordinates of the 29 loading points from left to right as $X = \{X_1, X_2, \cdots, X_{29}\}$, and the longitudinal coordinates of the midlines between adjacent loading points as $x = \{x_1, x_2, \cdots, x_{28}\}$. We have

$$x_i = (X_i + X_{i+1}) \times 0.5, \quad (i = 1, 2, \cdots, 28) \tag{3}$$

The counterweight value F for the ith loading point is

$$F_i = \begin{cases} 9 \times G_j(0, x_1) & , i = 1 \\ 9 \times G_j(x_i, x_{i+1}) & , i = 2, 3, \cdots, 28 \quad (j = 1, 2, \cdots, 22) \\ 9 \times G_j(x_{28}, 60) & , i = 29 \end{cases} \tag{4}$$

In the formula, $G_j(x_i, x_{i+1})$, $(x_i, x_{i+1} \in [0, 60])$ represents the total self-weight of all constructed components within the range of the centroid longitudinal coordinates (x_i, x_{i+1}) during the jth construction stage.

3.2.2. Counterweight Values of Each Loading Point at Different Construction Stages

The model bridge is divided into 20 construction stages. Manually calculating the counterweight loads of the 29 loading points at different construction stages would be a massive workload. In this study, the ANSYS APDL [31] programming language was utilized to develop a set of counterweight load calculation programs, which can automate the aforementioned task. The basic process is shown in Figure 9.

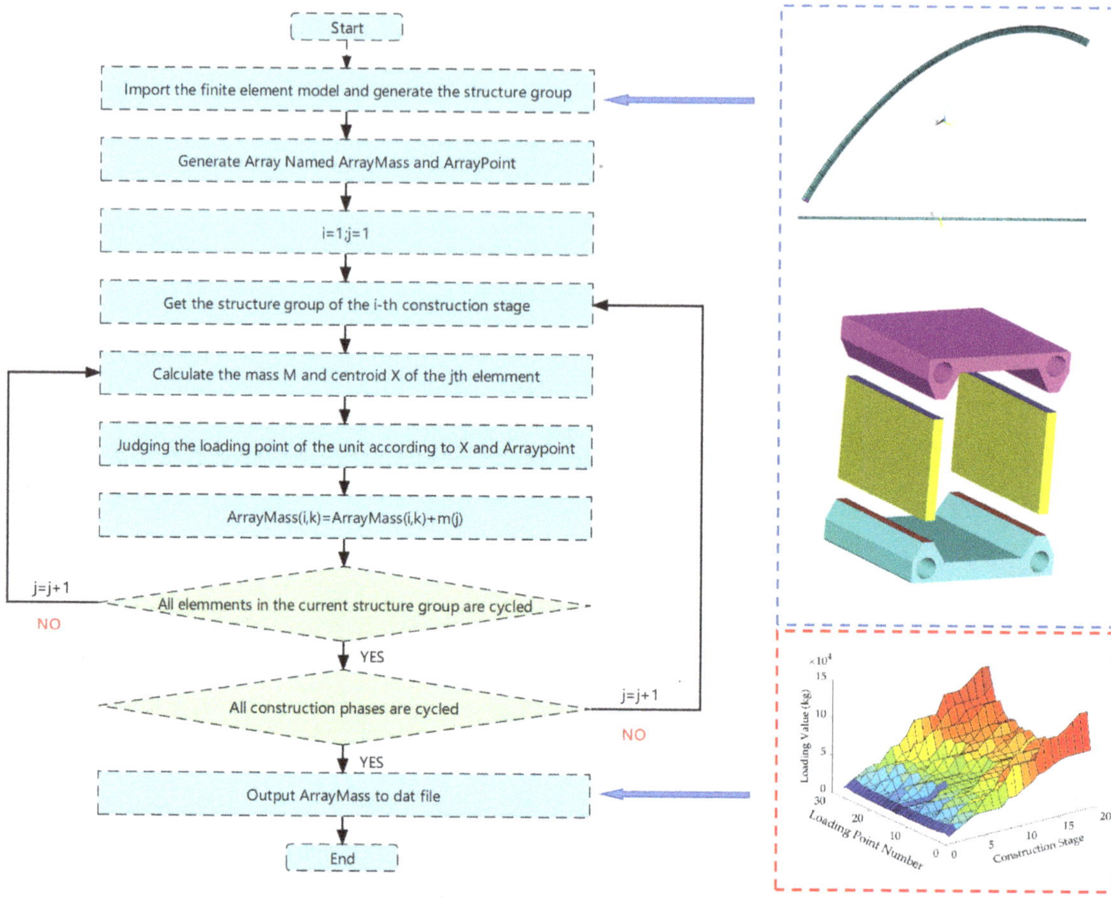

Figure 9. Flow chart of counterweight statistics at each loading point.

In Figure 9, ArrayMass is a 20 × 29 two-dimensional array, with the first dimension used to store construction stage information and the second dimension used to store loading point numbers; ArrayPoint is a one-dimensional array with 28 elements, and $x = \{x_1, x_2, \cdots, x_{28}\}$ is used to store information; i represents the construction stage, taking $i \in \{1 \le i \le 20 | i \in N^+\}$; and j represents the element number in the finite element model, taking $j \in \{1 \le j \le 6660 | j \in N^+\}$.

3.2.3. Optimization of Load Application Points' Weighting Values

In comparison to the original bridge, the model bridge design exhibits an increase of 12.30~30.27% in the cross-sectional areas of steel tubes, an increase of 7.42~23.77% in the chord members in the cross-sectional areas, a thickening in specific regions of the external concrete web, and additional thickening of the concrete near the steel tubes. Applying a nine-fold weighting value as per the model bridge would be unreasonable; thus, optimization of the weighting values for each load application point at various construction stages is required.

Given that the design of the model bridge, in comparison to the original bridge, only features scaling up without scaling down, the load value at each application point is determined based on the model bridge. These load values exceed the theoretical weighting values for the original bridge after scaling down, which prompts the introduction of the concept of a weighting reduction factor. This factor is defined as the ratio of the required weighting values for the actual model test design and the original bridge at a 1:10 scale without any modifications to component dimensions. During calculations, the following principles should be followed for the weighting reduction factor:

(1) Weighting calculations for components participating in the overall structural force, such as the main steel tubes and inner concrete of the rigid steel framework, should be nine times their actual dimensions.

(2) Weighting calculations for components participating in localized structural forces, such as the steel framework web members and top and bottom transverse links, should be nine times their actual dimensions after scaling down the original bridge components.

(3) As the rigid framework forms an arch in stages, the reduction factors at different construction stages for each load application point vary and should be calculated separately.

(4) In the model test, the weighting at each load application point is progressively applied as the construction stage advances. When calculating the weighting reduction factor, component parameters involved in the calculation should only be taken from newly poured components at the current construction stage, as the previous stage's weighting has been completed and should no longer be included in the early-stage weighting results.

Thus, in the ith construction stage, the weighting reduction factor $P_s(i,j)$ for the jth load application point can be calculated using the following formula:

$$P_s(i,j) = \frac{\sum \rho_1 A_1 + \sum \rho_2 A_2}{\sum \rho_0 A_0} \tag{5}$$

In this formula, $\sum \rho_0 A_0$ represents the sum of the product of the component area and the corresponding material density near the jth load application point of the model bridge in the ith construction stage; $\sum \rho_1 A_1$ represents the sum of the product of the overall force-bearing component area and the corresponding material density near the jth load application point of the model bridge in the ith construction stage; and $\sum \rho_2 A_2$ represents the sum of the product of the cross-sectional area and the corresponding material density of the original bridge's localized force-bearing components, which have been scaled down to 1:10, near the jth load application point in the ith construction stage.

4. Reliability Verification of the Loading System

4.1. Test of Loading System with a Single Set of Pulley Group

To verify the reliability and mechanical efficiency of the loading device, the loading system with a single set of pulley group was selected for the staged loading test, as shown in Figure 10.

In this system, the horizontal beam of the portal frame is used as a substitute for the main arch ring. Gradual loading is achieved by placing 40 cm × 40 cm × 2 cm steel plates in the counterweight box. During the test, the loading was divided into four stages, with each stage having loads of 0.2 t, 0.4 t, 0.6 t, and 0.8 t, respectively. After the loading was completed, the total mass of the counterweight box was 2.0 t. By reading real-time data from the load cell, the actual amplification factor of the pulley-group loading system can be obtained. The results are shown in Figure 11.

As shown in Figure 11, after each loading is completed, the load magnification factor of the single-pulley-group loading device requires approximately 10 min to reach a stable state. This is because the steel wire rope needs to overcome the resistance at the pulley to transmit the tension of the outermost steel wire rope layer by layer to the innermost steel wire rope. This process is referred to as the hysteresis effect. It is this effect that causes the load magnification factor to suddenly drop during each loading, as indicated by points A, B, C, and D in the figure. Moreover, after the completion of each loading level, the load magnification factor remains relatively stable around 6.6 times, indicating that the mechanical efficiency of the single-pulley-group loading system is approximately 94.29%, meeting the test requirements. Additionally, the influence of friction loss on the loading value can be eliminated based on the specific readings of the tension sensor.

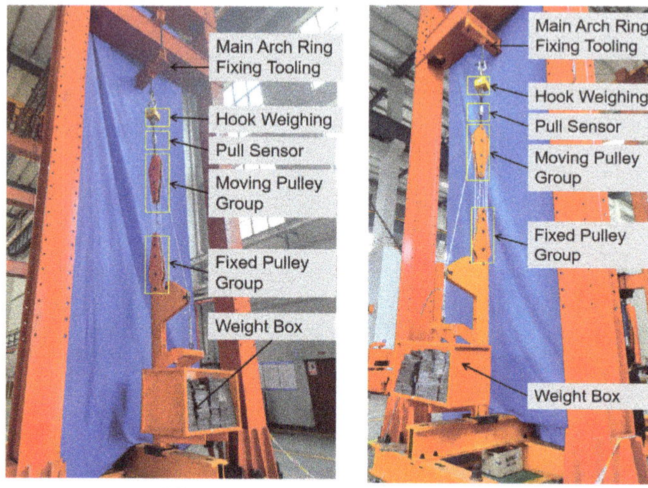

Figure 10. Test of loading system with a single set of pulley group. (**a**) Loading system with a single pulley set under viewpoint 1; (**b**) Loading system with a single pulley set under viewpoint 2.

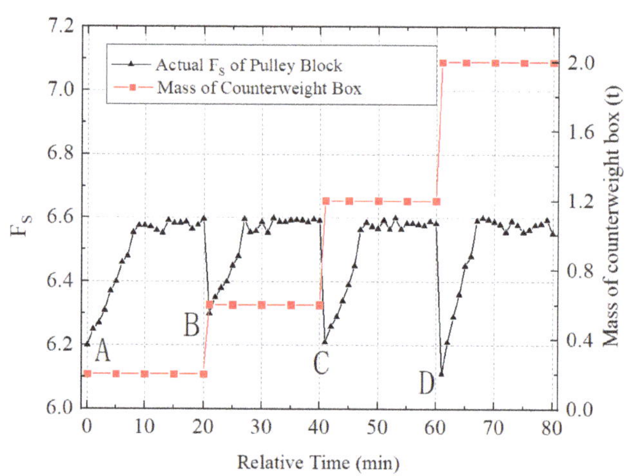

Figure 11. Time load curve of the loading test.

4.2. Array-Type Pulley-Group Loading System Test

To verify the reliability and stability of the array-type loading system in practical applications, selected loading points during the external concreting stage of the model test were chosen, and their design load values are presented in Table 2. The Donghua DHDAS data acquisition system was employed to collect real-time counterweight data of the relevant loading points, and the results are shown in Figure 12.

Table 2. Design loading value of selected loading points.

Number of Loading Points	Value of Theoretical Weight (kg)	Load Order
6#	1069	1
11#	982	2
15#	2320	3

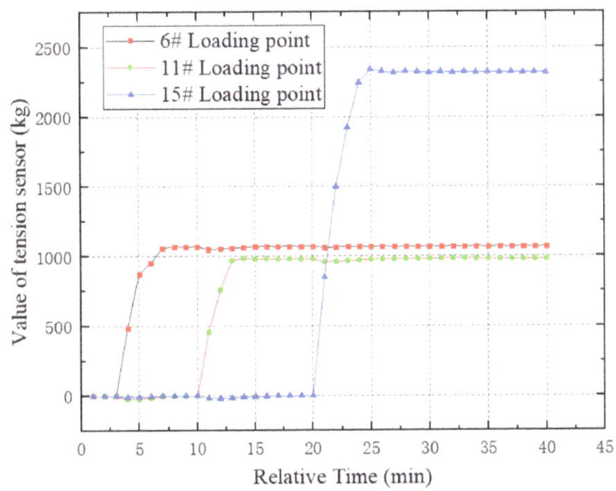

Figure 12. Time load curve of each loading point.

As shown in Figure 12, approximately 5 min after each loading point has completed loading, the values of the tension sensors can reach the designed load and remain relatively stable. Moreover, the load value of the *i*th loading point is affected by the *j*th loading point. This is because the counterweight of the latter causes a vertical deflection at the loading position of the former. This deflection is transmitted to the movable pulley group, causing the movable pulley group and the fixed pulley group to move closer together, which results in a certain degree of slack in the steel wire rope and a decrease in the total load value. However, after about 5 min, the load at each point returns to the original load due to the equal tension of the steel wire rope on both sides of the same pulley, demonstrating that the self-stabilizing characteristic of the system causes the mutual influence between the loading points to diminish over time. This confirms that the array-type pulley-group loading system exhibits good stability and high reliability.

4.3. Comparison of Finite Element Simulation Results

To verify the correctness of the load optimization algorithm for the array-type self-balancing pulley-group loading system, finite element models of the original bridge and the model bridge were established using the ANSYS APDL 18.0 software [30], as shown in Figures 13 and 14. In both cases, the rigid steel skeleton and internal concrete were simulated using Beam188 elements, while the external concrete was simulated using Shell181 elements. The displacement and rotation of both arch feet were constrained, and the construction process was simulated using birth and death elements. During calculations, only the self-weight load was considered for the original bridge, whereas the model test bridge considered not only the self-weight load but also the vertical load applied by each pulley-group loading device on the structure.

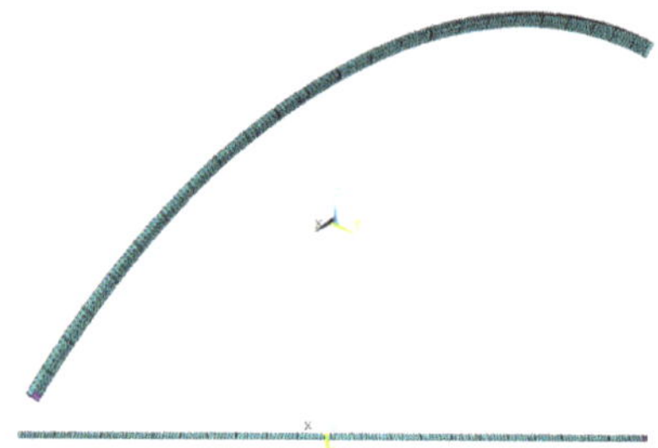

Figure 13. Finite element model of the main arch ring in the model test.

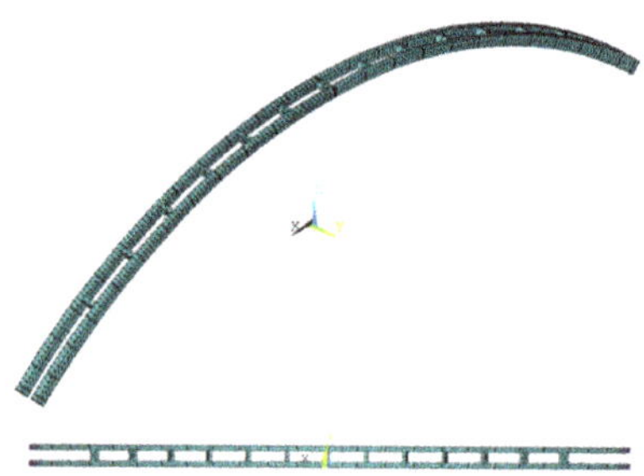

Figure 14. Finite element model of the main arch ring of the original bridge.

The stress values of the main steel tubes, internal concrete, and external concrete in the rigid steel skeletons of the original bridge and the model test bridge at key sections during the construction process were extracted and compared. Some of the results are shown in Figure 15.

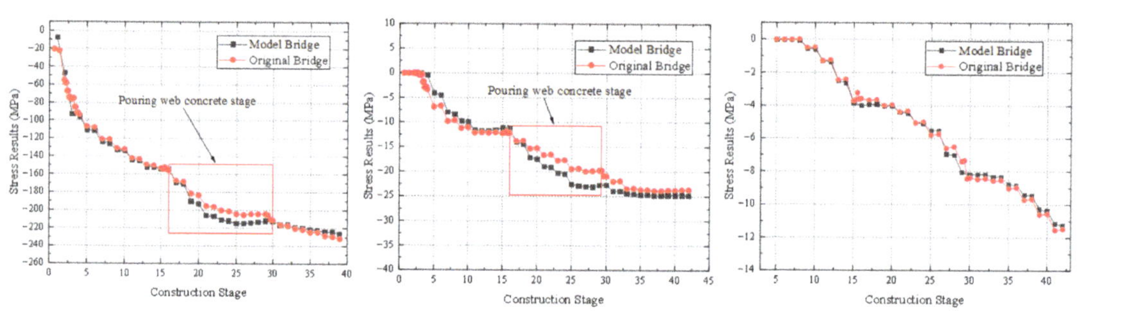

(**a**) (**b**) (**c**)

Figure 15. Comparison of the theoretical stress results of different components of the original bridge and the model test bridge. (**a**) stress results of the upper chords of 3/16-span sections; (**b**) stress results of the upper chord of 1/4 span sections; (**c**) stress results of the roof concrete of 1/4 Spans sections.

In Figure 15, the stress results of different components of the model bridge and the original bridge exhibit similar trends at various construction stages. During the web concrete pouring stage, the stress results of the model bridge are slightly larger than those of the original bridge. This is because the web concrete thickness of the original bridge, which is 45 cm, was scaled down to only 45 mm, making it difficult to arrange the reinforcement. For construction safety, the web concrete thickness in this area was increased to 55 mm, resulting in an increased self-weight load of the scaled-down web concrete and, ultimately, slightly higher stress results for the model bridge during this construction process.

In Figure 15a, the stress of the rigid steel skeleton has a larger difference during the 2–3# construction stages, while the maximum difference in the other construction stages is 8.33%. This difference arises from the fact that, for safety reasons, the model test did not apply loads after the rigid steel skeleton was connected; instead, the counterweights were applied after the internal concrete was poured. Considering that the focus of the model test is on the structural state during the external concrete process, and subsequent construction stages show good agreement between the two stress results, the stress difference during the internal concrete pouring stage can be ignored. In Figure 15b, the maximum stress deviation between the model bridge and the original bridge for the main arch ring, excluding the web concrete pouring stage, is 9.34%. As shown in Figure 15c, the stress trends of the bottom slab concrete construction stages for the model bridge's and original bridge's main arch ring are essentially the same, with a maximum stress difference of 0.66 MPa. These results indicate that the model bridge's main arch ring can adequately reflect the stress state of the original bridge during construction, verifying the correctness of the load optimization algorithm for the array-type, self-balancing pulley-group loading system.

5. Conclusions

A 1:10 scale model test was conducted based on the Tian'e Longtan Grand Bridge with a main span of 600 m. Considering the shortcomings of traditional loading methods in the test design process, an array-type, self-balancing pulley-group loading system was designed. This system was introduced, tested, optimized, and compared with the finite element simulation calculations. The main conclusions are as follows:

(1) A single set of loading devices based on the pulley-group design can theoretically amplify the dead weight of the counterweight box by seven times, which can be applied to the loading point. Multiple sets of loading devices arranged longitudinally along the bridge can form an array-type, self-balancing pulley-group loading system, which does not require repeated load adjustments, offers high precision in counterweight values, and occupies less space.

(2) A calculation method for the counterweight value at any loading point at different construction stages was proposed, and automation of load statistics was achieved using the ANSYS APDL language. Considering the differences in the detailed dimensions between the model bridge and the original bridge, the calculation principle of the counterweight reduction factor was proposed, thereby optimizing the counterweight calculation method when designing the arch bridge model test according to stress equivalence.

(3) A counterweight test of loading devices with a single set of a pulley group was conducted, revealing that due to the friction between the pulley and the steel wire rope, the actual load magnification factor of the loading devices is approximately 6.6 times, with a mechanical efficiency of about 94.29%.

(4) A full-bridge, array-type loading system counterweight test was carried out to analyze the changes in the load values at the 6th, 11th, and 15th loading points. The results show that with the array-type loading system, the actual load at each point can reach the design value. Moreover, the self-balancing feature of the loading system can eliminate the impact of vertical deformation of the structure on the loading accuracy of each point, confirming the high reliability of the system.

(5) Using the ANSYS APDL software, finite element models of the original bridge and the model test bridge were established. Based on the counterweight optimization method proposed in this study, the counterweight load required at each loading point of the model bridge was calculated and applied. The stress results of the rigid steel frame, the concrete inside the pipe, and the externally wrapped concrete at the key sections of the two bridges were compared. The results show that the stress results of different components of the model bridge and the original bridge have basically the same trend at each construction stage. During the stage of casting web concrete, the stress of the rigid steel frame and the concrete inside the pipe of the model bridge was

slightly higher than that of the original bridge due to the increased self-weight load caused by the increase in the thickness of the model bridge's web from the design value of 45 mm to 55 mm. At other key construction stages, the maximum relative errors in the stress results of the rigid steel frame and the concrete inside the pipe for the two bridges are 8.33% and 9.34%, respectively. The maximum absolute error of the bottom plate concrete is 0.66 MPa, verifying the correctness of the counterweight optimization algorithm of the array-type, self-balancing pulley-group loading system.

It is worth mentioning that the loading point of this experiment considered the position of the columns on the original bridge arch, which can meet the simulation loading requirements of the arch structure. Considering the long-term behavior and failure mode of the structure under full bridge load, it will be the next research direction.

Author Contributions: Conceptualization, Y.F. and J.Z.; methodology, Y.F. and C.L.; software, Y.F.; validation, J.X., J.Y. and S.W.; formal analysis, Y.F.; investigation, J.Z.; resources, C.L.; data curation, J.X.; writing—original draft preparation, J.Y.; writing—review and editing, S.W.; visualization, Y.F.; supervision, J.Y.; project administration, J.X.; funding acquisition, J.Z. All authors have read and agreed to the published version of the manuscript.

Funding: This work was supported by Science and Technology Project of Chongqing Municipal Transportation Bureau (Grant No. KJXM2021-0966); Guangxi key research and development plan project (Grant No. GuikeAB22036007); Independent research and development project of the State Key Laboratory of Mountain Bridge and Tunnel Engineering (Grant No. SKLBT-YF2103); Project of Science and Technology Program of Department of Transport, Hubei Province (Grant No. 2020-186-1-6) and Chongqing Jiaotong University Postgraduate Scientific Research Innovation Project (Grant No. CYB22232).

Data Availability Statement: The data presented in this study are available from the first and corresponding author upon request. The data are not publicly available due to the policy of the data provider.

Conflicts of Interest: The authors declare no conflict of interest. The sponsors had no role in the design, execution, interpretation, or writing of the study.

References

1. Zheng, J. New development trend of long span concrete arch bridges in China. *J. Chongqing Jiaotong Univ. (Nat. Sci. Ed.)* **2016**, *35*, 8–11.
2. Xin, J.; Zhou, J.; Zhou, F.; Yang, S.X.; Zhou, Y. Bearing Capacity Model of Corroded RC Eccentric Compression Columns Based on Hermite Interpolation and Fourier Fitting. *Appl. Sci.* **2019**, *9*, 24. [CrossRef]
3. Hu, Z.; Zhang, M.; Kong, X.; Wu, D. Static and dynamic similarity model of long-span steel truss arch bridge. *Chin. J. Highw.* **2014**, *27*, 82–89.
4. Zong, Z.; Zhou, R.; Huang, X.; Xia, Z. Seismic response study on a multi-span cable-stayed bridge scale model under multi-support excitations. Part I: Shaking table tests. *J. Zhejiang Univ.-Sci. A* **2014**, *15*, 351–363. [CrossRef]
5. Chen, L.; Wu, H.; Fang, Q.; Li, R. Full-scale experimental study of a reinforced concrete bridge pier under truck collision. *J. Bridge Eng.* **2021**, *26*, 05021008. [CrossRef]
6. Zheng, W.; Sheng, X.; Zhu, Z.; Luo, T.; Liu, Z. Experimental study on vibration characteristics of unit-plate ballastless track systems laid on long-span bridges using full-scale test rigs. *Sensors* **2020**, *20*, 1744. [CrossRef]
7. Li, Z.; Zhao, C.; Shu, Y.; Deng, K.; Cui, B.; Su, Y. Full-scale test and simulation of a PBL anchorage system for suspension bridges. *Struct. Infrastruct. Eng.* **2020**, *16*, 452–464. [CrossRef]
8. Chen, B.; Wei, J.; Lin, J. Experimental study on spatial stress of concrete filled steel tubular (single tube) single rib arch. *Eng. Mech.* **2006**, *23*, 99–106.
9. Liu, Z.; Wang, S.; Xin, J.; Zhou, J.; Li, Y. Experimental study on failure of single-pipe arch of concrete-filled steel catenary pipe. *J. Railw. Sci. Eng.* **2022**, 1–11. [CrossRef]
10. Wang, G.; Xia, Q.; Wang, Y.; Liu, C.; Sun, D. Research on static performance of double deck corrugated steel concrete composite arch. *J. Build. Struc.* **2021**, *42* (Suppl. S2), 358–364.
11. Liu, A.; Zhang, J.; Zhao, X.; Li, Y.; Mei, L. Zhongshan First Bridge Model Test and Theoretical Analysis. *China Highw. J.* **2005**, *18*, 75–79+83.
12. Li, Y.; Wang, Y. Static model experiment study on a concrete-filled steel tubular arch bridge. *Key Eng. Mater.* **2015**, *648*, 61–71. [CrossRef]

13. Zhang, S.; Zhao, R.; Jia, Y.; Wang, Y.; Xie, H. Model test research on long-span high-speed railway rigid skeleton concrete arch bridge. *J. Southwest Jiaotong Univ.* **2017**, *52*, 1088–1096.
14. Tang, X.; Xie, J.; Zhao, S.; Dong, W.; Pu, Q.; Yang, Y.; Weng, H. Experimental study on reinforced concrete arch model of Wanxian Yangtze River Bridge. *J. Southwest Jiao Tong Univ.* **1994**, *29*, 362–367.
15. Li, C.; Liu, J.; Jiang, L. Theoretical analysis of gravity distortion and quality distortion on scale test error. *J. Railw. Sci. Eng.* **2022**, 1–9. [CrossRef]
16. Srinivas, V.; Sasmal, S.; Ramanjaneyulu, K.; Ravisankar, K. Performance evaluation of a stone masonry–arch railway bridge under increased axle loads. *J. Perform. Constr. Fac.* **2014**, *28*, 363–375. [CrossRef]
17. Alampalli, S.; Frangopol, D.M.; Grimson, J.; Halling, M.W.; Kosnik, D.E.; Lantsoght, E.O.; Zhou, Y.E. Bridge load testing: State-of-the-practice. *J. Bridge Eng.* **2021**, *26*, 03120002. [CrossRef]
18. Lu, P.; Xu, Z.; Chen, Y.; Zhou, Y. Prediction method of bridge static load test results based on Kriging model. *Eng. Struct.* **2020**, *214*, 110641. [CrossRef]
19. Commander, B. Evolution of bridge diagnostic load testing in the USA. *Front. Built Environ.* **2019**, *5*, 57. [CrossRef]
20. Feng, S.; Xiao, F. The design of Longtan Roller Compacted Concrete gravity dam. In *Roller Compacted Concrete Dams*; Routledge: London, UK, 2018; pp. 597–602.
21. Li, X.; Wang, M.; Xiao, J.; Zou, Q.; Liu, D. Experimental study on aerodynamic characteristics of high-speed train on a truss bridge: A moving model test. *J. Wind Eng. Ind. Aerod.* **2018**, *179*, 26–38. [CrossRef]
22. Zhang, C.; Jiang, G.; Su, L.; Lei, D.; Liu, W.; Wang, Z. Large-scale shaking table model test on seismic performance of bridge-pile-foundation slope with anti-sliding piles: A case study. *Bull. Eng. Geol. Environ.* **2020**, *79*, 1429–1447. [CrossRef]
23. Gusella, F.; Orlando, M. Analysis of the dissipative behavior of steel beams for braces in three-point bending. *Eng. Struct.* **2021**, *244*, 112717. [CrossRef]
24. He, Y. *Preparation and Performance Study of C60 Fine-Stone Self-Compacting Concrete Encased in Steel Tube Arch*; Chongqing Jiaotong University: Chongqing, China, 2022.
25. Carpinteri, A.; Corrado, M.; Mancini, G.; Mancini, G.; Paggi, M. Size-Scale Effects on Plastic Rotational Capacity of Reinforced Concrete Beams. *ACI Struct. J.* **2009**, *106*, 887–896.
26. Zhang, S.; Zhao, R.; Zhan, Y.; Xu, T.; Yang, X. Model test research on influence of shrinkage and creep on deformation of high-iron concrete arch bridge. *J. Chin. Railway Soc.* **2016**, *38*, 102–110.
27. Li, S.; Xin, J.; Jiang, Y.; Wang, C.; Zhou, J.; Yang, X. Temperature-induced deflection separation based on bridge deflection data using the TVFEMD-PE-KLD method. *J. Civ. Struct. Health Monit.* **2023**, *13*, 781–797. [CrossRef]
28. Wolny, S. Dynamic loading of the pulley block in a hoisting installation in normal operating conditions. *Arch. Min. Sci.* **2009**, *54*, 261–284.
29. Liu, Y.; Jing, X.; Wan, Y. Structural Design and Simulation of Ejection Device with Pulley Block. In *2019 2nd International Conference on Mathematics, Modeling and Simulation Technologies and Applications (MMSTA 2019)*; Atlantis Press: Amsterdam, The Netherlands, 2019; pp. 26–29.
30. Reutov, A.A.; Kobishchanov, V.V.; Sakalo, V.I. Dynamic modeling of lift hoisting mechanism block pulley. *Procedia Eng.* **2016**, *150*, 1303–1310. [CrossRef]
31. ANSYS. *Structural Finite Element Analysis (FEA) Software*; ANSYS Inc.: Canonsburg, PA, USA, 2023.

buildings

MDPI

Article

Improvement in the Seismic Performance of a Super-Long-Span Concrete-Filled Steel-Tube-Arch Bridge

Dan Ye [1,2,*], Yijin Tong [1], Lijun Gan [1], Zhuoran Tang [1] and Ruijie Zhang [1]

[1] State Key Laboratory of Mountain Bridge and Tunnel Engineering, Chongqing Jiaotong University, Chongqing 400074, China; lambotong@163.com (Y.T.); 18996466751@163.com (L.G.); t1255981152@163.com (Z.T.); xjdzhangruijie@163.com (R.Z.)

[2] School of Tourism and Service Management, Chongqing University of Education, Chongqing 400065, China

* Correspondence: yedan@cque.edu.cn

Abstract: The applicability of current seismic-performance-improvement technologies needs to be studied. This research took a super-long-span CFST arch bridge with a total length of 788 m as the object on which to perform a non-linear time-history analysis and a seismic-check calculation according to the seismic response, so as to reveal the seismic weak points of the arch bridge. After the completion of the bridge's construction, we arranged and utilized the stayed buckle cables (SBCs) reasonably. The seismic performance of the super-long-span CFST arch bridge was improved through friction-pendulum bearings (FPBs) and SBCs. The research shows that FPBs can solve the problem of the insufficient shear resistance of bearings, and SBCs can address the problem whereby the compressive stress of the transverse connection of the main arch exceeds the allowable stress. Moreover, SBCs can increase the transverse stiffness of arch bridges and reduce their seismic responses. Finally, a combination of FPBs and SBCs was adopted to improve the overall seismic performance of the arch bridge and obtain the best seismic-performance-improvement effect.

Keywords: CFST arch bridge; super-long span; friction-pendulum bearing; stayed-buckle cable; seismic-performance improvement

Citation: Ye, D.; Tong, Y.; Gan, L.; Tang, Z.; Zhang, R. Improvement in the Seismic Performance of a Super-Long-Span Concrete-Filled Steel-Tube-Arch Bridge. *Buildings* **2023**, *13*, 1811. https://doi.org/10.3390/buildings13071811

Academic Editor: John Mander

Received: 9 June 2023
Revised: 27 June 2023
Accepted: 12 July 2023
Published: 16 July 2023

1. Introduction

The long-span deck-type concrete-filled steel tube (CFST) arch bridge has good terrain adaptability, while being economical and providing high levels of structural stiffness [1–6]. It is suitable for building in the mountainous area of southwest China. It is also one of the most cost-competitive bridge types in the 300-to-600-meter-span range. However, China is a country subject to frequent earthquakes, especially in the south-west region. All the earthquakes in recent years have caused severe losses.

Under the effects of earthquakes, the inertial force of a deck-type CFST arch bridge is mainly concentrated in the upper part, and excessive inertial force of the upper structure directly leads to the shear failure of the bearings between the upper and lower structures. The placement of an isolation device between the upper and lower structures can reduce the inertial force of the upper structure, which not only overcomes the problem of the insufficient shear resistance of the bearings, but also achieves the purpose of protecting the lower structure. The technique of base isolation originated in the 1960s and 1970s [7–9]. Its main purpose is to introduce a flexible bottom layer into the structure, so that the structure can be isolated from the ground motion induced by the earthquake, thus protecting the superstructure from earthquake damage. In the 1960s, Li Li, a Chinese scholar, proposed and studied the idea and method of using a sand layer and a rubber layer for isolation [10]. At the same time, Kelly, an American scholar, proposed the use of laminated rubber bearings as a method of isolation [11]. Later, Ali and others used the finite element method to conduct numerical simulations and calculated and evaluated the damping effect of lead–rubber bearings applied to long-span cable-stayed bridges [12]. Since then, foundation-isolation technology has developed rapidly; isolation technology has matured, the design theory has been refined, and the application of isolation technology has become increasingly widespread [13]. The friction-pendulum bearing (FPB) is a promising new type of seismic isolation device, which converts seismic energy into heat energy through friction-energy dissipation. Its unique circular sliding surface gives it a self-resetting function [14–16]; in recent years, research into this feature has increased. Many scholars [17–27] studied the seismic responses of FPBs applied to beam bridges and cable-stayed bridges, among others. The economical and structural efficiency of FPBs for the retrofitting of continuous beam bridges in the State of Illinois was studied by Dicleli et al. [17]. Almazán et al. [18] presented a physical model for frictional pendulum isolators (FPS) that was ready to be implemented in most commercial software, and the physical model was validated by studying the earthquake response of a continuous beam bridge. Murat et al. [19,20] proposed a new model for the FPS and examined the effects of modeling parameters on the response of a three-dimensional multi-span continuous (MSC) steel-girder-bridge model seismically isolated with the FPS. Jun Gao et al. [22] assessed the damping performance of frictional pendulum bearings in long-span cable-stayed bridge bearings. Zhiqiang Li et al. [23] explored vibration mitigation and isolation for a long-span cable-stayed bridge with double-arch pylons based on FPBs. Jing Li et al. [24] studied the parametric optimization of vibration mitigation and isolation scheme for a railway cable-stayed bridge. A global non-linear time-history analysis of the Benicia–Martinez Bridge, which is a long-span truss bridge, was conducted using ADINA by Mutobe et al. [25]. Ingham [26] presented a non-linear time-history analysis undertaken in support of the seismic retrofitting of the bridge, and the work included the installation of friction-pendulum-isolation bearings. Marin-Artieda et al. [27] proposed an XY-FPB that consisted of two perpendicular steel rails with opposing concave surfaces and a connector and applied it to a truss bridge. At present, isolation technology is mainly applied to the seismic resistance of medium- and small-span continuous beam bridges, cable-stayed bridges, and truss bridges in areas of high seismic intensity. However, the application of isolation technology in long-span CFST arch bridges requires further research.

The transverse connecting system of the main arch is a vulnerable component; because of the high level of stiffness of the lateral connection system, there a large internal force response and buckling instability occur under earthquake conditions [28,29]. Of the existing seismic performance-improvement technologies, buckling-restrained brace (BRB) may be the only type that can effectively address the buckling instability of the transverse connection system of the main arch [30–37]. However, compared with buildings, the super-long-span CFST arch bridge has the characteristics of large internal force and small displacement. As a displacement-type energy-dissipation device, the BRB is applied to a super-long-span CFST arch bridge, and it is very likely that the energy-dissipation effect is lower than the expected effect, making it necessary to find another way in which to improve the seismic performance of the transverse connection system of the main arch.

The cable-hoisting-and-inclined-pulling-and-fastening method (CHIPFM) is the most commonly used method of construction for long-span CFST arch bridges. The associated construction process is as follows. First, a temporary buckle tower and cable tower are built before the installation of the arch rib. At the same time, the arch rib is divided according to the hoisting capacity of the cable-hoisting system (CHS). Next, the CHS is used to hoist the arch rib to the designated position section by section for cantilever assembly. At the same time, each arch-rib section is buckled onto the buckle tower with the help of an anchor cable until the arch rib is closed. Finally, the CHS is removed. The stayed buckle cable (SBC) is the main item of equipment in the CHS. It is often used in such work; it is only a temporary facility, and it is removed after construction. The CHIPFM has been vigorously promoted and used in the construction of large-span DCFST arch bridges because of its mature construction technology, fast construction speed, convenient transportation of materials, strong spanning ability, and applicability to canyons and rivers with complex terrain and harsh environments [38–41]. Scholars have conducted significant research into the CHIPFM, to good effect [42–44]. The existing research mainly focuses on the calculation of the cable forces of SBCs and construction control [42–45]. However, relevant research on the impact of SBCs on the seismic performance of arch bridges has not been performed. As a temporary measure in the CHIPFM, SBCs complete missions after the arch rib is closed and cannot be reused, resulting in a waste of resources. Since SBCs show the characteristics of convenient installation, easy replacement after earthquake damage, and reasonable arrangement to increase the transverse stiffness of arch bridges, they can be retained after the completion of bridge construction to improve the transverse seismic performance of super-large-span arch bridges, providing a new way of improving the seismic performances of these bridges.

The existing research still lacks evidence that can guide the seismic-performance-improvement design of super-long-span CFST arch bridges. Therefore, this research took a super-long-span CFST arch bridge with a total length of 788 m as the object on which to perform a non-linear time-history analysis and a seismic-checking calculation according to its seismic response, so as to reveal the seismic weak points of the arch bridge. The seismic performance of the arch bridge was improved through FPBs and SBCs, making it possible to propose a more suitable seismic-performance-improvement technology for super-long-span concrete-filled steel-tube arch bridges. Finally, the seismic-performance-improvement effects of the FPBs, SBCs, and combination schemes were compared, and the optimal scheme was established.

2. Seismic-Response Analysis of Original Model

2.1. Project Overview and Modeling Method

The background bridge is a prestressed-concrete continuous rigid-frame bridge with a CFST arch bridge, with a length of 788 m. The construction-drawing model of the arch bridge was taken as the research object to improve its seismic performance. The main bridge is a CFST arch bridge with a span of 500 m, and the rise-span ratio is 1/4.76. The arch axis adopts a catenary, with an arch-axis coefficient of m = 2.0, and the bridge deck is horizontally arranged, with four lines of railway track. The main arch ring adopts a CFST transverse dumbbell-shaped four-limb truss-basket arch, with two arch ribs on the left and right. The main beam of the main bridge adopts 10 $\times$ 40.8-m steel-box continuous beams arranged on left and right sections. Spherical steel bearings are used to support the main beam. The left and right approaches are (51 + 66 + 66)-meter prestressed-concrete continuous rigid-frame bridges. The piers are variable-cross-section reinforced-concrete-frame piers, which show an inverted V shape when viewed horizontally.

The dynamic calculation model established by bridge-finite-element software is shown in Figure 1. The CFST arch rib was established by the double-element method. Apart from the spherical steel bearings, which feature elastic connection elements, the other components of the bridge were submitted to a beam-element-simulation analysis. The finite element-calculation mesh contained 2780 beam elements, and the number of nodes was 1442. According to the design data, the bridge site is on bedrock, so the interaction between the foundation and the structure did not need to be considered [44]. In the model, the pier bottom and arch foot of the approach bridge were set as the consolidation-constraint boundary.

The main pier and main beam of the approach bridge adopt consolidation, while the tops of each pier (column) of the whole bridge adopt four bearings. Specifically, the longitudinal movable bearings of model TJQZ-10000 were used at the approach-bridge side of columns 1 # and 10 #, and the fixed bearings of model TJQZ-7000 were used at the main bridge side. The vault adopted the fixed bearings of model TJQZ-12500. The tops of columns 2–10 # used the longitudinal movable bearings of model TJQZ-12500. As the bearings were arranged transversely and symmetrically, the layout plan of the main bridge bearings along the longitudinal direction of the bridge is shown in Figure 2.

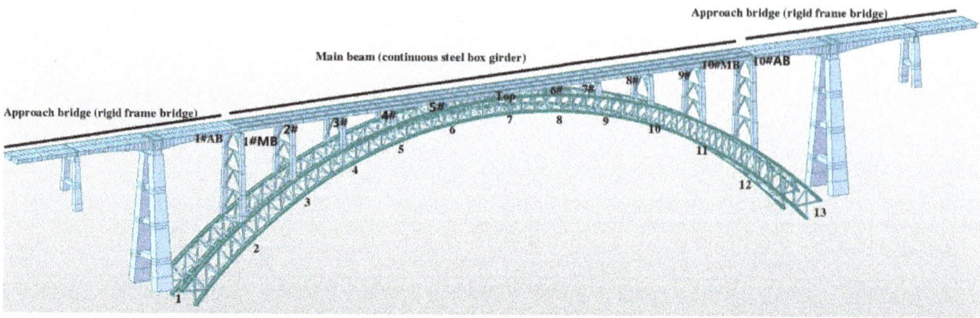

Figure 1. Finite element calculation model.

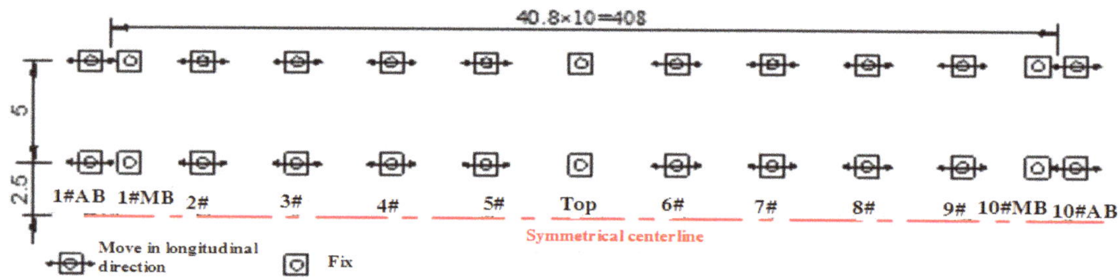

Figure 2. Layout plan of main bridge bearings (unit: m).

The bridge site of the project is a mountain-canyon landform, with high mountains and deep valleys, steep slopes, and a V-shaped river valley. The ground elevation on both banks of the river valley is 2700–4700 m, and the maximum elevation difference exceeds 2000 m. The topographic map of the site is shown in Figure 3. The intention of the project was to use CHIPFM to construct arch ribs.

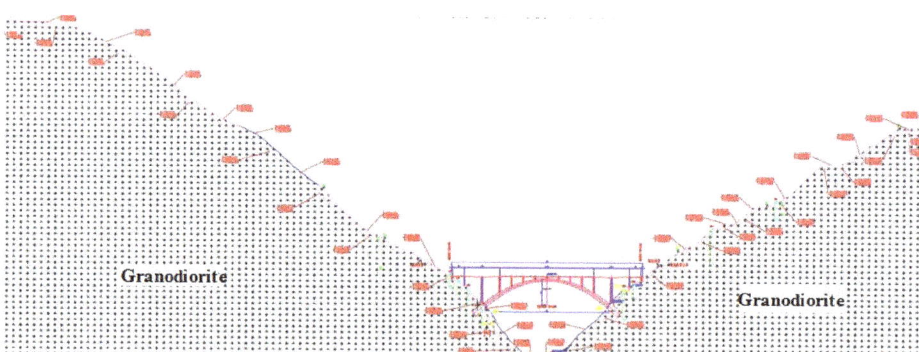

Figure 3. Topographic map of project site.

According to the seismic-safety-assessment report, the seismic fortification intensity of the bridge site is 7 degrees, the basic acceleration is 0.15 g, the site's characteristic period is 0.65 s, and the site is classified as Class I. Through the seismic analysis, the bridge was still in the elastic stage under Level-I earthquake excitation (E1), the internal force of the section passed the checking calculation, and the transverse seismic performance of the arch bridge was worse than the longitudinal seismic performance. Therefore, this paper focuses on the transverse seismic performance under Level-II earthquake excitation (E2) of the arch bridge. Three artificial seismic waves with a 50-year exceedance probability of 2% provided in the bridge's site-safety-assessment report were used for the non-linear time-history analysis. The peak accelerations of the three waves were 0.23 g, 0.21 g, and 0.27 g, respectively, after considering the terrain effect. In the non-linear time-history analysis, the transverse + vertical input mode was adopted for the ground motion, without considering the influence of the longitudinal earthquake excitation. The vertical ground-motion input was obtained by reducing the transverse seismic wave, and the reduction coefficient was 0.65. The acceleration-time-history curves of the three artificial seismic waves are shown in Figure 4. To ensure the clarity of the discussion, the envelope values of the calculated results of the three waves are discussed later.

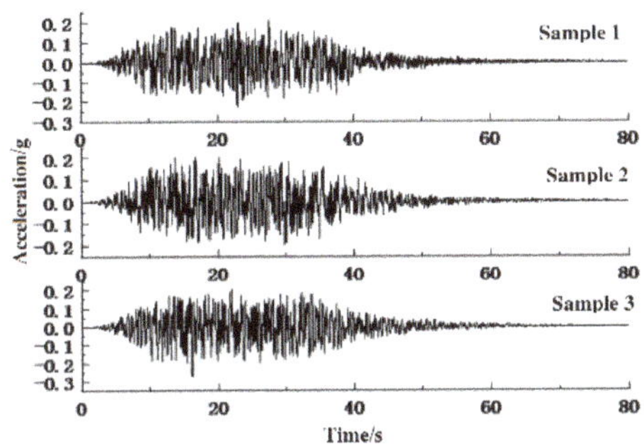

Figure 4. Time-history curve of seismic-wave acceleration.

2.2. *Analysis of Transverse Seismic Performances of Bearings of the Original Model*

For the through-arch basket-handle bridge, the inclination angle of the arch rib did not increase constantly. Geometrically, the inclination angle was limited by the span, the width and height of the bridge deck, the sagitta, and other factors at the crown of the arch, as shown in Figure 5. The relationship between the inclination angle of the arch rib and these parameters was derived as follows.

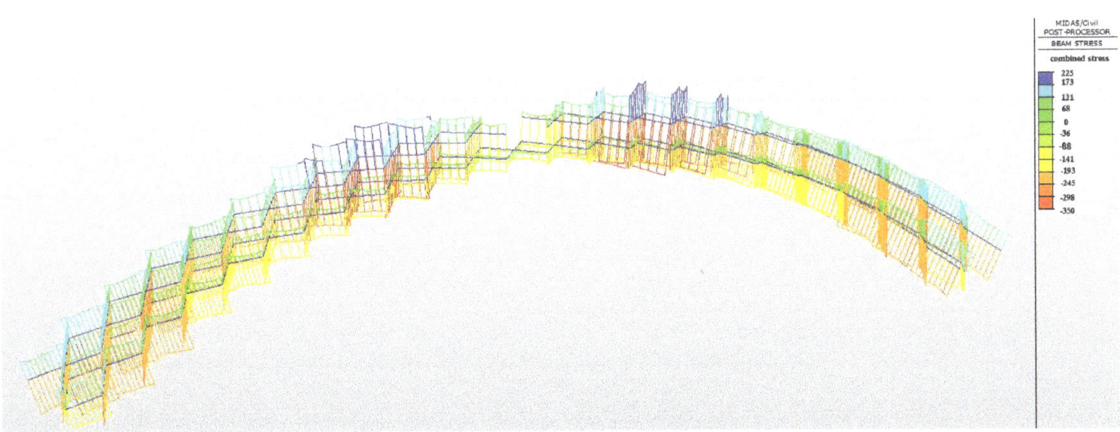

Figure 5. Axial-stress-envelope diagram of horizontal diagonal bracing between arch ribs of the original model.

The bearings, as the key components connecting the substructure and the superstructure, may be the seismic weak points of the deck-arch bridge. The safety of the bearings was evaluated through the capacity–demand ratio (CDR). The CDR method evaluates the seismic safety of bridges by comparing the seismic demands and capacities of each key component in the bridge-structure system [46,47]:

$$R_Q = Q_c / (Q_{sd}^{\max} + Q_{dd}) \tag{1}$$

$$R_\Delta = \Delta_c / (\Delta_{sd}^{\max} + \Delta_{dd}) \tag{2}$$

where R_Q and R_Δ are the CDR of bearings' shear resistance and deformation, respectively; and Q_c, $Q_{sd}^{\max}$, and Q_{dd} are the shear bearing capacity, seismic shear demand, and dead-load shear demand on the bearings, respectively; Δ_c, $\Delta_{sd}^{\max}$, and Δ_{dd} are the deformation capacity of the bearings, the seismic-deformation demand, and the dead-load deformation-demand, respectively. Formula (1) is applicable to the fixed direction of the spherical steel bearings, and Formula (2) is applicable to the FPBs and the moving direction of the spherical steel bearings. The shear capacity of the fixed direction of the spherical steel bearings is taken as 30% of the designed vertical-bearing capacity.

When the CDR is greater than 1, the system is safe. When the CDR is less than 1, the system is not safe. As the top of each pier (column) of the bridge adopts four bearings, the maximum seismic-shear force in each row of bearings was used for a checking calculation when analyzing the transverse seismic performance of the bearings. The CDRs of the bearings of the main bridge are listed in Table 1.

Table 1. CDRs of main bridge bearings.

Location of Bearings	1#AB	1#MB	2#	3#	4#	5#	top	6#	7#	8#	9#	10#MB	10#AB
Shear bearing capacity/kN	3000	2100	3750	3750	3750	3750	3750	3750	3750	3750	3750	2100	3000
Shear demand/kN	3792	3244	2765	3069	3280	1415	3717	1359	3263	3169	2768	3321	3745
CDR	0.79	0.65	1.36	1.22	1.14	2.65	0.99	2.76	1.15	1.18	1.35	0.63	0.8

Table 1 shows that the CDR of the bearings of the vault was 0.99, and the CDRs of the bearings of columns 1# and 10# were 0.79, 0.65, 0.63, and 0.80, respectively, which were less than 1; the CDR of the bearings at the tops of the other columns was greater than 1. This shows that the shear capacity of the arch crown and the 1# and 10# spherical steel bearings was insufficient, and that seismic measurements in the transverse direction needed to be undertaken. It is worth noting that these findings were affected by the high and low columns on one side of the side pier of the approach bridge; there was no symmetry in the curves with respect to the arch crown.

2.3. Stress-Seismic-Response Analysis of the Transverse Connection System of the Main Arch in the Original Model

Due to the high level of rigidity of the transverse connection system of the main arch, stability is easily lost due to the large internal forces imposed during an earthquake. Therefore, the stress seismic response of the transverse connection system of the main arch was assessed.

Through this analysis, it was found that the transverse connection system of the main arch with the highest stress in the original model took the form of horizontal diagonal bracing. As shown in Figure 5, the maximum combined tensile stress on the horizontal diagonal braces was 225 MPa, and the maximum combined compressive stress was 350 MPa. The stress capacity of the members in the elastoplastic state under seismic action was calculated by using the basic allowable stress method and considering the improvement of various factors. According to the basic allowable stress method, the allowable tensile stress of the horizontal diagonal brace was 350 MPa, and the allowable compressive stress was 297 MPa. The compressive stress on the horizontal diagonal brace of the original model was greater than the allowable compressive stress; that is, the horizontal diagonal brace of the main arch entered the elastic–plastic stage, in which it was likely to buckle and lose stability, and transverse-vibration-reduction measures needed to be implemented.

2.4. Seismic Internal Force Response Analysis of Arch Rib Section

The seismic performance of the CFST arch bridge first depends on the seismic performance of the CFST main arch ring, so this paper focuses on the internal force of the arch-rib section. As the left and right arch ribs are symmetrical and their internal forces are similar, to save space, only the internal-force response of the right arch rib section was analyzed. When studying the axial force and the out-of-plane bending moment and in-plane bending moment of the arch-rib section, the maximum and minimum internal forces were considered and expressed in maximum and minimum working conditions, respectively. When analyzing the shear force on an arch rib section, only the most unfavorable situation was considered. The calculated results of internal-force seismic response of the arch rib are shown in Figure 6.

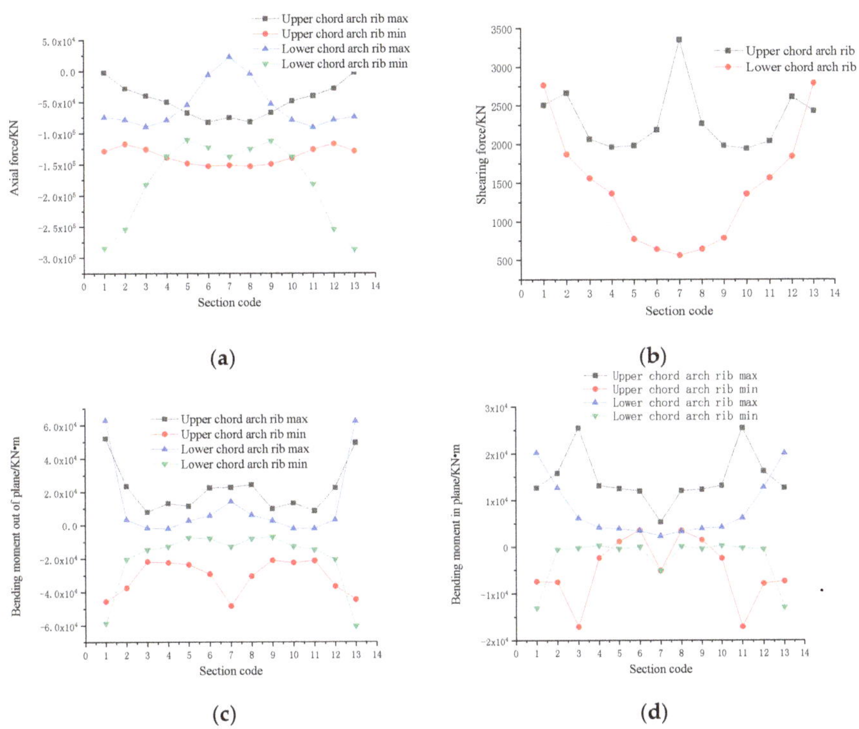

Figure 6. Internal-force envelope of an arch-rib section of the original model. (**a**) Axial-force demand. (**b**) Shear demand. (**c**) Out-of-plane bending-moment demand. (**d**) In-plane bending -moment demand.

Figure 6a illustrates that the axial-force difference between the arch foot of the upper-chord arch rib and the crown section of the upper-chord arch rib was small, and that the axial-force distribution was relatively uniform; the axial force of the arch-foot section of the lower-chord arch rib was greater than that of the arch-crown section of the lower-chord arch rib. It can be seen from Figure 6b that, except at the arch foot, the shear force on the upper-chord arch-rib section was greater than that on the lower-chord arch-rib section, and the shear force at the arch crown of the upper-chord arch rib was much larger than that at the arch crown of the lower-chord arch rib. Figure 6c,d indicate that the bending moment in the upper-chord arch -rib section was greater than that -n the lower-chord arch-rib section, except at the arch foot.

Transverse seismic conditions mainly affect the out-of-plane bending moment of arch-bridge structures, so we focus on the out-of-plane bending performance here. The axial force affects the flexural capacity of the member, and the out-of-plane bending moment of the section determines the flexural demand of the section. In this paper, a CDR method is used to evaluate the out-of-plane bending performance of the arch-rib section:

$$R_M = M_c^{min}/(M_e^{max} + M_d^{max}) \quad (3)$$

In Formula (3), R_M is the CDR of the bending resistance of the arch rib, M_c^{min} is the bending capacity of the section, and M_e^{max} and M_d^{max} are the bending-moment demands of the section under the combined working conditions of seismic and dead loads, respectively. The fiber model in XTRACT software was used to calculate the bending capacity of the section. The steel-structure material adopted the strain-hardening model, and the confined concrete used a Mander model.

The CDRs of the out-of-plane bending of key sections of the main arch rib were calculated, and the results are shown in Table 2.

Table 2. CDR of out-of-plane bending of arch-rib section of the original model.

Section Code	1	2	3	4	5	6	7	8	9	10	11	12	13
Upper chord	7.5	9.7	18.7	15.4	15.4	13.2	8.0	12.7	17.2	15.3	19.0	9.8	7.9
Lower chord	10.4	28.5	26.9	29.4	43.8	40.0	13.7	36.7	47.1	29.4	27.1	28.2	8.6

The CDRs of key arch sections under the most unfavorable working conditions are shown in Table 2. The CDRs of the upper-chord arch-rib section were much smaller than those of the lower-chord arch-rib section, and the CDRs of the arch crown and arch foot were smaller than those of the other sections. Therefore, the arch-rib sections of the upper chord, the arch crown of the lower chord, and the arch foot are the weak points in this arch bridge. The CDRs of the upper-chord arch-rib section, the arch crown of the lower chord, and the arch foot of the lower chord should be increased through the implementation of seismic measures.

3. Seismic-Performance-Improvement Schemes

3.1. The FPBs Scheme

The analysis in Section 2.4 indicated that the shear resistance of the spherical steel bearings on the arch crown and the tops of columns 1 # and 10 # was insufficient; that is, the anti-seismic calculation of the bearings failed, and transverse anti-seismic measures needed to be taken. The seismic isolation's design using FPBs is mainly discussed in this paper.

The FPBs adopt lag-system simulation, and the simplified calculation model for FPBs is shown in Figure 7. The main design parameters of each FPB obtained through repeated optimization calculations and verifications are presented in Table 3.

To verify the reliability of the FPB parameters, a testing company was entrusted to conduct a pseudo-static test on the FPB-7000, and the measured hysteresis curve of the FPB-7000 was obtained. The test results were compared with the finite element simulation results, as shown in Figure 8; the hysteresis curve of the FPBs in the finite element analysis was very close to the experimental results, which verified that the parameters of the FPBs in this paper were set accurately and that the mechanical properties of the FPBs were simulated accurately.

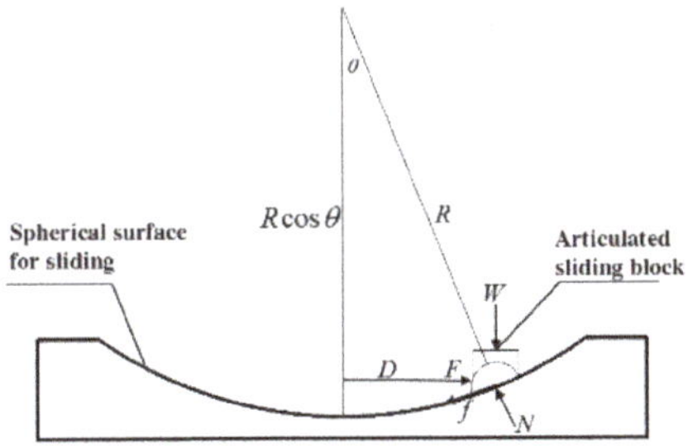

Figure 7. Simplified calculation model for FPB.

Table 3. Design parameters of FPBs.

Position	Model and Specification	Vertical Load (kN)	Yield Displacement (mm)	Design Displacement (mm)	Slideway Radius (m)	Static Friction Coefficient	Dynamic Friction Coefficient	Shear Equivalent Stiffness (kN/m)	Initial Stiffness (kN/m)
1#AB/10#AB	FPB-10000	10,000	2	300	5	0.04	0.03	3000	150,000
1#MB/10#MB	FPB-7000	7000	2	300	5	0.04	0.03	2100	105,000
Top	FPB-12500	12,500	2	300	5	0.04	0.03	3750	187,500

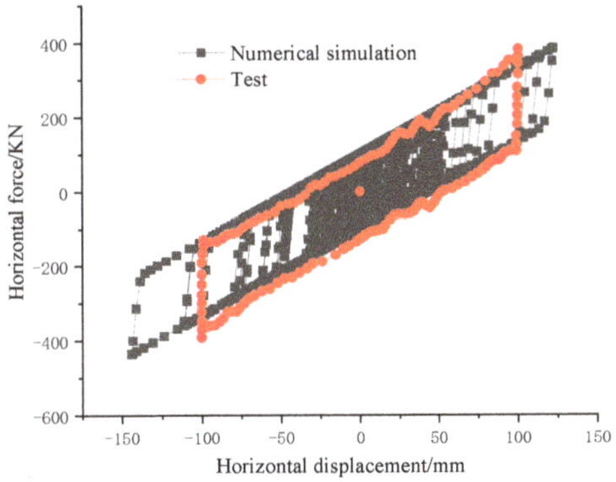

Figure 8. Hysteretic curve of FPBs.

To study the effect of the FPBs on the seismic performance of the long-span deck CFST arch bridge and the influence of its layout on the seismic performance of the bridge, a variety of layout schemes were developed, and the isolation effects were compared to obtain the optimal scheme. Due to space constraints, the results are presented here directly. Considering the excessive displacements of the FPBs under E2 excitation in the full bridge-isolation scheme, the optimal plan involves replacing only the spherical steel bearings with shear failure in the original plan; that is, the spherical steel bearings at the top of the vault and on the side column are replaced by FPBs. The bearing-layout scheme is summarized in Table 4, where TJQZ is a spherical steel bearing. The spherical steel bearings passing the shear-resistance-checking calculation in Table 1 are retained in this scheme. This scheme is called Model 1. It should be noted that because the main beams at the junction of the main approach bridge are not connected, the main beams at the junction of the main approach bridge suffer a high level of relative displacement under earthquake excitation. To prevent the relative displacement between the main beams from leading to dislocation between the tracks, which would affect traffic safety, the structural measures that limit the relative displacement in the transverse direction were adopted between the main beams of the main approach bridge, and the elastic connection was used in the finite element model to simulate the displacement-limiting effect of the structural measures.

Table 4. Layout scheme of the bearings in Model 1.

Location of Bearings	1#AB	1#MB	2#	3#	4#	5#	Top	6~10#
Type of bearings	FPB-10000	FPB-7000	TJQZ-12500	TJQZ-12500	TJQZ-12500	TJQZ-12500	FPB-12500	Symmetrical arrangement

3.2. The SBC Scheme

It can be seen from Section 2.2 that the compressive stress of the horizontal diagonal brace of the transverse connection system of the main arch exceeds the allowable compressive stress, which may lead to buckling instability. Therefore, it is necessary to implement seismic lateral measures to improve the seismic performance of this arch bridge in the transverse direction.

In this project, CHIPFM is proposed for the arch-rib construction. After an arch-rib construction is completed, the buckle tower and all SBCs are removed. The SBCs used can only assist in the construction, and the SBCs removed after the completion of the construction cannot assist in the construction of other arch bridges again. The SBCs are discarded, causing a certain level of waste. Therefore, the SBCs removed after the construction were reused in the present work. After the completion of the bridge's construction, the SBCs are not removed directly, but remain as the "fuses" to support the arch bridge's seismic resistance. The specific method is as follows: the end of SBC connected with the arch rib (called the A end) does not need to be moved, but the other end of the SBC (called the B end) is anchored in the canyon bedrock, and the arch bridge's transverse stiffness is increased through the reasonable arrangement of SBCs, so as to improve the arch bridge's transverse seismic performance. Moreover, SBCs are easy to replace and can be replaced directly in case of damage after an earthquake. The schematic diagram of the SBC scheme is shown in Figure 9.

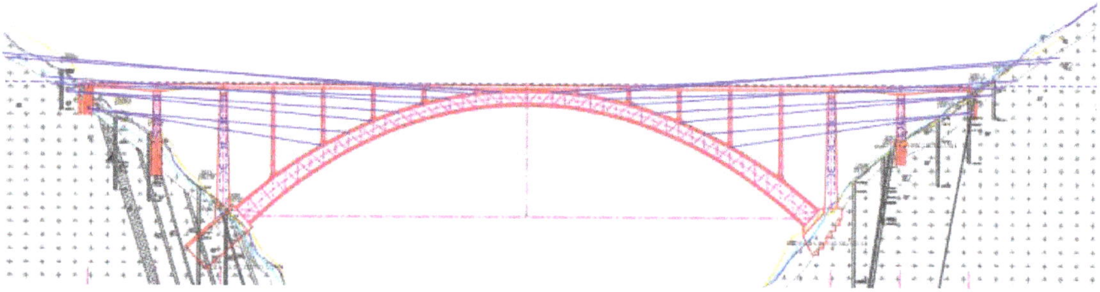

Figure 9. Schematic diagram of SBC scheme.

The seismic-performance-improvement effects of several SBCs layout schemes were compared, and the best scheme was obtained, as illustrated in Figure 10. This scheme is called Model 2. Compared with the construction-layout scheme, the SBCs near the arch foot in Model 2 are removed, leaving only the SBCs of the $\frac{1}{4}$-arch-to-$\frac{3}{4}$-arch-rib section. The nodes (called the A end) connecting the remaining SBC and the arch rib remain unchanged, but the other end of the SBC (called the B end) needs to be adjusted, as follows. The B ends of the six SBCs at the top of each side are moved 50 m to the lateral outer side, and the B ends of the other SBCs are moved 30–40 m in the vertical downward direction. The B ends of the SBCs move a certain distance in the vertical downward direction to prevent the SBCs from bearing the self-weight of the structure as far as possible, so as not to change the structural system of the original model, while endowing the arch0bridge structure with greater lateral stiffness and better seismic-performance improvements.

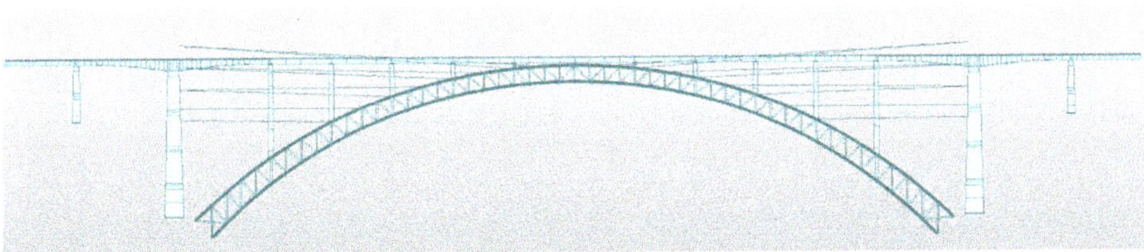

(a)

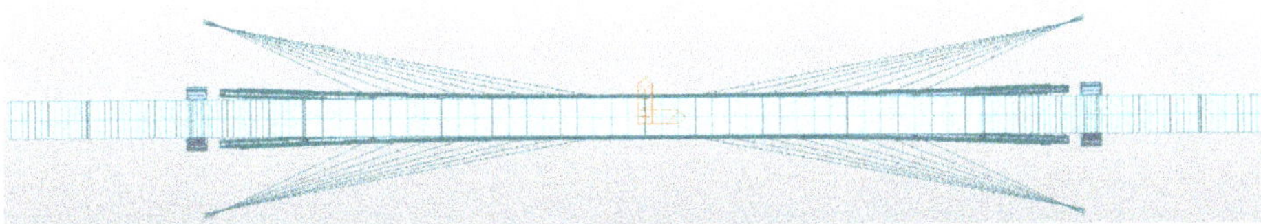

(b)

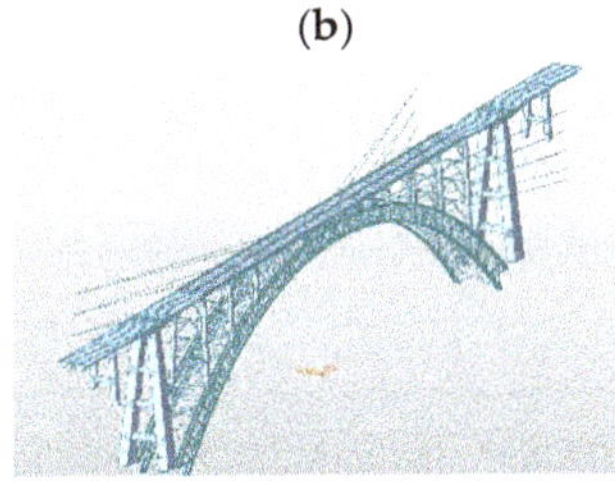

(c)

Figure 10. Layout of SBC. (**a**) Front view. (**b**) Top view. (**c**) Stereogram.

3.3. Combination of SBCs and FPBs

The combined scheme of SBCs and FPBs was adopted to improve the seismic performance of the arch bridge. In this scheme, FPBs are only set on column 1# and column 10#. The bearing layout is shown in Table 5, and the reader is referred to Section 3.2 for details of the SBC layout: this scheme is called Model 3.

Table 5. Layout of the bearings in Model 3.

Location of Bearings	1#AB/10#AB	1#MB/10#MB	2#	3#	4#	5#	top	6~10#
Type of bearing	FPB-10000	FPB-7000	TJQZ-12500	TJQZ-12500	TJQZ-12500	TJQZ-12500	TJQZ-12500	Symmetrical arrangement

4. Results and Analysis

4.1. Analysis of Natural Vibration Characteristics

The first three natural periods of all the models are shown in Table 6. Due to the influence of the fixed bearings of columns 2#–9#, the first three natural periods in Model 1 demonstrated little change compared with the original model. Compared with the original model, the first-order natural periods of vibration of Model 2 and Model 3 varied greatly. The first-order natural vibration period of Model 2 decreased by 12.2%, and the first-order natural vibration period of Model 3 decreased by 12%, which was closely related to the increased lateral stiffness when using the SBCs.

Table 6. Comparison of the first three natural periods of models.

Natural Vibration Period/s	First-Order	Second-Order	Third-Order
Original model	3.913	2.424	2.296
Model 1	3.935	2.479	2.424
Model 2	3.437	2.304	2.290
Model 3	3.444	2.304	2.290
Reduction Rate	**First-Order**	**Second-Order**	**Third-Order**
Model 1	−0.6%	−2.3%	−5.6%
Model 2	12.2%	5.0%	0.3%
Model 3	12.0%	5.0%	0.3%

4.2. Comparative Analysis of Overall Displacement Response

The overall transverse displacement responses of all the models are shown in Table 7. The maximum overall displacement in all the models occurred on the main beam at the corresponding position of the vault. Compared with the original model, the overall displacement envelope of Model 1 changed little. Compared with the original model, the overall displacement envelope of Model 2 changed greatly, with the maximum displacement decreasing by 28.6% and the minimum displacement decreasing by 36.2%. The SBCs greatly reduced the seismic-displacement response of the arch-bridge structure; this was largely related to the increase in the transverse stiffness of the arch bridge structure by the SBCs. Compared with the original model, the overall displacement envelope in Model 3 changed greatly, with the maximum displacement decreasing by 30.3% and the minimum displacement decreasing by 40.2%. Model 3 had the highest rate of reduction in overall displacement response of all the models.

Table 7. Comparison of overall displacement responses of Models.

Displacement Responses/mm	Original Model	Model 1	Model 2	Model 3
Maximum	535	534	382	373
Minimum	−577	−577	−368	−345
Reduction Rate	**Original Model**	**Model 1**	**Model 2**	**Model 3**
Maximum	0.0%	0.2%	28.6%	30.3%
Minimum	0.0%	0.0%	36.2%	40.2%

4.3. Seismic-Performance Check of Bearings

The safety-factor CDR was obtained by using the CDR method to evaluate the safety status of the bearings under the E2 earthquake excitation of transverse bridge direction. The CDRs of the bearings were calculated by using Formula (1) and Formula (2). The safety factor needed be greater than 1 to pass the design check. The CDRs of the bearings at the top of each column are listed in Table 8.

Table 8. CDRs of bearings of all models.

Location of Bearings	1# AB	1# MB	2#	3#	4#	5#	top	6#	7#	8#	9#	10#MB	10# AB
Original Model	0.79	0.65	1.36	1.22	1.14	2.65	0.99	2.76	1.15	1.18	1.35	0.63	0.8
Model 1	1.16	1.16	1.03	1.42	1.16	1.2	4.11	1.21	1.14	1.38	1.02	1.44	1.44
Model 2	0.87	0.64	1.44	1.19	1.13	2.86	1.2	2.76	1.13	1.16	1.44	0.62	0.8
Model 3	1.16	1.16	1.14	1.38	1.21	2.75	1.34	2.83	1.20	1.35	1.18	1.76	1.79

As shown in Table 8, after replacing the spherical steel bearings at column 1# with the FPBs in Model 1, the CDR of the adjacent spherical steel bearings (corresponding to the position of column 2#) decrease. The same rule can be observed at the vault. However, since the shear resistance of the spherical steel bearings at columns 2 # and 5# in the original model has a certain margin, the CDR value is still greater than 1 after it decreases. Similarly, the bearings of columns 3 #–9 # passed the design check. Table 8 demonstrates that after replacing columns 1 # and 10 # with the FPBs in Model 1, the displacement demand of the FPBs under E2 earthquake excitation is less than the design displacement of the FPBs (300 mm), and the CDR value is greater than 1. Therefore, FPBs passed the seismic-calculation check.

As shown in Table 8, the shear CDR of the spherical steel bearings of the arch crown increased from 0.99 to 1.2 when using the SBC in Model 2. That is, after the seismic response of the arch crown in Model 2 reduced, the shear demand of the spherical steel bearing in the arch crown also reduced. However, the shear CDRs of the spherical steel bearings at the corresponding positions of columns 1 # and 10 # remained below 1. The shear-check calculations for the spherical steel bearings at the corresponding positions of columns 1 # and 10 # still failed, and transverse-vibration-reduction measures should be taken.

As shown in Table 8, the CDR of all the bearings in Model 3 is greater than 1. Therefore, it can be concluded that all the bearings in the combination scheme passed the seismic-design checks.

4.4. Comparative Analysis of Stress Response of the Transverse Connection System of the Main Arch

Through analysis, it was found that the transverse connection system of the main arch with the highest stress of all the models was that of the horizontal diagonal bracings, and the stress-response-envelope values of the horizontal diagonal bracings are shown in Table 9.

Table 9. The stress response envelope values of the horizontal diagonal bracings.

Index	Allowable Stress	Original Model	Model 1	Model 2	Model 3
Maximum tensile stress	330	225	189	170	141
Maximum compressive stress	297	350	316	290	261

Compared with the original model, the maximum tensile stress of the horizontal diagonal bracings in Model 1 decreased by 16%, and the maximum compressive stress of the horizontal diagonal bracings in Model 1 reduced by 10%. However, the horizontal diagonal bracings in Model 1 still bear a stress that is greater than the allowable compressive stress, which may lead to buckling instability.

Compared with the original model, the maximum tensile stress on the horizontal diagonal bracings in Model 2 decreased by 25%, and the maximum compressive stress reduced by 17%. The maximum tensile stress and maximum compressive stress of the horizontal diagonal bracings in Model 2 do not exceed the allowable stress, and the check was thus passed. Therefore, the application of SBCs to the CFST arch bridge can reduce the stress response of the transverse connection system of the main arch.

Compared with the original model, the maximum tensile stress on the horizontal diagonal bracings in Model 3 decreased by 38%, and the maximum compressive stress on the horizontal diagonal bracings in Model 3 reduced by 26%. The maximum tensile stress and maximum compressive stress on the horizontal diagonal bracings in Model 3 do not exceed the allowable stress, and the check on the transverse connection system of the main arch was passed. Compared with the other schemes, the combination scheme had the highest seismic-reduction rate and the best seismic-performance-improvement effect.

4.5. Comparative Analysis of Internal-Force Response of Arch Rib

The CDR is an index to visually express the internal-force response of a section. The CDRs of the out-of-plane bending of the arch rib in the models are shown in Figure 11, and the lifting factor of the safety factor of the arch rib in the models is shown in Table 10, where positive values mean the CDRs increase and negative values suggest that the CDRs decrease.

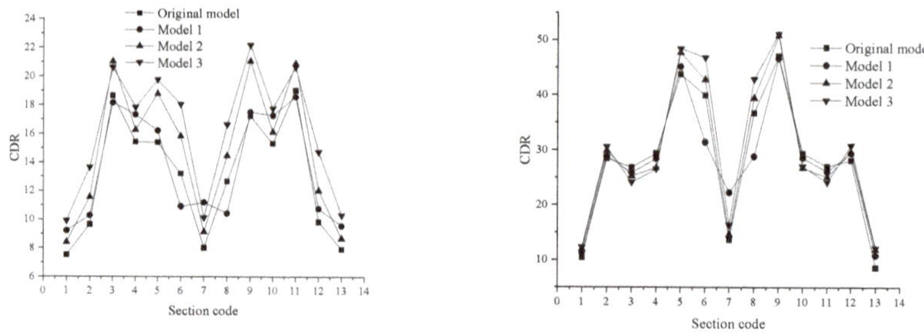

(a) (b)

Figure 11. CDR comparison of out-of-plane bending of arch-rib sections of models. (**a**) Upper chord-arch-rib section. (**b**) Lower-chord arch-rib section.

Table 10. CDR-lifting factors for the out-of-plane bending moment.

Section Code		1	2	3	4	5	6	7	8	9	10	11	12	13
Upper chord	model 1	22%	6%	−3%	12%	5%	−17%	39%	−18%	2%	13%	−2%	9%	21%
	model 2	12%	20%	13%	5%	22%	20%	14%	14%	22%	5%	10%	22%	9%
	model 3	32%	42%	11%	16%	28%	37%	26%	31%	29%	16%	8%	50%	30%
Lower chord	model 1	8%	1%	−4%	−3%	3%	−21%	63%	−21%	−1%	−3%	−4%	5%	26%
	model 2	15%	5%	−6%	−9%	9%	7%	7%	7%	8%	−9%	−7%	5%	39%
	model 3	18%	7%	−10%	−10%	11%	17%	20%	17%	9%	−8%	−11%	10%	41%

Figure 11 and Table 10 show that after the FPB was set, the CDRs of the corresponding sections (section codes 1, 2, 7, 12, and 13) greatly improved, regardless of the upper or lower chord, but the CDRs of the adjacent sections (section codes 3, 6, 8, and 11) decreased to a certain extent. That is, the CDRs at the corresponding arch-rib location where the FPB was set rose, but this had an adverse effect on the arch-rib section at the adjacent location of the FPB. In particular, after the FPB was set at the vault, the CDRs of the arch0rib sections at adjacent locations decreased by at least 17%.

As shown in Figure 11 and Table 10, the CDRs of the upper-chord arch-rib section improved by at least 5%, and the CDRs of the 77% section increased by more than 10%. The CDRs of most of the sections of the lower-chord arch rib increased. The arch-foot and arch-crown sections of the upper-chord arch rib and the lower-chord arch rib were the seismic weak points of the arch rib. The CDRs of the arch-rib weak points were improved through the reasonable arrangement of the SBCs.

Figure 11 and Table 10 demonstrate that the CDR of the upper-chord arch-rib section was greatly improved. The CDRs of all the sections increased by more than 8%, and the CDRs of 92% of the sections increased by more than 11%. The CDRs of most of the sections of the lower-chord arch rib increased. Compared with Models 1 and 2, the CDRs of arch-rib weak points can be better improved through the reasonable arrangement of SBCs and FPBs.

5. Conclusions

Taking a super-long-span CFST with a total length of 788 m as the research object, a non-linear time-history analysis was conducted under E2 excitation in the cross-bridge direction, plus dead-load conditions, and the seismic-design-check calculation was performed according to the seismic response. As a result, the seismic weak points were revealed and the seismic-performance-improvement scheme was explored with a view to proposing a better seismic-performance0improvement scheme for this long-span CFST arch bridge.

(1) The seismic-response analysis of the original model shows that the shear resistance of spherical steel bearings is insufficient and that the compressive stress of the horizontal diagonal brace in the transverse connection of the main arch exceeds the allowable compressive stress. The arch-rib section of the upper chord, the arch crown of the lower chord, and the arch foot of the lower chord are the weak points of this arch bridge.
(2) The seismic-response-analysis results for Model 1 (the FPBs scheme) show that the FPB layout and the setting parameters of the model are reasonable, and the seismic checks for all the bearings were passed. However, the horizontal diagonal brace still failed to pass the seismic-design checks.
(3) The seismic-response analysis of Model 2 (the SBCs scheme) showed that the horizontal diagonal bracing of Model 2 passed the design checks. The first natural vibration period of Model 2 decreased by 12%, indicating that the SBCs increased the transverse stiffness of the arch bridge. The internal-force response of the upper-chord arch-rib section was greatly reduced, and the internal force of the arch-crown and arch-foot sections of the lower-chord arch-rib section reduced to a certain extent. However, the shear-checking calculations of the spherical steel bearings at the corresponding positions of columns 1 # and 10 # still failed
(4) The seismic-response analysis of the combined scheme of SBCs and FPBs (Model 3) showed that the horizontal braces and bearings in the transverse connection of the main arch passed the seismic-design checks. The maximum tensile stress of the horizontal brace decreased by 38%, and the maximum compressive stress decreased by 26%. The internal-force responses of most of the sections of the arch rib in Model 3 reduced to varying degrees, and the CDR of the upper-chord arch rib-section greatly improved. The CDRs of all the sections increased by more than 8%, and the CDRs of 92% of the sections increased by more than 11%. Compared with Models 1 and 2, Model 3 had the best seismic-performance-improvement effect.

Author Contributions: D.Y., methodology and finite element analysis; Y.T., drawing and translation; L.G., review and finite element analysis; Z.T., editing and finite element analysis; R.Z., methodology and concepting. All authors have read and agreed to the published version of the manuscript.

Funding: This study was supported by the Key research and development plans of Guangxi Province (grant no. Guike AB22036007), the Talent Program Project of Chongqing (grant no. cstc2022ycjh bgzxm0133), the Natural Science Foundation of Chongqing (grant no. CSTB2022TIAD-KPX0205), The Natural Science Foundation of China (grant no. 52008064); The Science and Technology Research Program of Chongqing Municipal Education Commission of China (grant no. KJQN202000737); The Postdoctoral Research Program of Chongqing (grant no. 2022CQBSHTB3082); The Postgraduate Research and Innovation Project of Chongqing Jiaotong University (grant no. CYB23247).

Data Availability Statement: Data available on request due to restrictions.

Conflicts of Interest: The authors declare no conflict of interest.

References

1. Lu, Z.; Guo, C.; Li, G. Air void and ring gap effect on CFST arch bridges dynamic performance. *J. Constr. Steel Res.* **2020**, *177*, 106418. [CrossRef]
2. Geng, Y.; Ranzi, G.; Wang, Y.-T. Out-of-plane creep buckling analysis on slender concrete-filled steel tubular arches. *J. Constr. Steel Res.* **2018**, *140*, 174–190. [CrossRef]
3. Tang, Q.; Xin, J.; Jiang, Y.; Zhou, J.; Li, S.; Fu, L. Fast identification of random loads using the transmissibility of power spectral density and improved adaptive multiplicative regularization. *J. Sound Vib.* **2022**, *534*, 117033. [CrossRef]
4. Xin, J.; Zhou, C.; Jiang, Y.; Tang, Q.; Yang, X.; Zhou, J. A signal recovery method for bridge monitoring system using TVFEMD and encoder-decoder aided LSTM. *Measurement* **2023**, *214*, 112797. [CrossRef]
5. Zeng, Y.; He, H.; Qu, Y.; Sun, X.; Tan, H.; Zhou, J. Numerical Simulation of Fatigue Cracking of Diaphragm Notch in Orthotropic Steel Deck Model. *Materials* **2023**, *16*, 467. [CrossRef]
6. Zeng, Y.; Qiu, Z.; Yang, C.; Haozheng, S.; Xiang, Z.; Zhou, J. Fatigue experimental study on full-scale large sectional model of orthotropic steel deck of urban rail bridge. *Adv. Mech. Eng.* **2023**, *15*, 16878132231155271. [CrossRef]

7. Aydin, E.; Öztürk, B.; Kebeli, Y.E.; Gültepe, G. An Experimental Study on the Effects of Different Pendulum Damper Designs on Structural Behavior. In *Seismic Isolation, Energy Dissipation and Active Vibration Control of Structures*; WCSI 2022. Lecture Notes in Civil Engineering; Springer: Cham, Switzerland, 2023; Volume 309, pp. 240–253. [CrossRef]
8. Ozturk, B.; Cetin, H.; Aydin, E. Optimum vertical location and design of multiple tuned mass dampers under seismic excitations. *Structures* **2022**, *41*, 1141–1163. [CrossRef]
9. Makris, N. Seismic isolation: Early history. *Earthq. Eng. Struct. Dyn.* **2018**, *48*, 269–283. [CrossRef]
10. Li, L. *Isolation and Damping Technology*; Beijing Seismological Publishing House: Beijing, China, 1989; pp. 5–32.
11. Kelly, J. Aseismic Base Isolation. *Shock. Vib. Dig.* **1985**, *17*, 3–14. [CrossRef]
12. Ali, H.-E.M.; Abdel-Ghaffar, A.M. Modeling of Rubber and Lead Passive-Control Bearings for Seismic Analysis. *J. Struct. Eng.* **1995**, *121*, 1134–1144. [CrossRef]
13. Deringöl, A.H.; Güneyisi, E.M. Single and combined use of friction-damped and base-isolated systems in ordinary buildings. *J. Constr. Steel Res.* **2020**, *174*, 106308. [CrossRef]
14. Zayas, V.A.; Low, S.S.; Mahin, S.A.; Zayas, S.S.L.V.A. A Simple Pendulum Technique for Achieving Seismic Isolation. *Earthq. Spectra* **1990**, *6*, 317–333. [CrossRef]
15. Su, L.; Ahmadi, G.; Tadjbakhsh, I.G. A comparative study of performances of various base isolation systems, part I: Shear beam structures. *Earthq. Eng. Struct. Dyn.* **1989**, *18*, 11–32. [CrossRef]
16. Constantinou, M.; Mokha, A.; Reinhorn, A. Teflon Bearings in Base Isolation II: Modeling. *J. Struct. Eng.* **1990**, *116*, 455–474. [CrossRef]
17. Dicleli, M.; Mansour, M. Seismic retrofitting of highway bridges in Illinois using friction pendulum seismic isolation bearings and modeling procedures. *Eng. Struct.* **2003**, *25*, 1139–1156. [CrossRef]
18. Almazán, J.L.; De la Llera, J.C. Physical model for dynamic analysis of structures with FPS isolators. *Earthq. Eng. Struct. Dyn.* **2003**, *32*, 1157–1184. [CrossRef]
19. Eröz, M.; DesRoches, R. Bridge seismic response as a function of the Friction Pendulum System (FPS) modeling assumptions. *Eng. Struct.* **2008**, *30*, 3204–3212. [CrossRef]
20. Eröz, M.; DesRoches, R. The influence of design parameters on the response of bridges seismically isolated with the Friction Pendulum System (FPS). *Eng. Struct.* **2013**, *56*, 585–599. [CrossRef]
21. Brito, M.B.; Ishibashi, H.; Akiyama, M. Shaking table tests of a reinforced concrete bridge pier with a low-cost sliding pendulum system. *Earthq. Eng. Struct. Dyn.* **2018**, *48*, 366–386. [CrossRef]
22. Gao, J.; Lin, X. Analysis of Damping Performance of Frictional Pendulum Bearings in Long Span Cable Stayed Bridge Bearings in Long Span Cable-Stayed Bridge. *Bridge Constr.* **2020**, *50*, 55–60. (In Chinese)
23. Li, Z.; Wang, H.; Wang, D. Study of Vibration Mitigation and Isolation for Long-Span Cable-Stayed Bridge with Double Arch Pylons Based on FPBs. *World Bridges* **2018**, *46*, 36–40. (In Chinese)
24. Li, J.; Qian, Y.; Yang, H.; Jin, C. Study of Parametric Optimization of Vibration Mitigation and Isolation Scheme for Railway Cable-Stayed Bridge. *Bridge Constr.* **2019**, *49*, 45–50. (In Chinese)
25. Mutobe, R.; Cooper, T. Nonlinear analysis of a large bridge with isolation bearings. *Comput. Struct.* **1999**, *72*, 279–292. [CrossRef]
26. Ingham, T.J. Analysis of the Million Dollar Bridge for seismic retrofit. *Comput. Struct.* **2003**, *81*, 673–679. [CrossRef]
27. Marin-Artieda, C.C.; Whittaker, A.S.; Constantinou, M.C. Experimental Study of the XY-Friction Pendulum Bearing for Bridge Applications. *J. Bridg. Eng.* **2009**, *14*, 193–202. [CrossRef]
28. Xiaoli, L.; Zou, Y.; Wang, D. Seismic damage characteristics and research of arch bridges under strong earthquake. *World Earthq. Eng.* **2018**, *34*, 33–43. (In Chinese)
29. Bruneau, M. Performance of steel bridges during the 1995 Hyogo-ken Nanbu (Kobe, Japan) earthquake. *Can. J. Civ. Eng.* **1996**, *23*, 678–713. (In Canadian) [CrossRef]
30. Usami, T.; Lu, Z.; Ge, H. A seismic upgrading method for steel arch bridges using buckling-restrained braces. *Earthq. Eng. Struct. Dyn.* **2005**, *34*, 471–496. [CrossRef]
31. El-Bahey, S.; Bruneau, M. Buckling restrained braces as structural fuses for the seismic retrofit of reinforced concrete bridge bents. *Eng. Struct.* **2011**, *33*, 1052–1061. [CrossRef]
32. Bazaez, R.; Dusicka, P. Cyclic behavior of reinforced concrete bridge bent retrofitted with buckling restrained braces. *Eng. Struct.* **2016**, *119*, 34–48. [CrossRef]
33. Dong, H.; Du, X.; Han, Q.; Hao, H.; Bi, K.; Wang, X. Performance of an innovative self-centering buckling restrained brace for mitigating seismic responses of bridge structures with double-column piers. *Eng. Struct.* **2017**, *148*, 47–62. [CrossRef]
34. Jin, S.; Ai, P.; Zhou, J.; Bai, J. Seismic performance of an assembled self-centering buckling-restrained brace and its application in arch bridge structures. *J. Constr. Steel Res.* **2022**, *199*, 107600. [CrossRef]
35. Shi, Y.; Zhong, Z.; Qin, H.; Han, J.; Sun, Z. Toggle buckling-restrained brace systems and a corresponding design method for the seismic retrofit of bridge bents. *Eng. Struct.* **2020**, *221*, 110996. [CrossRef]
36. Wang, Y.; Ibarra, L.; Pantelides, C. Effect of incidence angle on the seismic performance of skewed bridges retrofitted with buckling-restrained braces. *Eng. Struct.* **2020**, *211*, 110411. [CrossRef]
37. Zheng, J.; Wang, J. Concrete-Filled Steel Tube Arch Bridges in China. *Engineering* **2017**, *4*, 143–155. [CrossRef]
38. Bai, J.; He, J.; Li, C.; Jin, S.; Yang, H. Experimental investigation on the seismic performance of a novel damage-control replaceable RC beam-to-column joint. *Eng. Struct.* **2022**, *267*, 114692. [CrossRef]

39. Chen, B.-C.; Wang, T.-L. Overview of Concrete Filled Steel Tube Arch Bridges in China. *Pract. Period. Struct. Des. Constr.* **2009**, *14*, 70–80. [CrossRef]
40. Han, L.-H.; Li, W.; Bjorhovde, R. Developments and advanced applications of concrete-filled steel tubular (CFST) structures: Members. *J. Constr. Steel Res.* **2014**, *100*, 211–228. [CrossRef]
41. Gao, X.; Liu, L.J.; Yao, S.K.; Yang, J.W.; Li, Y. Commonly Encountered Damages in Cable Members of CFST Arch Bridge and Detection Methods. *Key Eng. Mater.* **2012**, *619*, 71–80. [CrossRef]
42. Hong, X.; Renzhong, T.; Yaodong, C. Calculation of SBC forces for concrete-filled steel tubular arch bridge with long span in process of lifting. *J. Zhejiang Univ. Eng. Sci.* **2004**, *38*, 5. (In Chinese)
43. Qian, Z.H.; Jian-ting, Z.H.; Hu, M.A.; Xiao-gang, L.I.; Lan, Z.H. Improved algorithm of cable force for one time cable tensioning on steel tube arch ribs with segmental hoisting. *J. Traffic Transp. Eng.* **2020**, *20*, 92–101. [CrossRef]
44. Zhou, Q.; Zhou, J.; Chen, J.; Feng, Y. Improved algorithm of cable force optimization of one-time tensioning for arch bridge. *J. Fuzhou Univ. Nat. Sci. Ed.* **2019**, *47*, 412–416.
45. Wang, H.; Yang, Y.D.; Li, A.Q.; Qiao, J.D.; Chen, Z.Q. Influences of soil-pile-structure interaction on seismic response of long span CFST arch bridge. *J. Southeast Univ. Nat. Sci. Ed.* **2005**, *3*, 433–437. (In Chinese)
46. Priestley, M.J.N.; Seible, F.; Calvi, G.M. *Seismic Design and Retrofit of Bridges*; John Wiley & Sons, Inc.: New York, NY, USA, 1996.
47. Susantha, K.; Ge, H.; Usami, T. Uniaxial stress–strain relationship of concrete confined by various shaped steel tubes. *Eng. Struct.* **2001**, *23*, 1331–1347. [CrossRef]

MDPI
St. Alban-Anlage 66
4052 Basel
Switzerland
www.mdpi.com

Buildings Editorial Office
E-mail: buildings@mdpi.com
www.mdpi.com/journal/buildings

www.ingramcontent.com/pod-product-compliance
Lightning Source LLC
Chambersburg PA
CBHW040837250926
51850CB00007BA/1336
9783036584485